T0310170

**Environmentally Conscious
Fossil Energy Production**

Environmentally Conscious Fossil Energy Production

Edited by
Myer Kutz and Ali Elkamel

WILEY

JOHN WILEY & SONS, INC.

This book is printed on acid-free paper. ∞

Copyright © 2010 by John Wiley & Sons, Inc. All rights reserved

Published by John Wiley & Sons, Inc., Hoboken, New Jersey
Published simultaneously in Canada

No part of this publication may be reproduced, stored in a retrieval system, or transmitted in any form or by any means, electronic, mechanical, photocopying, recording, scanning, or otherwise, except as permitted under Section 107 or 108 of the 1976 United States Copyright Act, without either the prior written permission of the Publisher, or authorization through payment of the appropriate per-copy fee to the Copyright Clearance Center, 222 Rosewood Drive, Danvers, MA 01923, (978) 750-8400, fax (978) 646-8600, or on the web at www.copyright.com. Requests to the Publisher for permission should be addressed to the Permissions Department, John Wiley & Sons, Inc., 111 River Street, Hoboken, NJ 07030, (201) 748-6011, fax (201) 748-6008, or online at www.wiley.com/go/permissions.

Limit of Liability/Disclaimer of Warranty: While the publisher and the author have used their best efforts in preparing this book, they make no representations or warranties with respect to the accuracy or completeness of the contents of this book and specifically disclaim any implied warranties of merchantability or fitness for a particular purpose. No warranty may be created or extended by sales representatives or written sales materials. The advice and strategies contained herein may not be suitable for your situation. You should consult with a professional where appropriate. Neither the publisher nor the author shall be liable for any loss of profit or any other commercial damages, including but not limited to special, incidental, consequential, or other damages.

For general information about our other products and services, please contact our Customer Care Department within the United States at (800) 762-2974, outside the United States at (317) 572-3993 or fax (317) 572-4002.

Wiley also publishes its books in a variety of electronic formats. Some content that appears in print may not be available in electronic books. For more information about Wiley products, visit our web site at www.wiley.com.

Library of Congress Cataloging-in-Publication Data:

Environmentally conscious fossil energy production / edited by Myer Kutz and Ali Elkamel
  p. cm
 Includes index.
 ISBN 978-0-470-23301-6 (cloth)
 1. Fossil fuels—Environmental aspects. 2. Renewable energy sources. 3. Global warming—Prevention. 4. Sustainable engineering. I. Kutz, Myer. II. Elkamel, Ali.
 TP318.E58 2010
 665.5028′6—dc22

2009020808

Printed in the United States of America

10 9 8 7 6 5 4 3 2 1

Contents

Contributors

Sabah A. Abdul-Wahab
Sultan Qaboos University
Sultanate of Oman

M. Rafiqul Awal
Texas Tech University
Lubbock, Texas

Terry Carter
Petroleum Resources Centre
Ministry of Natural Resources
Ontario, Canada

Eric Croiset
University of Waterloo
Ontario, Canada

Peter Douglas
University of Waterloo
Ontario, Canada

Ali Elkamel
University of Waterloo
Ontario, Canada

R. Larry Grayson
The Pennsylvania State University
University Park, Pennsylvania

Haslenda Hashim
University of Waterloo
Ontario, Canada

K. Mohammed
University of New Brunswick
Fredericton, New Brunswick, Canada

J. Guillermo Ordorica-Garcia
University of Waterloo
Ontario, Canada

B. V. Reddy
University of Ontario Institute
  of Technology
Oshawa, Ontario, Canada

M. R. Riazi
Kuwait University
Safat, Kuwait

Ahmed Shafeen
Natural Resources Canada
Ottowa, Ontario, Canada

Preface

Many readers will approach this series of books in Environmentally Conscious Engineering with some degree of familiarity with, knowledge about, or even expertise in, one or more of a range of environmental issues, such as climate change, pollution, and waste. Such capabilities may be useful for readers of this series, but they aren't strictly necessary, for the purpose of this series is not to help engineering practitioners and managers deal with the *effects* of man-induced environmental change. Nor is it to argue about whether such effects degrade the environment only marginally or to such an extent that civilization as we know it is in peril, or that any effects are nothing more than a scientific-establishment-and-media-driven hoax and can be safely ignored. (Authors of a plethora of books, even including fiction, and an endless list of articles in scientific and technical journals, have weighed in on these matters, of course.) On the other hand, this series of engineering books does take as a given that the overwhelming majority in the scientific community is correct and that the future of civilization depends on minimizing environmental damage from industrial, as well as personal, activities. At the same time, the series does not advocate solutions that emphasize only curtailing or cutting back on these activities. Instead, its purpose is to exhort and enable engineering practitioners and managers to reduce environmental impacts, to engage, in other words, in Environmentally Conscious Engineering, a catalog of practical technologies and techniques that can improve or modify just about anything engineers do, whether they are involved in designing something, making something, obtaining or manufacturing materials and chemicals with which to make something, generating power, transporting people and freight, handling materials anywhere in the chain between manufacturing operations, warehousing, and distribution, or handling and transporting both municipal and dangerous wastes.

Increasingly, engineering practitioners and managers need to know how to respond to challenges of integrating environmentally conscious technologies, techniques, strategies, and objectives into their daily work, and, thereby, find opportunities to lower costs and increase profits while managing to limit environmental impacts. Engineering practitioners and managers also increasingly face challenges in complying with changing environmental laws. So companies seeking a competitive advantage and better bottom lines are employing environmentally responsible methods to meet the demands of their stakeholders, who now include not only owners and stockholders, but also customers, regulators, employees, and the larger, even worldwide community.

Engineering professionals need references that go far beyond traditional primers that cover only regulatory compliance. They need integrated approaches centered on innovative methods and trends in using environmentally friendly processes, as well as resources that provide a foundation for understanding and implementing principles of environmentally conscious engineering. To help engineering practitioners and managers meet these needs, I envisioned a flexibly connected series of edited books, each devoted to a broad topic under the umbrella of Environmentally Conscious Engineering.

The intended audience for the series is practicing engineers and upper-level students in a number of areas—mechanical, chemical, industrial, manufacturing, plant, power generation, transportation, and environmental—as well as engineering managers. This audience is broad and multidisciplinary. Practitioners work in a variety of organizations, including institutions of higher learning, design, manufacturing, power generation, transportation, warehousing, waste management, distribution, and consulting firms, as well as federal, state, and local government agencies. So what made sense in my mind was a series of relatively short books, rather than a single, enormous book, even though the topics in some of the smaller volumes have linkages and some of the topics might be suitably contained in more than one freestanding volume. In this way, each volume is targeted at a particular segment of the broader audience. At the same time, a linked series is appropriate because every practitioner, researcher, and bureaucrat can't be an expert on every topic, especially in so broad and multidisciplinary a field, and may need to read an authoritative summary on a professional level of a subject that he or she is not intimately familiar with but may need to know about for a number of different reasons.

The **Environmentally Conscious Engineering Series** is comprised of practical references for engineers who are seeking to answer a question, solve a problem, reduce a cost, or improve a system or facility. These books are not a research monographs. The purpose is to show readers what options are available in a particular situation and which option they might choose to solve problems at hand. I want these books to serve as a source of practical advice to readers. I would like them to be the first information resource a practicing engineer reaches for when faced with a new problem or opportunity—a place to turn to even before turning to other print sources, even any officially sanctioned ones, or to sites on the Internet. So the books have to be more than references or collections of background readings. In each chapter, readers should feel that they are in the hands of an experienced consultant who is providing sensible advice that can lead to beneficial action and results.

The six earlier volumes in the series have covered mechanical design, manufacturing, materials handling, materials and chemicals processing, alternative energy production, and transportation. The seventh series volume, **Environmentally Conscious Fossil Energy Production**, has linkages to some of those earlier volumes, particularly alternative energy production and transportation. While this volume might be deemed the most controversial in the **Environmentally Conscious Engineering Series**, it deals with technologies that cannot be avoided for the intermediate future. After all, energy demand is growing worldwide, both in developed and in developing countries. While conservation methodologies and alternative energy sources are coming on line, fossil energy production will continue to be used to meet human needs and desires. The contributors to this volume, especially Ali Elkamel, who originated the idea for the volume and contributed three of the chapters together with colleagues, all recognize that the fossil energy sector currently produces unacceptable amounts of greenhouse gas emissions responsible for global climate change and must be carefully managed and regulated lest it create environmental damage in other ways, including pollution from spills occurring during oil ocean transport. Contributors have provided chapters that provide solutions for mitigating some of the most devastating problems that the fossil energy sector can create.

I asked the contributors, located not only in North America, but also in the Middle East, to provide short statements about the contents of their chapters and why the chapters are important. Here are their responses:

M. Rafiqul Awal (Texas Tech University in Lubbock, Texas), who contributed the chapter on **Environmentally Conscious Petroleum Engineering**, writes, "The book opens with a chapter on environmental consciousness in the petroleum sector. True, the petroleum industry was *not* environmentally savvy in the first half of its century-old global existence, partly because the world itself was not environmentally conscious. But spurred by rising global ecological awareness and government regulation in the Western hemisphere after World War II, the petroleum industry has taken cognizance of its responsibilities. Consequently, not only has it *plugged holes* via technological and human resources development, but most oil companies in the product distribution and marketing sector, as well as oilfield operators, have taken proactive roles in planning their operations with almost zero tolerance for environmental damage."

Ali Elkamel (University of Waterloo in Waterloo, Ontario, Canada), who co-edited this book and along with J. Guillermo Ordorica-Garcia, Peter Douglas, and Eric Croiset, contributed the chapter on **Carbon Management and Hydrogen Requirements in Oil Sands Operations**, writes, "In this chapter, energy requirements associated with producing synthetic crude oil (SCO) and bitumen from oil sands are modeled and quantified, based on current commercial production schemes. These energy requirements are expressed in terms of amounts of hot water, steam, power, hydrogen, diesel fuel, and process fuel for upgrading processes. The production schemes considered are: a) mined bitumen, upgraded to SCO; b) thermal bitumen, upgraded to SCO and c) thermal bitumen, diluted.

Additionally three distinct bitumen upgrading methods were modeled and incorporated into schemes a) and b). In addition to energy demands, greenhouse gas (GHG) emissions associated with supplying energy required to produce bitumen and SCO are also illustrated. Calculations are illustrated on two distinct situations. The first is a base case in which all the energy is produced using current technology, in the year 2003. The second situation is a future production scenario, where energy demands are computed for SCO and bitumen production levels corresponding to the years 2012 and 2030."

R. Larry Grayson (The Pennsylvania State University in University Park, Pennsylvania), who contributed the chapter on **Environmentally Conscious Coal Mining**, writes, "Following a description of the historical environmental performance of the U.S. coal industry, this chapter details the extent of current environmental regulations and environmental achievements of recent decades. Through regulation and enlightened management, improvements have been impressive. Impacts on air, land, and water have been mitigated dramatically. Today the global coal industry, through major companies, has embraced sustainable development principles, just as governments have, and future improvements will address remaining and emerging environmental, health, and safety related issues in a proactive way."

Sabah A. Abdul-Wahab (Sultan Qaboos University in Muscat, Sultanate of Oman), who contributed the chapter on **Maritime Oil Transport and Pollution Prevention**, writes, "This chapter addresses three case studies on maritime oil transport and pollution prevention. In the first case study, an innovative, real-time, and environment friendly in-situ device for early detection of oil pollution in seawater was developed. The device is very important for early detection of oil pollution in seawater and hence minimizes damage caused. It can contribute to the development of advanced practices in oil spill detection. In the second case study, an innovative technique to detect fissures in ship structures at an early stage was presented. Ihe design concept can be used in operational pipelines, such as those used to transport gas or oil. In the third case study, the ground-level concentration of nitrogen oxides (NO_x) released from ships in berth was predicted. This prediction will help to identify ships that cause emission problems in port and also will help ships' crews to take corrective actions to reduce and control."

M. R. Riazi (Kuwait University in Kuwait), who contributed the chapter on **Accidental Oil Spills Behavior and Control**, writes, "This chapter is devoted to oil spill occurrence, behavior and control. Oil is transmitted from producing countries to consuming countries mainly through tankers. The products of crude oil such as gasoline and fuel oil may also be transferred from refinery sites to the market. Oil spills occur mainly during such activities as a result of collision, accidents or human errors. In addition, during offshore production of oil, significant amounts of oil may be released over the sea surface causing significant damage to the environment and threatening the lives of humans and living organs in aquatic system. It is important to prevent such accidents but once they have occurred to use an appropriate clean-up technique. In this chapter, the nature of oil spills, their occurrence, prevention and clean-up methods are discussed. Selection of the right clean up technique mainly depends on the nature of a spill and its behavior. Predictive models are presented to simulate spill behavior and its impact especially from ecological and toxic points of view."

Ahmed Shafeen (Natural Resources Canada, Ottowa, Ontario, Canada), who along with Terry Carter, contributed the chapter on **Geological Sequestration of Greenhouse Gases**, writes, "The recent rapid rise in the CO_2 content of the atmosphere and its causal linkage to global warming has politicians and the scientific community scrambling to identify a solution. Of all the methods proposed, geological sequestration of CO_2 is generally acknowledged to have the best potential to mitigate CO_2 emissions into the atmosphere in the large quantities necessary to have an effective near to medium-term impact. This chapter provides an overview of this promising technology."

Ali Elkamel (University of Waterloo in Waterloo, Ontario, Canada), who co-edited this book and along with J. Guillermo Ordorica-Garcia, Peter Douglas, and Eric Croiset, contributed the chapter on **Clean-Coal Technology: Gasification Pathway**, writes, "In this chapter, relevant information concerning power plant technology and carbon dioxide capture techniques is presented. The chapter covers current gasification techniques, including commercial gasification processes. The advantages of integrated gasification combined cycle (IGCC) power plants, stages of development, and selected commercial IGCC projects are presented. These are plants that employ synthetic gas from coal as a fuel. Details concerning the production of synthetic gas from coal are reviewed as well as the performance of IGCC plants. The more recent integrated coal gasification fuel cell combined cycle (IGFC) is also discussed. The chapter ends with a detailed classification of all major CO_2 capture technologies."

Ali Elkamel (University of Waterloo in Waterloo, Ontario, Canada), who co-edited this book and along with Ali Elkamel, Haslenda Hashim, Peter L. Douglas, and Eric Croiset, contributed the chapter on **An Integrated Approach for Carbon Mitigation in the Electric Power Generation Sector,** writes, "This chapter presents the modeling of fleet-wide energy planning that can be used to determine the optimal structure necessary to meet a given CO_2 reduction target while maintaining or enhancing power to the grid. The modeling framework discussed here incorporates power generation as well as CO_2 emissions from a fleet of generating stations (hydroelectric, fossil fuel, nuclear and wind). The modeling is based on mixed integer programming (MIP) and is used to optimize an existing fleet as well as recommend new additional generating stations, carbon capture and storage, and retrofit actions to meet a CO_2 reduction target and electricity demand at a minimum overall cost. The framework is illustrated by a case study that deals with the energy supply system operated by Ontario Power Generation (OPG) for the province of Ontario, Canada."

B.V. Reddy (University of Ontario Institute of Technology in Oshawa, Ontario, Canada), who along with K. Mohammed, contributed the chapter on **Energy and Exergy Analyses of Natural Gas-Fired Combined Cycle Power Generation Systems**, writes, "The present chapter provides details on thermodynamic analysis of natural gas fired combined cycle power generation systems, on methodology to conduct energy analysis, on the role of operating parameters such as gas turbine inlet temperature and gas turbine pressure ratio. But energy analysis alone has limitations in providing details on the overall performance improvement for the power generation unit. Exergy analysis helps identify sources of irreversibilities and losses in components and in the overall system and improves performance. The present chapter provides details on how to conduct exergy analysis for a natural gas fired combined cycle power generation system with reheat and supplementary combustion chambers."

That ends the contributors' comments. I would like to express my heartfelt thanks to all of them, and especially to Ali Elkamel, my esteemed co-editor, for having taken the opportunity to work on this book. Their lives are terribly busy, and it is wonderful that they found the time to write thoughtful and complex chapters. I developed the book because I believed it could have a meaningful impact on the way many engineers approach their daily work, and I am gratified that the contributors thought enough of the idea that they were willing to participate in the project. Thanks also to my editor, Bob Argentieri, for his faith in the project from the outset. And a special note of thanks to my wife Arlene, whose constant support keeps me going.

Myer Kutz
Delmar, NY

CHAPTER 1

ENVIRONMENTALLY CONSCIOUS PETROLEUM ENGINEERING

M. Rafiqul Awal
Bob L. Herd Department of Petroleum Engineering
Texas Tech University
Lubbock, Texas

1   INTRODUCTION

Petroleum (from Greek *petro* — rock, and *oleum* — oil) has been known to mankind in two forms: oil and natural gas. Oil includes crude oil and various refined products from crude oil: gasoline, diesel, and heavier oils like heating oil, furnace and bunker oil, and tar. Petroleum is second to coal in terms of energy reserves in earth's subsurface structures and natural source of useful energy. But its use for energy began quite late: 1850s, after the first commercial oil well flowed from Colonel Drake's well in Titusville, Pennsylvania (1859). Limited use for heat and light existed before that in various parts of the world, where natural seepage of crude oil created surface pools. Oil seeps were used in ancient times and production of oil took place in a number of countries long before Drake's well produced oil. Shortly after Drake's well came into production, several other wells went online in places such as in Canada and Europe. Some of them became oil centers in their own right.

The use of petroleum became widespread with the advent of *kerosene*, a liquid fuel that was initially made from coal (in 1846 by Canadian geologist Abraham Gesner) but could be made from petroleum (in 1848 by Scottish chemist James Young). Kerosene steadily replaced oil from *sperm whale* (which almost became extinct due to its odorless prized oil by 1950) as an affordable illuminant in North America and found a great demand in Europe and elsewhere on the globe (Pees 1989 and Pees and Stewart 1995).

However, the boom and industrial advance that began near Titusville on August 27, 1859 (*Drake Day*), gathered momentum so quickly and enlarged so greatly that a veritable industrial explosion took place (Owen 1975).

From early twentieth century onward, civilization has been marked by oil. Industrial production in general, and its agricultural model in particular, are largely dependent on fossil fuels. Oil-based civilization has given birth to an urban model where the automobile becomes the determining factor in its design.

With the spread of usage of petroleum, environmental pollutions also ensued. An entirely new range of highly contaminating products derived from oil has been generated. These products are highly contaminating not only in their production process but also in their disposal (Oil Watch).

In spite of all of that, all the predictions point toward an increase in energy demand for the next decade, and oil will play a prevailing role. In the United States, for example, by 2025, even with expected dramatic gains in efficiency, total energy *consumption* is forecasted to increase by 36 percent, petroleum by 39 percent, natural gas by 40 percent, coal by 34 percent, electricity by 49 percent, and renewable energy by 38 percent, according to the Energy Information Administration (2005). Figure 1 shows that in order to meet the demand for energy, the role of petroleum is unchallengeable—a cold fact that only emphasizes the need for ever-more consciousness in the extraction, transportation, and use of petroleum.

It took many decades before *petroleum engineering* emerged as a formal engineering profession, having crystallized into a science in the late 1920s. In the meantime, oil became the lifeline of world's naval powers, leading to fierce global competition to own and monopolize the new *black gold*. The greatest victim was perhaps the *environment*, taking spills of all scales, before institutionalized consciousness dawned in 1924 (Owens 1975). Before that, several decades saw rampant drilling and waste disposal at site, and as fierce competition to capture the global oil market ensued between the Dutch conglomerate Royal Shell Oil Company (headed by Sir Henri Wilhelm August Deterding, b.1866, d.1939), and the American monopolist, the Standard Oil Company of New Jersey (headed by John D. Rockefeller, b. 1839, d.1937), oil spills, blowouts, and prolonged burning of oil wells and pits became a commonplace (Hanighen 1934).

With the advent of environmental awareness in the post–WWII era, regulations for environmental impact assessment by government and certified agencies have reduced environmental pollution in all sectors of the petroleum industries. But accidents causing widescale damage to the environment for months are still not uncommon. For example, the *Exxon Valdez* tanker grounded in 1989 off Alaskan shores, spilling more than 11 million gallons of oil (see Figure 2) (Kvonvolden

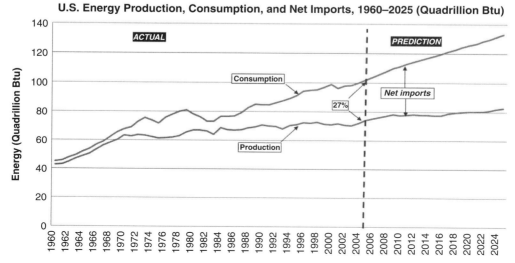

Figure 1 Total energy production and consumption, 1970–2025 (quadrillion Btu) (EIA 2005)

Figure 2   Massive oil spill from *Exxon Valdez* oil tanker (1989) in Alaska's North Slope (Kvonvolden 2004)

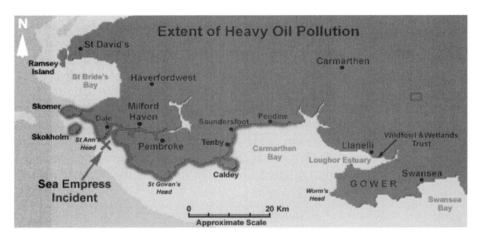

Figure 3   Oil spill from the *Sea Empress* spread to vast coastal areas of southwest coast of Wales in February 1996 (Dyrynda and Symberlist 2002)

2004). Owing to quick spreading of oil over seawater and currents, tides and wind effect, oil spills have the potential to pollute thousands of square miles of water, as happened when the *Sea Empress* oil spill occurred off the coast of England in 1996 (Dyrynda and Symberlist 2002). Figure 3 shows the approximate extent of coastline that was affected by heavy oil pollution during February and March of 1996. The heaviest and most persistent oiling affected shores to the south and east of the point of grounding.

The meteoric growth of oil was spurred by a series of new industrial inventions, in addition to cleaner devices for illumination. These included the automobile

(1895), electric generator (1896), motorized maritime vessels and submarines (1911), and the advent aircraft (1906) for public and goods transport. Such needs could then be met by occurrence of oil in shallow reservoirs, cheap and quick mobility through cross-country pipelines, and a rapid growth of refining and petrochemical technologies that prompted thousands of useful products for everyday life, from paving of roads to producing soaps, plastics, cosmetics, and medicines.

Before embarking on the various types of environmental damages that arise or may arise from petroleum, it is worthwhile to explore the vast world of petroleum. The petroleum industry is divided into two broad divisions: the *upstream* and the *downstream* petroleum industry.

The former is often referred to as the *exploration and production* (E&P) industry (comprising various geological and geophysical exploration operations), followed by drilling activities and long-term well operations.

The latter include such operations as refining, petrochemical processing, and pipeline transportation of various consumer products, including natural gas.

The physical gap between the upstream and downstream is usually butted onshore by crude oil and gas transportation pipelines—from the oil fields to the refinery. In case of offshore E&P, both pipelines (short distance of a few hundred miles, except the Trans Alaskan Pipeline System (TAPS), and some Russian pipelines, which span several thousand miles) and maritime vessels (tankers,

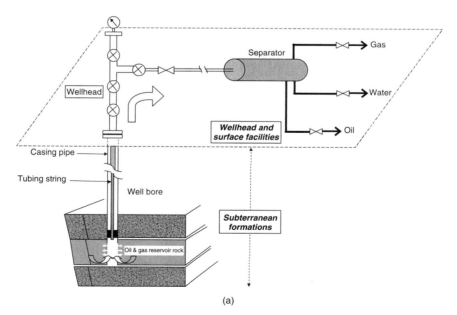

(a)

Figure 4 (a) Typical petroleum production system (Guo et al. 2007) (b) Sketch showing sub–sea wells, pipelines—all on sea floor, and steel structures used for drilling or production (Guo et al. 2007) (c) From source to sink—oil and gas production and consumption encompasses a vast topography on land, sea, and air

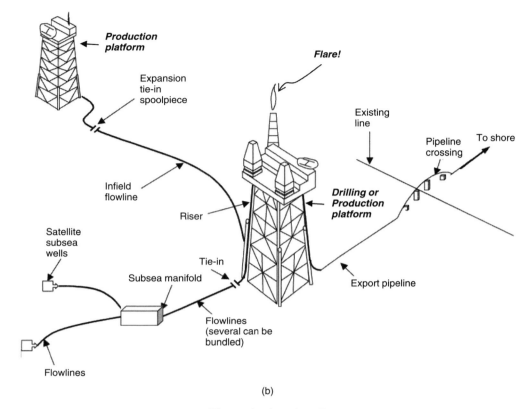

(b)

Figure 4 (*continued*)

barges, etc.) are used to carry produced hydrocarbons for export/import or to the refineries. Figure 4 depicts the comprehensive picture of the entire petroleum industry. Actually, the complete spatial domain of petroleum, from source to consumption points, is much more extensive if transportation and mass storage of various petroleum products for the end market are included.

In its long journey, from deep in Earth's interior (2,000 to 25,000 feet) to end use as fuel or petrochemical, and of course after its transformation by combustion and/or disposal, petroleum significantly affects the environment. Figure 5 shows a typical petroleum production system, highlighting a widely distributed contact with the environment—both subsurface and above ground.

As shown in Figure 5, the *reservoir* contains oil and/or gas, and highly saline water in porous, permeable and/or fractured networks (natural and/or manmade) media. The oil and/or natural gas flows along with water from the reservoir rock into the *wellbore*. The wellbore is protected by a steel *casing pipe* throughout the life of the well. To minimize the corrosion of the casing pipe, almost 90 percent of the casing's interior is sealed off with a *packer* system, and the reservoir fluids are transported through a *tubing string* inside the casing.

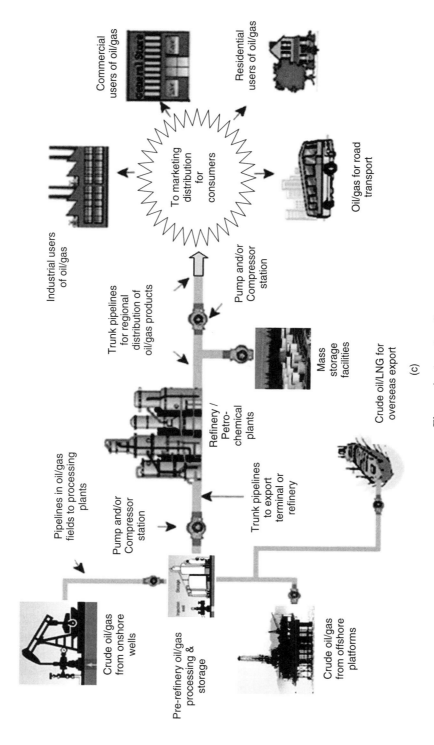

Commercial users of oil/gas

Residential users of oil/gas

Industrial users of oil/gas

To marketing distribution for consumers

Oil/gas for road transport

Trunk pipelines for regional distribution of oil/gas products

Pump and/or Compressor station

Mass storage facilities

Crude oil/LNG for overseas export

Pipelines in oil/gas fields to processing plants

Refinery / Petro-chemical plants

Pump and/or Compressor station

Trunk pipelines to export terminal or refinery

Crude oil/gas from onshore wells

Pre-refinery oil/gas processing & storage

Crude oil/gas from offshore platforms

(c)

Figure 4 (*continued*)

7

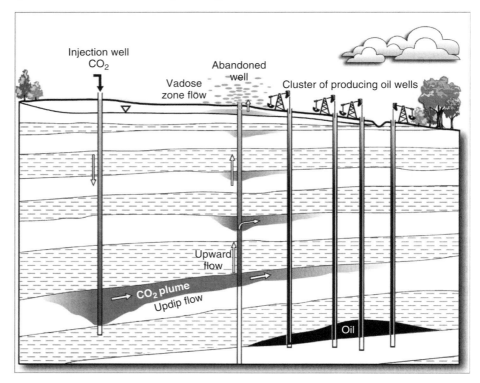

Figure 5 Clusters of production and injection wells used for petroleum extraction pose numerous environmental threats, which need continued development in petroleum engineering, and environmental awareness by government agencies and oil operators

From a number of wellheads, called the *Christmas tree*, surface flow lines carry oil and gas to processing plants, called *gathering stations*, which are located in the oilfield itself. After initial separation of oil from associated gas and water, the partially stabilized oil is pipelined to a *central tank farm* (CTF), again in the oilfield. At the CTF, the crude oil is processed to specifications required by refineries or sales for export. Because of the huge quantity of crude oil that must be processed for such requirements as desalting, dewaxing, sulfur removal (especially for crude oils in Saudi Arabia and Brazil, which contain over 10,000 ppm of sulfur), and deemulsification, the CTFs generate huge amount of wastes, mostly solids and liquids.

The final processing stage for the crude oil is the refinery and petrochemical plants, where the maximum pollution occurs—most notably in the form of atmospheric emissions that can be felt from miles away due to the odor of airborne chemical species, such as oxides of nitrogen and sulfur.

In the entire oil and gas chain, other sources of pollution are the local and trunk pipelines, oil tankers, and huge storage tank farms (Figures 6 and 7).

(a)   (b)

Figure 6   Petroleum transportation by (a) oilfield pipelines and (b) cross-country pipelines leave a large environmental footprint (Merriam-Webster)

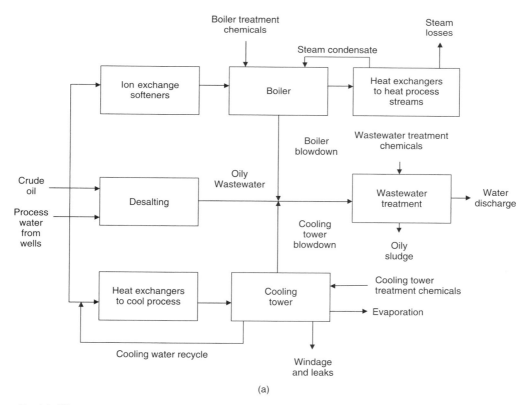

(a)

Figure 7   (a) Wastewater treatment at oilfield central tank farm and refinery (Allen and Shonnard 2002) (b) Petroleum refining, showing chemical processes that converts crude oil into various usable products (Allen and Shonnard 2002)

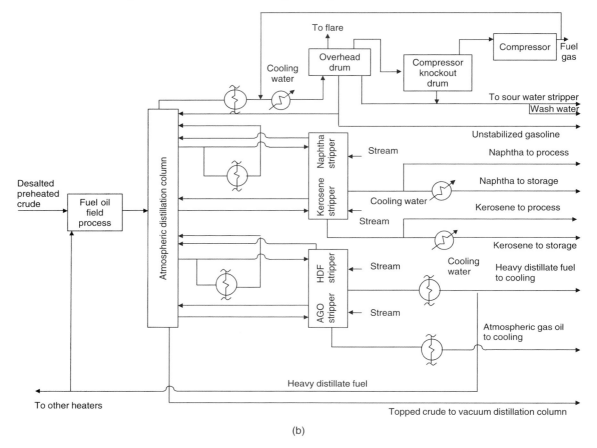

(b)

Figure 7　(*continued*)

Because oil and gas are so visible to the public eye—and so are oil sheens in water surface and odors of various gases emitted by refineries—there is a common perception that petroleum is one of the worst polluters. Figure 8 busts this perception.

A global awareness was initiated by the United Nations, which culminated in the Kyoto Protocol (1998) for greenhouse gas emission control. Although the Kyoto Protocol has been ratified into an international law (2004), practical measures taken by various countries are all but satisfactory.

The scope of this chapter is limited to documenting environmental consciousness in petroleum engineering, both within academia and throughout the larger industry.

The presentation is organized by dividing petroleum engineering into its distinct components, and then addressing environmental issues at various subfields. The historical developments in each subfield are followed by suggested actions that need to be taken.

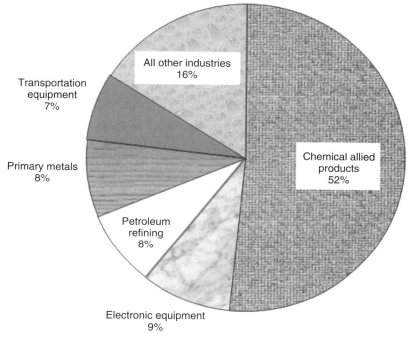

(a) Industrial Hazardous Waste Generation

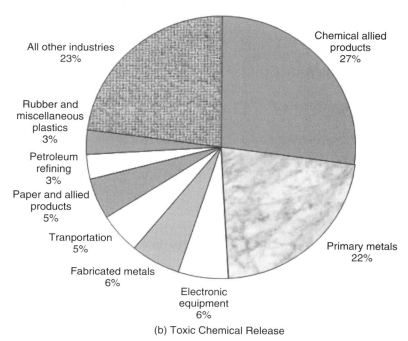

(b) Toxic Chemical Release

Figure 8 Petroleum refining has relatively low pollution signature: (a) industrial hazardous waste generation; (b) Toxic chemical release in the United States (by industry sector) (Allen and Shonnard 2002)

2 EXTENT OF ENVIRONMENTAL IMPACT OF PETROLEUM

The spread of petroleum also increased environmental pollution—in land, sea, and air—as production and distribution surpassed progress in containing oil spills during drilling and production in the oil and gas fields and during surface transportation—to stabilizing plants (known as central tank farms), to refineries through trunk lines, and ultimately to every nook and corner of the civilized world for bulk and consumer distributions.

At this point, it is prudent to define some of the terms related to the environment in order to avoid confusion and misinterpretation. The following definitions are taken from California Department of Environment (Therkelsen 1973):

- *Environment* is all physical and biological features of an area, such as the topography, geology, soil, hydrology, climate, plants, and wildlife.
- An *environmental problem* is anything that interferes with, alters, or has some generally detrimental effect on all or any of the environmental components.
- *Ecology* is the study of the relationship between an organism, plant or animal, and its environment.
- *Oil fields* generally constitute an environment composed of the physical features of the field, the organisms present, and all the factors associated with oil exploration but mainly oil production, such as drilling and production equipment, access roads, and waste disposal facilities.

However, satisfactory prevention and solutions to petroleum-induced environmental damages should recognize that these problems are complex due to the involvement of many interrelated factors, all of which need to be considered. They include water and air quality, land use, wildlife and habitat requirements, aesthetic values and recreational uses, and regional economics.

World oil production continues to rise, from 8.5 million tonnes (1 tonne equals about 7 barrels) in 1985 to 11.7 million tonnes in 2000 (EIA 2005). In that same time, the number of offshore oil and gas platforms rose from a few thousand to approximately 8,300 fixed or floating offshore platforms. Current world average production of petroleum is about 90 million barrels per day, with 1.5 to 2 percent annual growth (RCEP 2000).

2.1 Petroleum Pollution in Land and Sea

Historically, oil and gas exploration and production of petroleum have represented a significant source of spills. Table 1 shows how much of it goes to the sea during maritime transportation and carried as effluent from offshore rigs, production platforms, and city sewerage.

Petroleum release in marine waters occurs from four major sources:

1. Natural seeps
2. Releases that occur during the extraction of petroleum

Table 1   Natural and Anthropogenic Release of Petroleum in the Sea

	North America			Worldwide		
	Best	Min	Max	Best	Min	Max
		Estimate			Estimate	
Natural Seeps	160	80	240	600	200	2000
Extraction of Petroleum	3.0	2.3	4.3	38	20	62
Platforms	0.16	0.15	0.18	.86	0.29	1.4
Atmospheric Deposition	0.12	0.07	0.45	1.3	0.38	2.6
Produced waters	2.7	2.1	3.7	36	19	58
Transportation of Petroleum	9.1	7.4	11	150	120	260
Pipeline Spills	1.9	1.7	2.1	12	6.1	37
Tank Vessel Spills	5.3	4.0	6.4	100	93	130
Operational Discharges(cargo washings)	Na	na	na	36	18	72
Coastal facility spills	1.9	1.7	2.2	4.9	2.4	15
Atmospheric Deposition	0.01	trace	0.02	0.4	0.2	1
Consumption of Petroleum	84	16	2000	480	130	6000
Land Based (River and Runoff)	54	2.6	1900	140	6.8	5000
Recreational Marine Vessel	5.6	2.2	9	Nd	nd	nd
Spills(Non-Tank Vessels)	1.2	1.1	1.4	7.1	6.5	8.8
Operational Discharges(Vessels >100GT)	0.10	0.03	0.30	270	90	810
Operational Discharges(Vessels <100GT)	0.12	0.03	0.30	Nd	nd	nd
Atmospheric Deposition	21	9.1	81	52	23	200
Jettisoned aircraft fuel	1.5	1.0	4.4	7.5	5.0	22
Total(in thousands of tonnes)	260	110	2300	1300	470	8300

"Oil in the Sea III: Inputs, Fates, and Effects," report from National Research Council, http://dels.nas.edu/dels/rpt_briefs/oil_in_the_sea_final.pdf.

3. Transportation of petroleum products

4. Consumption of petroleum products

Natural seeps contribute the highest amount of petroleum to the marine environment, accounting for 45 percent of the total annual load to the world's oceans and 60 percent of the estimated total load to North American waters. The last three include all significant sources of *anthropogenic* (manmade) petroleum pollution. Releases from extraction and transportation of petroleum represent less than 10 percent of inputs from human activity. Chronic releases during consumption of petroleum, which include urban runoff, polluted rivers, and discharges from commercial and recreational marine vessels, contribute up to 85 percent of the anthropogenic load to North American waters (NRC). These releases can pose significant risks to the sensitive coastal environments, where they most often occur.

The major sources of petroleum discharges in the transportation sector include pipeline spills, tank vessel spills, operational discharges from cargo washings,

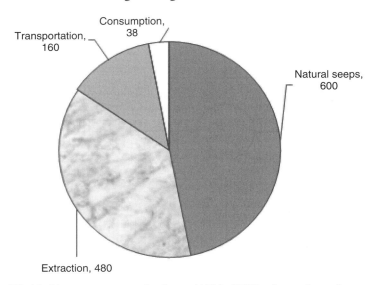

Figure 9 Worldwide average annual release (1990–1999) of petroleum into sea, in tones (1 tonne = 7 barrels) (Source: "Oil in the Sea III: Inputs, Fates, and Effects," report from National Research Council, http://dels.nas.edu/dels/rpt_briefs/oil_in_the_sea_final.pdf.)

and coastal facilities spills. Although the devastating impact of spills has been well publicized with images of oil-covered shores and wildlife (see, for example, Figure 9), releases from the transport of petroleum now amount to less than 13 percent worldwide.

As a matter of fact, transportation-related spills are down for several reasons. In the United States, the enactment of the Oil Pollution Act of 1990 placed increased liability on responsible parties, and other regulations required the phase-out of older vessels and the implementation of new technology and safety procedures. By 1999, approximately two-thirds of the tankers operating worldwide had either double hulls or segregated tank arrangements—a vast improvement over older single-hull ships. Operational discharges from cargo washing are now illegal in North America, a law that is rigorously enforced. However, there still remains a risk of spills in regions with less stringent safety procedures or maritime inspection practices. On the contrary, pollution from urban areas is expected to rise. At current trends, 60 percent of U.S. population will live along the coast, and worldwide two-thirds of the urban centers with populations of 2.5 million or more are near coastal areas, by the year 2010 (NRC).

2.2 Petroleum Pollution in the Atmosphere

Since only about 6 percent of petroleum is used as noncombustion chemicals, the remaining 94 percent ends as airborne pollution. Perhaps airborne pollution has overshadowed all other forms petroleum pollutions; initially due to lack of

Table 2   Amount of CO_2 Emitted per Million Btu of Energy from Various Fuels

Refined Products from Crude Oil	Pounds CO_2 per Million Btu
Aviation gasoline	152.7
Distillate fuel (No. 1, No. 2, No. 4 Fuel Oil and Diesel)	161.4
Jet fuel	156.3
Kerosene	159.6
Liquefied petroleum gases (LPG)	139.1
Motor gasoline	156.4
Petroleum coke	225.1
Residual fuel (No. 5 and No. 6 fuel oil)	173.9
Natural and other fuel gases	
Methane	115.3
Landfill gas	115.3
Flare gas	120.7
Natural gas (pipeline)	117.1
Propane	139.2
Coal	
Anthracite	227.4
Bituminous	205.3
Subbituminous	212.7
Lignite	215.4

(Energy Information Administration 2000)

combustion efficiency, then exponential consumption in power generation and mass transport systems. In fact, global warming is attributed to anthropogenic (i.e., manmade) CO_2 release in the atmosphere, in which combustion of petroleum has taken a seat close to coal. Table 2 shows a comparison of CO_2 emitted by various fuels.

During the past decade, however, improved production technology and safety training of personnel have dramatically reduced both blowouts and daily operational spills. Today, accidental spills from platforms represent about 1 percent (NRC).

3   ENVIRONMENTAL IMPACT OF WASTES

Reis (1996) has given a comprehensive review of various wastes produced in connection with production and transportation of petroleum.

The primary measure of the environmental impact of petroleum wastes is their toxicity to exposed organisms. The *toxicity* of a substance is most commonly reported as its concentration in water that results in the death of half of the exposed organisms within a given length of 96 hours (4 days), although other times have been used (Reis 1996). Common test organisms include *Mysid* shrimp or *Sheepshead Minnows* for marine waters and fathead minnows or rainbow trout for freshwaters. The concentration that is lethal to half of the exposed population during the test is called LC_{50}. High values of LC_{50} mean that high concentrations of the substance are required for lethal effects to be observed, and this indicates a low toxicity. A related measure of toxicity is the concentration at which half of the exposed organisms exhibit sublethal effects; this concentration is called EC_{50}. Another measure of toxicity is reported as *No Observable Effect Concentration* (NOEC), the concentration below which no effects are observed.

Other effects of hydrocarbons include stunted plant growth if the hydrocarbon concentration in contaminated soil is above about 1 percent by weight. Hydrocarbons can also impact higher organisms that may become exposed following an accidental release. For example, marine animals that use hair or feathers for insulation can die of hypothermia if coated with oil. Coated animals can also ingest fatal quantities of hydrocarbons during washing and grooming activities (Reis 1996).

The high dissolved-salt concentration of most produced water can also impact the environment. Typical dissolved-salt concentrations affect the ability of plants to absorb water and nutrients from soil. They can also alter the mechanical structure of the soil, which disrupts the transport of air and water to root systems. The LC_{50} values for dissolved salt concentrations for freshwater organisms are on the order of 1,000 ppm (Reis 1996).

The toxicity of drilling muds varies considerably, depending on their composition. Toxicities (LC_{50}) of water-based muds containing small percentages of hydrocarbons can be a few thousand ppm. The LC_{50} of polymer muds, however, can exceed one million, which means that fewer than 50 percent of a test species will have died during the test period. The toxicity of heavy metals found in the upstream petroleum industry varies widely. The toxicity of many heavy metals lies in their interference with the action of enzymes, which limits or stops normal biochemical processes in cells. General effects include damage to the liver, kidney, or reproductive, blood forming, or nervous systems. With some metals, these effects may also include mutations or tumors, Heavy-metal concentrations allowed in drinking water vary for each metal, but are generally below about 0.01 mg/L. The heavy metals in offshore drilling fluid discharges normally combine quickly with the naturally abundant sulfates in seawater to form insoluble sulfates and precipitates that settle to the sea floor. This process renders the heavy metals inaccessible for bioaccumulation or consumption. The various chemicals used during production activities can also affect the environment. Their toxicities vary considerably, from highly toxic to essentially nontoxic.

4 PROVENANCE OF ENVIRONMENTAL DAMAGE

We now identify the sources of environmental damages encountered during petroleum exploration, production, and transportation. Reis (1996) has given a broad description on the subject, which are summarized in this section. Then we switch our discussion to some potential sources of threats to environmental that have yet to be addressed.

In the upstream petroleum industry, drilling and production are two major operations that can potentially impact the environment by generating a significant volume of wastes, as well as leakage and emissions. Based on more in-depth understanding of these processes, operations can be improved in order to minimize or eliminate the adverse environmental impacts.

Drilling is the process in which a hole is made in the ground to allow subsurface hydrocarbons to flow to the surface. The wastes generated during drilling are the rock removed to make the hole (as cuttings), the fluid used to lift the cuttings, and various materials added to the fluid to change its properties to make it more suitable for use and to condition the hole.

Production is the process by which hydrocarbons flow to the surface to be treated and used. Water is often produced with hydrocarbons and contains a variety of contaminants. These contaminants include dissolved and suspended hydrocarbons and other organic materials, as well as dissolved and suspended solids. A variety of chemicals are also used during production to ensure efficient operations.

During drilling and production activities, a variety of air pollutants are emitted. The primary source of air pollutants are the emissions from internal combustion engines, with lesser amounts from other operations, fugitive emissions, and site remediation activities.

4.1 Drilling

The drilling rig has a huge environmental footprint—both above ground and underground (see Figure 10). In the early days, uncontrolled flow of oil and gas during drilling, called gushers, were common at drilling sites, and the spills often exceeding millions of gallons of oil played havoc on the environment (see Figure 11). Although modern drilling technology has advanced, human errors still cause similar uncontrolled flow, called a well blowout (see Figure 12). Well blowouts are described in detail in Section 7.1.

The process of drilling oil and gas wells also generates a variety of different types of wastes. Some of these wastes are natural byproducts of drilling through the earth (e.g., drill cuttings) and some come from materials used to drill the well (e.g., drilling fluid and its associated additives).

During drilling, the drilling fluid tends to penetrate the permeable rock strata because an overpressure is always maintained to prevent blowout from high-pressure gas or oil strata. Special care is taken to avoid groundwater contamination while drilling under the water table. Environmental regulation requires setting and cementing a steel casing pipe (surface casing) to isolate the freshwater bearing strata from drilling fluid contamination, as drilling is continued until target depth.

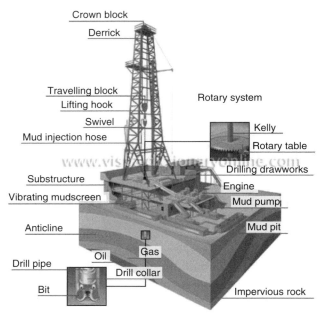

Figure 10 The drilling rig has a huge environmental footprint—both above ground and underground

Figure 11 A *gusher* (uncontrolled flow of oil and gas after oil-bearing strata is ruptured by drill bit) used to be common site at most successful drilling, circa 1900 (Energy Industry Photos)

Figure 12 Accidental *well blowouts* are still not uncommon, despite quantum progress in oil well drilling technology (Energy Industry Photos)

The drilling fluid is usually a water-based system of numerous solid additives—mainly bentonite, barites, hematite, antioxidants (oxygen scavengers), corrosion inhibitors, biocides, lubricants, pH-control chemicals, polymers, and salts (sodium and potassium chlorides). Diesel oil is the fluid base when an oil-base drilling fluid is required for drilling through water-sensitive shale strata, thus worsening the sensitivity to the environment—both surface and subsurface.

The situation is further aggravated when the drilling fluid returns to the surface, carrying cuttings of subsurface rock strata that contain heavy metals as well as radioactive materials.

Heavy metals can enter drilling fluids in two ways: Many metals are naturally occurring in most formations and will be incorporated into the fluid during drilling. Other metals are added to the drilling fluid as part of the additives used to alter the fluid properties. Heavy metals naturally occur in most rocks and soils, although at relatively low concentrations. Although the concentrations of the major elements will vary from carbonate to siliceous rocks, the concentration of the trace elements, including heavy metals, is probably representative of rocks and soils of many other areas. Another significant source of heavy metals in drilling fluid is the thread compound (pipe dope) used on the pipe threads when making up a drill string.

Drilling Fluid Treatments

For maintaining drilling fluid properties as well as minimizing pollution, elaborate treatment process is involved at the rig site. The basic equipment includes shale-shakers, de-sanders and de-silters, de-gassers, a battery of centrifugal and cyclones for unwanted solids removal, and chemical treaters. After separating the solids from the mud, a significant volume of liquid is normally retained with the cuttings. Volumetric measurements from offshore platforms have shown that the total volume of liquids with the cuttings after discharge can be 53 to 73 percent (Wojtanowicz 2008a,b,c). In some cases, further dewatering of the solids may be required before disposal, in which case advanced separation methods are used. One difficulty with using advanced technology for improved separations at a drill site is the high cost of equipment rental. The expenditure for this equipment can be easier to justify if a good economic model for their benefits is used.

Drilling Solid Waste Treatments

The most common method for the disposal of drilling wastes for onshore wells is in on-site reserves pits. The contents of reserves pits vary, depending on the drilling mud and the types of formations drilled. Reserves pits, however, can cause local environmental impact, particularly older pits that contain materials that are currently banned from such disposal or that were not constructed according to current regulations. The environmental impact of modern reserves pits is minimal.

Regulations for the design and monitoring of reserves pits during and after drilling can vary significantly with location. Unlined pits are most commonly used for freshwater mud systems, while pits lined with an impermeable barrier are used for salt- or oil-based mud systems. Following the completion of drilling of the well, the pits are eventually dewatered, covered with a few feet of soil, and abandoned. For offshore applications, steel tanks are used as reserves pits. The solids, after being separated from the mud, are typically discharged into the sea, where they settle to the bottom around the drilling rig. In some areas, however, regulations require that any waste mud and cuttings be transported to shore for disposal.

4.2 Petroleum Production Operations

The production of oil and gas generates a variety of wastes. The largest waste stream is produced water, with its associated constituents. Reis (1996) has given an excellent review of both production process and the wastes that are generated during production.

Production operations involve principally flowing oil and/or gas from a distributed system of wells spread over thousands of acres of land, or cluster of wells in offshore locations—both sea floors and various types of fixed platforms or floating structures. The produced fluids are transported to central processing plants, called gathering stations via field pipelines. Large oil fields also have a central tank farm that receives partially stabilized fluids from the satellite gathering stations. These gathering stations and the central farms are designed as plants to separate undesirable liquids (brine) and solids (sand, wax, sulfur, etc.)

Produced Water

The largest volume waste stream in the upstream petroleum industry is produced water. For mature oil fields, the volume of produced water can be several orders of magnitude greater than the volume of produced oil. The environmental impact of produced waters arises from its chemical composition. Produced water contains dissolved solids and hydrocarbons (dissolved and suspended), and is depleted in oxygen.

The most common dissolved solid is salt (sodium chloride). Salt concentration in produced water ranges between a few parts per thousand to hundreds of parts per thousand (ppt). For comparison, seawater contains 35 parts per thousand.

In addition to salt, many produced waters also contain high levels of calcium, magnesium, and potassium, with lower amounts of aluminum, antimony, arsenic, barium, boron, chromium, cobalt, copper, gold, iron, lead, magnesium, manganese, nickel, phosphorus, platinum, radon, radium, silicon, silver, sodium, strontium, tin, uranium, and vanadium (Reis 1996).

The current U.S. Environmental Protection Agency (US EPA) limits for the discharge of hydrocarbons in water for the "best available technology" are 29 mg/L on a monthly average and 42 mg/L for a daily maximum. Like all regulatory targets, these numbers are subject to change.

The concentrations of dissolved hydrocarbons in produced water depend on the solubility of the hydrocarbon. For discharges in the Gulf of Mexico, dissolved hydrocarbon concentrations for phenols, benzene, and toluene were found to be between 1,000 and 6,000 micrograms/L, while the concentrations of high-molecular-weight hydrocarbons was considerably lower (Wojtanowicz 2008a,b,c).

Produced water is invariably oxygen depleted. If discharged, oxygen-depleted water can impact fauna, requiring dissolved oxygen for respiration. Oxygen depletion can be a problem for discharge in shallow estuaries and canals, particularly if the produced water forms a layer along the bottom because of its higher density. This dense layer would be isolated from the atmosphere, limiting its contact with oxygen. Oxygen depletion is normally not a problem for discharge in deep water or in high-energy environments because of rapid dilution of the produced water in the surrounding environment.

Production Chemicals

Many extraneous chemicals enter the produced water stream for a variety of problems in oilfield operations. The most common problems are emulsions, corrosion, scale, microbial growth, suspended particles, foams, and dirty equipment. A variety of chemicals are often added to the water to avoid those problems.

Produced water often consists of an oil-in-water emulsion. Chemicals are commonly used to lower the electrostatic forces on the oil droplets to allow them to coalesce into larger droplets. Common chemicals used for this purpose include surfactants, alcohols, and fatty acids.

Produced water can be very corrosive to production equipment. Corrosion is caused primarily by the presence of dissolved oxygen, carbon dioxide, and/or hydrogen sulfide gases. Although produced water is initially oxygen depleted, oxygen can enter the produced fluid stream as a result of agitation during pumping or by atmospheric diffusion in holding tanks and surface impoundments. The oxygen content of water can be minimized by designing the system to exclude oxygen contact with the water.

Carbon dioxide and hydrogen sulfide can occur naturally in the formation and be produced with the water. Carbon dioxide forms carbonic acid, which lowers the pH and increases the corrosiveness of the water. Hydrogen sulfide corrosion can occur as a result of bacterial action on sulfates and is more often a surface or near-surface phenomenon.

Complex inorganic salts like *sodium chromate*, *sodium phosphate*, and *sodium nitrite* are also effective in slowing oxygen corrosion, particularly in high pH environments. Zinc salts of organic *phosphonic* acids and *sodium molybdate* have also been used for corrosion control. Zinc-based inhibitors are less toxic than chromates and should be used if possible. Organic anionic inhibitors, such as *sodium sulfonates* and *sodium phosphonates*, are also used in cooling waters and antifreeze. Current regulations may limit the use of some corrosion inhibitors. Hydrogen sulfide can be removed from produced fluids with a zinc scavenger. *Zinc carbonate* is widely used. This chemical reacts with hydrogen sulfide, producing insoluble zinc sulfide (Reis 1996).

For water injection systems, oxygen causes the largest problems with corrosion. Oxygen can be removed from water by stripping it with an inert gas, such as natural gas, steam, or flue gas, by vacuum deaeration, or by chemical treatment. Oxygen scavengers include sodium sulfite, *sodium bisulfite*, *ammonium bisulfate*, *sulfur dioxide*, *sodium hydrosulfite*, and *hydrazine* (Reis 1996).

Microbial growth in produced water can produce hydrogen sulfide gas by the chemical reduction of sulfates. In addition to causing corrosion, the presence of the bacteria themselves can affect production operations. To minimize these problems, biocides are often added to the produced water to inhibit microbial growth. Biocides used include *aldehydes*, quaternary ammonium salts, and amine acetate salts. Chlorine compounds are also used (Reis 1996).

Well Stimulation by Acidizing

Well stimulation is a widely practiced operation, which introduces thousands of gallons of chemical fluids from surface into the well. Most of the reactive chemicals are spent in situ, but the unspent portion and a host of other chemicals are flown back to surface after the operation. These chemical-laden fluids must be disposed of.

There are two most common forms of stimulation: acidizing and hydraulic fracturing. This section will explain acidizing; the next will discuss hydraulic fracturing.

Acids are used to dissolve acid-soluble materials in the rock matrix around the wellbore. These acid-soluble materials can include formation rocks and clays, as well as other chemicals that entered the rock matrix during drilling. A variety of inorganic and organic acids can be used, depending on the formation. These acids include *hydrochloric, formic, acetic, and hydrofluoric* acid. Other chemical ingredients are also added to optimize the process.

The most widely used acid is hydrochloric acid. Its main application is in low-permeability carbonate reservoirs. The major reaction products produced during acidizing are carbon dioxide, calcium chloride, and water. Spent acid returned from a well has high chloride content.

Hydrofluoric acid is used to stimulate wells in sandstone formations. It is normally used in a mixture of hydrochloric or formic acids, and is used primarily to dissolve clays and drilling fluids. The reaction products are various forms of fluorosilicates. Like hydrochloric acid, it is highly corrosive.

Formic acid is a weak organic acid that is used in mixtures during stimulation. Formic acid is commonly used as a preservative. It is relatively noncorrosive and can be used at temperatures as high as 400°F.

Acetic acid is used to dissolve carbonate materials, either separately or in combination with hydrochloric or formic acid. It is a slowly reacting acid that can penetrate deep into the formation and is useful for high-temperature applications. Reaction products are calcium, sodium, or aluminum acetates.

Commonly used additives include salts, alcohols, aromatic hydrocarbons, and other surfactants. Gelling agents such *xanthan gum* and hydroxyethyl cellulose, alcohols, acrylic polymers, aliphatic hydrocarbons, and amines, are also used. Retarders such as *alkyl sulfonates*, *alkyl amines*, or *alkyl phosphonates* are also used to reduce the reaction rate by forming hydrophobic films on carbonate surfaces.

During production, the spent acid returning to the surface may become emulsified with crude oil. These emulsions can be stabilized by the fines released during acidizing. To prevent such emulsions from forming, demulsifiers (surfactants) can be used. Common demulsifiers include organic amines, salts of quaternary amines, and *polyoxyethylated alkylphenols*. Glycol ether can be used as a mutual solvent for both spent acid and oil (Reis 1996).

Well Stimulation by Hydraulic Fracturing of Wells

During hydraulic fracturing, fluids are injected at a rate high enough so that the fluid pressure in the wellbore exceeds the tensile strength of the formation, rupturing the rock. The most commonly used base fluid for hydraulic fracturing is water. Various hydrocarbons can also be used as a base fluid, particularly where surface freezing may occur. Acid is also occasionally used when a combination of acidizing and hydraulic fracturing is desired. Liquefied gases, such as carbon dioxide or liquefied petroleum gas, can also be used, particularly to fracture gas wells. The use of a liquid base fluid in gas wells can reduce the gas production rate by lowering the gas-relative permeability.

After fracturing, the fluid pressure in the fracture drops when the well is placed back on production. To keep the fracture open during production, solids are injected with the base fluid to fill the fracture and prop it open. Materials used for proppants include sand, alumina pellets, and synthetic ceramic beads.

To lower the pump size requirement to fracture the rock, water-soluble polymers such as guar or xanthan gum, cellulose, or acrylics are used as viscosifiers. These polymers are frequently cross-linked with metal ions like boron, aluminum, titanium, antimony, or zirconium to further enhance their viscosity. To viscosify the oil-based fracture fluids, aluminum phosphate esters are commonly used. Surfactants are also occasionally used to create a liquid-air foam or oil-water emulsion to be used as the fracture fluid. To prevent degradation of many gels at high temperatures, stabilizers like methanol and sodium thiosulfate can be added.

Most polymers and cross-linkers need pH buffers: acetic, adipic, formic, or fumeric acids for low pH and sodium bicarbonate or sodium carbonate for high pH. For water-sensitive formations containing hydratable clays, clay stabilizing chemicals, such as sodium chloride, potassium chloride, calcium chloride, and ammonium chloride are used, which are returned to surface when the well is placed on production. But permanent stabilizers (such as quaternary amines, zirconium oxychloride, or hydroxyl-aluminum) bond to the clay surfaces to stabilize remain in the formation and are not removed with produced fluids.

The list of chemicals added is not closed yet. To suppress the pressure drop in the well tubing, high-molecular-weight polymers can be added as friction reducers to the fracture fluid. Alternatively, a cross-linking polymer that has a slow gelling time can be added. Bactericides such as glutaraldehyde, chlorophenates, quaternary amines, and iso-thiazoline are often added to protect these polymers from bacterial degradation. As if those chemicals are not enough, the process uses fluid-filtration control or diverter additives such as silica flower, granular salt, carbohydrates, and proteins for water-based fluids and organic particulates such as wax, pellets, or naphthalene granules.

After a fracture has been created, breakers are used to lower the gel viscosity so the fracture fluids can be easily removed from the fracture and not inhibit subsequent production. Common breakers for water-based fracture fluids are peroxydisulfates (Reis 1996).

Water Flooding and Enhanced Oil Recovery Operations
Most oil fields deplete in terms of pressure in a few years, after which field-wide water flooding and then chemical flooding (surfactants followed by polymer solutions) are used for many years in order to sustain commercial production rates and increase recovery. These operations generate wastes that greatly impact the environment.

These wastes include wastewater from cooling towers, water softening wastes, contaminated sediments, scrubber wastes, used filter media, various lubrication oils, and site construction wastes. Cooling towers are used for a variety of processes during oil and gas production. The cooling water used in these towers often contains chrome-based corrosion inhibitors and pentachlorophenol biocides.

In many areas, produced water is reinjected into the reservoir to assist hydro-carbon recovery. Unfortunately, the level of dissolved solids, particularly hard-ness ions (calcium and magnesium), is often too high to be used because they readily precipitate and can plug the formation. Thus, before produced water can be reinjected, it must be softened to exchange the hardness ions with softer ions, such as sodium.

The most common way to soften produced water is sulfonic acid, carboxylic acid. Strong acid resins can be regenerated simply by flushing with a concentrated solution of sodium chloride. Weak acid resins, however, must be regenerated by flushing with a strong acidlike hydrochloric or sulfuric and then neutralizing it with sodium hydroxide.

For producing heavy and viscous crude oils, thermal methods are used, either steam injection or direct in situ combustion by injecting hot air. In such cases, lease crude is burned, and the combustion gases may need to be scrubbed to remove pollutants like sulfur dioxide. One way to remove sulfur dioxide from combustion gases is to bubble it through aqueous solutions containing caustic chemicals like sodium hydroxide or sodium carbonate. Sulfur dioxide dissolves into water, forming sulfuric acid, which is neutralized by the caustic. Another form of scrubber uses various amines. In cold climates, such as Alaska's North Slope, methanol is used for freeze protection of equipment (Reis 1996).

4.3   Refinery and Petrochemical Industries

Various production chains that are based on the conversion of hydrocarbons into chemical products make up what is known as the petrochemical industry. This is one of the keystones of industry and technology from the twentieth and beginnings of the twenty-first centuries. The petrochemical industry has made possible the development of many products that today are considered normal and indispensable, such as computers, textiles, unbreakable toys, and a large quantity of other products that do not exist in nature and that did not exist before the mass use of oil. The belief that these oil-derived products are what assure an acceptable quality of life make it seem impossible to live without them.

The huge variety of end products of the petrochemical industry can be classi-fied into five main groups: plastics, synthetic fibers, synthetic rubber, detergents, and nitrogen fertilizers. The common name for plastics comes from its prop-erty of deformability in relation to plasticity (elasticity) under the influence of heat, pressure, or both. There are three important plastic families: thermo-plastics, thermo-resistant plastics, and polyurethanes. Thermoplastics constitute approximately 50 percent of the consumer plastic of the world. They include photographic films, plastic bags, pipes, furniture, construction material, toys, electronics, PVCs, valves, flowers, boots, and more.

The thermo-resistant plastics are used in electronics, decorative panels, and domestic utensils, for example. Polyurethane plastics are products with transpar-ent glass appearance or extra-light foams. Synthetic fibers include polyamides for fine lingerie, carpets, curtains, and swimsuits. Polyesters are used in suits, ties, water-resistant clothing, and carpets. Acrylic fibers substitute for wool. Synthetic rubber is the principal supplier of the automobile industry, since it is the fundamental element of tires. It is also used in some of its varieties in shoes and materials for terraces and roofs. Detergents are products soluble in water whose property is the ability to modify liquid surface tension, reducing or eliminating contained dirt. Its main uses are in the home in the form of powder, or liquids.

Industrial fertilizers for agriculture include sulfuric acid. The phosphates and synthesized ammoniums have placed in circulation a variety of chemical fertilizers. Via the petrochemical industry, the supply of hydrogen at a low cost has promoted the mass use of ammonium products as nitrogen that can be assimilated in its three variants: nitrates, sulfates and urea, and the infinite number of complex fertilizers.

However, the petrochemical industry has also created a great variety of agro-toxins (e.g., herbicides, fungicides, insecticides, etc.), which generate a large quantity of contaminants. The products themselves—as opposed to natural products—are not biodegradable. In addition, the secondary products involved in the production of the agro-toxins create contaminant byproducts as well. The local populations that live with the area of influence of petrochemical plants face serious health problems, due to the presence of the contaminants generated by the industry.

Among the contaminants typical of the industry are the *polyaromatic hydrocarbons* (PAH), considered as the most toxic hydrocarbons, together with the *monoaromatics*. Once PAHs are liberated into the aquatic environment, the degradation via microorganisms is often very slow, which leads to its accumulation in sediments, soils, aquatic and land plants, fish, and invertebrates. PAHs affect human health as well: Individuals develop cancer if exposed to a mixture of these components via inhalation or touch for prolonged periods of time. The *alkalibenzines* are very resistant to degradation and can accumulate in sediments. In toxic terms, the acute exposure to these products can cause depression to the central nervous system, leading to alterations in speech. The heavy metals (which include lead, mercury, zinc and copper) all are toxic to humans as to wild life (Reis 1996; Wojtanowicz 2008a,b,c).

Oil is therefore more than just energy. Via the petrochemical industry, five million different products can be obtained. It transformed the twentieth century. It also made us a civilization dependent on oil and on the transnationals that control oil exploitation and the petrochemical industry (RCEP 2000).

5 ENVIRONMENTAL COMPLIANCE

Historically, environmental compliance of the oil and gas industry (also referred to as hydrocarbon or petroleum industry) has evolved from one of regulation to self-motivation. Initial awareness to petroleum-related environmental pollution was slow, but concern for oil patch workers' safety and the need for conservation resulted in improved methods. Although the petroleum industry has often reacted to new regulations by changing operational practices the minimum amount required to meet the letter of the regulations, with time, several factors have led to environmentally responsible, cost-effective practices:

- Complex and rapidly changing regulatory environment
- Mutual education between regulators and the petroleum industry
- Increased public awareness of environmental concerns

Scenes like that shown in Figure 13 were common in oil fields and refineries until a few decades ago, when government environmental agencies around the developed countries put stringent laws in place. Despite regulations and improved technology in all areas of the petroleum industry, major disasters still occur due to human errors (e.g., the *Exxon Valdez* oil tanker accident in Alaska, 1989), natural disaster (e.g., Shell's oil spill at Kholmsk, Sakhalin Island, in 2004; Figure 14), and war (e.g., the Kuwait–Iraq War, 1991).

Five drivers behind environmental awareness and compliance are as follows:

1. *Conservation of oil.* The Oil Conservation Board in the United States was formed in December 19, 1924, under the Department of the Interior, to investigate the oil industry, recommend appropriate remedial action,

Figure 13 *Environmental consciousness?* A scene from the past, *before* Petroleum Engineering was born (circa 1935). Photo shows Standard Oil Refinery No. 1 in Cleveland, Ohio, 1899 (Standard Oil Company 2009)

Figure 14   Oil spills rarely occur, but when they do, environment takes a heavy toll. Photo shows Shell's Oil Spill at Kholmsk, Sakhalin Island (ECA Watch)

and cooperate with industry and state agencies in promoting resource conservation (Oil Watch).

2. *Health, safety, and environment (HSE).* Such practices aimed at improving technology at various phases of petroleum—exploration, production and distribution are motivated partly by operating companies' concern for their bottom line and for public awareness leading to government regulations. A shift in about 1990 to downhole oil-water separation and in situ underground disposal of brine is a prime example of oil field operating cost control that also pays a rich dividend to the environment (Wojtanowicz 2008a,b,c).

3. *Superior technology.* The Industrial Revolution has set in a perpetual penchant for improving efficiency and smaller footprints in everything manmade.

4. *Standard of living.* Man's quest for a higher standard of living has led to modifying dirty fossil fuels into cleaner forms (e.g., coal-to-liquid, underground coal gasification, reducing underground oil to cleaner gases by in situ microbial reduction, and use of solar panels to operate oil field instruments).

5. *Global warming and climate control.* The unusual rise of global atmospheric temperature over land masses since 1990 is attributed to, among other things, *anthropogenic* CO_2 released by fossil fuel combustion. The twentieth century's last two decades were the hottest in 400 years and

possibly the warmest for several millennia, according to a number of climate studies. And the United Nations' Intergovernmental Panel on Climate Change (IPCC) reports that 11 of the past 12 years are among the dozen warmest since 1850 (National Geographic 2007).

This partnership requires cooperation, teamwork, commitment, credibility, and trust among all parties involved in the exploration for and production of oil, including operating company managers, engineers, geologists, contractors, subcontractors, work crews, regulators, courts, and legislators.

Companies conforming to standards set by the American Petroleum Institute (API) have formally adopted a set of principles toward ensure environmental responsibility. These principles are known as the *Guiding Principles for Environmentally Responsible Petroleum Operations* (American Petroleum Institute 2009).

6   ENVIRONMENTAL REGULATIONS ON FLUID AND SOLIDS DISCHARGE

Two types of muds that are normally used in drilling operation are water-based muds (WBMs) and oil-based muds (OBMs). WBMs are by far the most commonly used muds, both onshore and offshore. WBMs are widely used in shallow wells and often in shallower portions of deeper wells, but are not effective in deeper wells. The uses of WBMs generate 7,000 to 13,000 bbl of waste per well. Depending on the depth and diameter of the well, about 1,400 to 2,800 bbl of that amount are drill cuttings (McMordie 1980). WBMs use water as their base fluid and do not contain any oil. WBMs are very economical and easy to dispose of because they can be fully biodegraded and are considered as having very low toxicity. In many countries, both WBMs and cuttings are discharged on site to the ocean.

During the past 30 years, OBMs have been developed and refined to overcome the limitation of WBMs applications. OBMs have been the mud of choice for a range of special situations, including high temperatures, hydratable shales, high angle and extended-reach well, high-density mud, and drilling trough to salt. Wells drilled with OBMs normally produced lower waste volume than those drilled with WBMs because very little slumping or caving in of the walls of the hole occurs. Also, the mud is reconditioned and reused rather than discharged at the end of the well. Only the drill cuttings will be expelled into the ocean. The average volume of OBM waste is estimated at 2,000 to 8,000 bbl per well. The base fluids of OBM are normally either diesel or mineral oil, even though nowadays many other types of low-toxicity oil are developed. Because they contain oil, OBMs waste cannot be discharged on site under the regulation of many countries (McMordie 1980).

Frac-packing has emerged as a preferred method of completing offshore wells in unconsolidated formations such as those prevalent in the Gulf of Mexico.

Fracturing fluids are a complex mixture of polymers and additives that must be balanced to achieve such properties as high viscosity, proper formation compatibility, appropriate breakers, and proper concentration to achieve the desired break times for cleanup. Some jobs may require up to a dozen additives.

The *frac-pack* operation manufactures and pumps large volumes of stimulation fluids from the fracturing stimulation vessel into the well. Most operations require at least two work string volumes of fracturing fluid to be reversed out and safely disposed. Greater than 3 miles (4.8 km) from shore, discharging the fluids offshore minimizes costs, risks, and logistics and deck space requirements. To qualify as offshore-disposable, the fluids must meet stringent criteria. No discharge of free oil means that waste streams cannot be discharged when they would cause a film or sheen on or discoloration of the surface of the receiving water.

In recent years, the development of completion fluids for use in offshore wells has taken on a new component—namely, that of satisfying new and pending environmental requirements. The fluid system must have a *mysid shrimp toxicology* (MST) of more than 30,000 ppm, and biodegradability of minimum 60 percent in 28 days, according to the Organization of Economic Cooperation and Development (OECD) (Baycroft 2005; Baycroft et al. 2005).

The new fluid includes chemical technology that meets all environmental goals, improves overall fluid performance, and maintains pricing levels within 5 percent of older fluids. *Offshore disposability is a significant advantage that reduces operational costs*. The new fluid system has a hexane extractable materials (HEM) value for oil and grease that averages less than 10 mg/l, along with a MST of over 30,000 ppm, which meets the toxicity standard for water-based drilling mud. Additionally, biodegradability meets or exceeds the OECD requirement (Baycroft 2005; Baycroft et al. 2005).

Various governments have set environmental compliances on drilling and well completion fluids and solids discharge in seawater. For example, the Indonesian government requires for both WBMs and OBMs 96 h LC_{50} at 30,000 ppm SPP (suspended particulate phase) to shrimp (Soegianto et al. 2008). The LC_{50} is a standard test (set by the US EPA: 96-hour Acute Static LC_{50}. EPA Method 58 FR 34592, *Mysidopsis bahia*) to determine the concentration of the substance which will prove lethal to 50 percent of a test population of the marine organism in 96 hours (Burke and Veil 1995).

The chapter appendixes show major U.S. environmental regulations pertaining to oilfield operations.

7 MAJOR ACHIEVEMENTS IN ENVIRONMENTALLY CONSCIOUS PETROLEUM ENGINEERING

The petroleum industry has developed various equipment, technologies, and operational procedures since the time of Colonel Drake, albeit at a slow pace until the 1950s, to cope with various hazardous situations and problems. With growing public awareness for pollutions, of which oil spills are most conspicuous, and consequent government regulations, the oil industry has eventually taken a proactive role in remedying and mitigating environmental impacts and in handling inadvertent spills when they occur. From basic concerns for safety of man and equipment in the petroleum chain, the industry has evolved to take environmental safety as a stimulus to its bottom line.

7.1 Barriers in Drilling

During drilling, a major source of surface contamination occurs when a *well blowout* occurs. A blowout is an uncontrolled flow of oil and/or gas that happens if the drilling fluid fails to contain the high pressure encountered during drilling through a known or unknown oil- or gas-bearing rock formation.

In order to prevent blowout, two barriers are used: a heavy drilling fluid, where heavy minerals like barium sulfate are used, and a high-pressure control device called the blowout preventer (BOP), which is installed at the surface between the surface casing and the rig floor.

7.2 Barriers in Production/Injection (Cementing)

The problem of hydraulic integrity of well annular seals has been addressed through both regulatory and technological measures. The two areas of regulatory initiatives to control annular integrity are *drilling permit regulations* and *injection permit regulations*. Drilling regulations focus mostly on the integrity of the surface casing. Typically, drilling permits require the surface pipe to be entirely cemented to protect freshwater sands from oil and gas zones (see Figure 15). In addition, typical drilling regulations may specify minimum footage for surface pipe, minimum *waiting-on-cement* (WOC) time, minimum volume of cement slurry to be used, minimum length of cement sheath above the top producing zone and at the salt–fresh groundwater interfaces, and the minimum testing requirements after completion (pressure test or *cement-bond log* tests (CBL)).

Currently, no quantitative requirements exist to verify a potential annular flow between well casing and formations. For production casing, drilling permits are not very specific about the verification of annular integrity, even though this integrity is most important in effectively isolating upper zones from produced hydrocarbons and brines.

Subsurface injection permits require an operator to provide evidence of the hydrodynamic integrity of the well's annular seal. However, no direct standardized tests for such integrity exist (Wojtanowicz 2008a,b,c). Usually, permit decisions are based on indirect evidence of the well's integrity, such as CBL, electric logs, the driller's log, and geological cross-plots, which indicate to the regulatory agency that no unusual environmental risk is involved. Typical generic criteria for wells injecting oilfield brines address three issues:

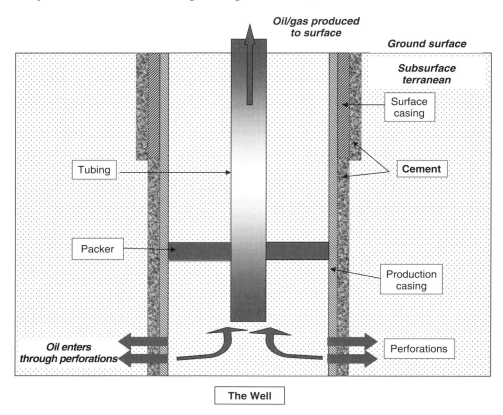

Figure 15 Cemented casing string for protecting the production and injection well as the primary barrier to zonal communication

1. Length of casing
2. Mechanical integrity (pressure) test procedure (wellhead pressure, test duration, maximum pressure drop) and its frequency (usually before the operation, then every five years)
3. Minimum distance to any abandoned well (usually 0.4–0.8 km).

A permit is also required for the annular injection of solid drilling waste, the common method of on-site disposal during drilling operations.

In the area of subsurface brine injection, the permit issue revolves around reliable techniques to prevent the stream of brine from migrating freely into the environment. The three main criteria are the *internal mechanical integrity* of the borehole installation (IMI), the *external* integrity of annular seals (EMI), and the integrity of the confining layer. The IMI practices of pressure testing casing as well as monitoring the annular pressure during injection are the most typical field technologies (Wojtanowicz 2008a,b,c).

7.3   Cuttings Reinjection

Subsurface disposal of solid waste has evolved from downhole injection of solids-free liquids combined with the well stimulation technique of hydraulic fracturing to the new technology of subsurface injection of slurrified solids (fluids having various concentrations of solids, from less than 1 percent to over 20 percent by volume). Slurry injection operations are batch processed, where drill cuttings are mixed with waste mud and water in the mixing/processing tanks, sent to a holding tank, and then injected downhole (Wojtanowicz 2008a,b c).

In the early 1980s, high-permeability annular injection of small volumes of drill cuttings became an environmentally sound alternative for on-site disposal of drilling waste, particularly in the Gulf Coast area. Later, slurry fracture injection technology was developed for disposal of drill cuttings from oil-based muds in Alaska and the North Sea, and for NORM (naturally occurring radioactive materials) disposal. In the mid-1990s, the first large commercial facility with dedicated injection wells began operation. This was followed by large-scale injection operations in Alaska and Gulf of Mexico (Wojtanowicz 2008a,b,c).

At present, annular injection is available for routine use offshore, with several different service companies providing a range of operations and engineering support. An example of continuing evolution of the technology was documented in a study on commingled drill cuttings and produced water injection. Also, slurry fracture injection has been used for disposal of oilfield wastes other than drilling mud and cuttings, such as produced sand, sediment from tank bottoms, unset cement, and unused fracture sand. However, the most common sources of waste injected are from ongoing drilling operations and from mud and cuttings stockpiled in tanks or stored in earthen pits (Wojtanowicz 2008a,b,c).

7.4 CO_2 Sequestration

The oil industry has use CO_2 as a means for enhanced oil recovery since the 1950s. Initially CO_2 was injected into the oil reservoir with water in dissolved form. In the 1960s the CO_2 injection became more popular when more oil could be recovered by achieving CO_2 miscibility with oil under high pressure. At high pressures a far greater supply of CO_2 is required, thereby creating a better CO_2 sink.

The oil industry experience with bulk CO_2 injection proved beneficial to the world when awareness of global warming and consequent demand for CO_2 sequestration in suitable underground geological structures grew in the late 1980s. For underground storage, both oil and gas reservoirs and aquifers are potential storage sites.

Oil and gas reservoirs have proven their storage capacities for hydrocarbons over millions of years, and are thus considered as safe storage sites for CO_2. In recent years, millions of tonnes of CO_2 have been injected into oil reservoirs for profitable EOR, and no serious leaks to the atmosphere have been reported.

This widespread use of CO_2 injection in the oil industry, where CO_2 often is transported over land hundreds of miles in pipes, also proves that CO_2 injection is a method that has been accepted as safe by society. Currently, more than 400 million tonnes CO_2 per year are injected (Holt et al. 2000).

The theoretical storage capacity of CO_2 in petroleum reservoirs may be estimated from the carbon density of hydrocarbons and CO_2. This is summarized in Table 3, where average densities of the reservoir fluids are estimated. From the table it is seen that almost one third of the CO_2 formed by combustion of oil can be stored in the oil reservoir volumes. Correspondingly, more than all the CO_2 formed by combustion of gas can be stored in the gas reservoirs. The world's petroleum resources are estimated at 141×10^{12} kg oil and 145×10^{12} Sm^3 (standard cubic meter) gas, respectively. By assuming some global average values regarding temperature, pressure, and formation factor for the world's petroleum reservoirs, the total storage capacity for CO_2 is estimated to 520×10^{12} kg. This mass of CO_2 corresponds to 68 percent of the CO_2 formed by combustion of the petroleum reserves (Holt et al. 2000).

CO_2 has for a long time been injected into oil reservoirs. Through a large number of projects, CO_2-injection has proven its effectiveness as an injection fluid for enhanced oil recovery. In a review of the performance of 25 CO_2 injection projects, pilots and field scale, the average incremental oil recovery obtained was 13 percent of original oil in place. Of 60 miscible CO_2 projects in North America, all were evaluated by the operators as successful, promising, or "too early to tell." The possible worldwide use of CO_2 for improved oil recovery purposes is summarized in Table 4, based on screening criteria mainly related to oil gravity and reservoir depth. Figure 16 shows that the deposition costs include sequestering from a gas turbine flue gas, compression, pipeline transport, and injection. The lowest cost estimate is based on separation and compression costs, with a 4.2 USD/ton addition for 500 km offshore pipeline transport and an equal addition for injection costs (Holt et al. 2000).

From Figure 15, we can see that the value of CO_2 can—in the best case, and for realistic high oil prices—only cover parts of the deposition costs. However, the use of CO_2 for oil recovery is the only large-scale use of CO_2 that offers any value at all. The first CO_2 deposition projects should therefore concentrate on the use of CO_2 in connection with oil production. CO_2 injection into depleted gas

Table 3   Carbon Density of Various Fossil Fuels, as Indicator of CO_2 Sequestration Volumes (Holt et al. 2000)

Fluid	Density (kg/Sm^3)	Mass Fraction of C	Carbon Density (kg/Sm^3)
Oil (CH_2)	704	12/14	604
Gas (C1.2H4.4)	178	14.4/18.8	136
CO_2	689	12/44	188

Table 4 Estimated Worldwide Potential Oil Recovery and CO_2 Utilization (Holt et al. 2000)

Region	Potential Oil		Potential CO_2 Utilization	
	10(9)m(3)	10(9) Tonnes	10(9)m(3)	10(9) Tonnes
Middle East	22	24	57	39
Western Hemisphere	4.6	5	15	10
Africa	2.1	2	5.8	8
East Europe and CIS	1.7	2	4.4	3
Asia Pacific	1.4	1	4.4	3
Western Europe	0.6	1	1.5	1
World Total	33	35	89	61

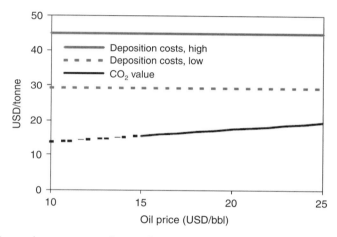

Figure 16 Separation, transportation, and storage costs for CO_2, and the value of CO_2 for improved oil recovery versus oil price (Holt et al. 2000)

reservoirs will have only marginal or no effect on the economy of the exploitation project. Since the gas reservoirs have proven their ability for long-term gas storage, the use of depleted gas reservoirs should be considered, especially in cases where infrastructure from the gas production still could be used.

In the calculation of recovery potential, separate recovery factors were used for miscible and immiscible projects. Table 4 contains both the potential incremental oil due to CO_2 injection and the CO_2 consumption. It is seen that more CO_2 is injected than the corresponding quantities of oil produced—apparently, on average, 2.7 volumes of CO_2 per volume of oil produced. This factor is, however, reduced to approximately 2.0 when the formation volume factor of oil is taken into account. Of this, one volume displaces the oil produced. The rest of the CO_2 may displace mobile water in the reservoir, or go into other parts of the formation (Holt et al. 2000).

7.5 Decommissioning Offshore Platforms

This is relatively new problem encountered in the petroleum industry. The offshore oil and gas industry had its beginnings in the Gulf of Mexico in the late 1940s. Now there are more than 7,000 drilling and production platforms located on the continental shelves of 53 countries. Some of these structures have been installed in areas of deep water and treacherous climates, and consequently, structure designs have adapted to withstand the environmental conditions of these areas. Some typical designs are shown in Figure 17. In the North Sea, which is an area that experiences some extreme environmental conditions, more than 200 structures have been installed. This combination of deep waters and extreme storm forces dictates large structures, some with component weights that exceed 50,000 tonnes (Day 2008).

Now, as oil and gas fields begin to deplete their reserves, the concern has turned to the removal and disposal of these structures at the end of their producing lives. A primary objective during the decommissioning is to protect the marine environment and the ecosystem by proper collection, control, transport, and disposal of various waste streams. Decommissioning is a dangerous phase

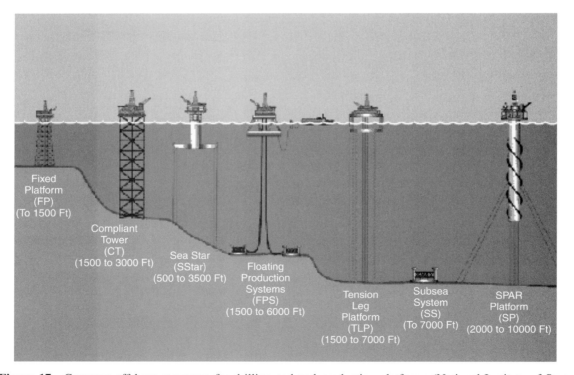

Figure 17 Common offshore structures for drilling pad and production platforms (National Institute of Standards and Technology, U.S. Minerals Management Service, Gulf of Mexico Region, Offshore Information, October 1999: http://www.atp.nist.gov/eao/grc04-863/chapt4.htm)

of the abandonment operation and creates the possibility of environmental pollution. Platform decommissioning will result in large amounts of waste liquids and solids. Many platforms will have chemical treatment additives, as well as possible toxic/hazardous materials such as methanol, biocides, antifoams, oxygen scavengers, corrosion inhibitors, paints, and solvents, some of which may cause damage to the marine environment if accidentally discharged (Day 2008).

Estimates indicate that the cost of some removals may exceed the cost of the original installation. Innovative removal and disposal techniques must be developed to limit costs and minimize the impact on the environment. International law provides the basic foundation of the legal requirements for the removal and disposal of offshore structures. The removal of installations was addressed by the 1958 *Geneva Convention on the Continental Shelf*, which stated that any installations that are abandoned or disused must be entirely removed. However, several parties to the Convention were soon adopting some form of local standards to allow for partial or nonremoval. The more widely accepted statement of international law is contained in the *United Nations Convention on the Law of the Sea* (UNCLOS), which allows for partial removal and has been widely accepted as it appears to represent customary international law in relation to abandonment. The *International Maritime Organization* (IMO) guidelines were issued using UNCLOS as a basis. These guidelines state that if the structure exists in less than 75 m of water and weighs less than 4,000 tonnes, it must be totally removed. Structures installed after January 1988 will have a water depth criterion of 100 m, forcing the owner to plan for the eventual abandonment in the initial design. If the removal is done partially, the installation must maintain a 55 m clear water column. There are exceptions in the guideline that allow for nonremoval, such as if the structure can serve a new use, including enhancement of a living resource; if the structure can be left without causing undue interference with other uses of the sea; or where removal is technically not feasible or an unacceptable risk to the environment or personnel. If the installation is to remain in place, it must be adequately maintained to prevent structural failure (Day 2008).

In general, the entire abandonment process of offshore platforms consists of seven steps:

1. *Well abandonment.* This involves the permanent plugging and abandonment of nonproductive well bores.
2. *Acquiring environmental permits.*
3. *Engineering.* An abandonment plan must be developed.
4. *Decommissioning.* All process equipment and facilities are shut down. Waste streams and associated activities are removed to ready the platform for a safe and environmentally sound demolition.
5. *Structure removal.* The deck or floating production facility is removed from the site, followed by removal of the jacket, bottom tether structures, or gravity base.

6. *Disposal.* This phase involves the disposal, recycle, or reuse of platform components onshore or offshore.

7. *Site clearance.* This includes final cleanup of sea-floor debris.

Disposal Methods

Once a platform or portions of a platform have been removed, the structure must be disposed of. Some disposal options include the following:

- *Onshore transport.* Generally, topside deck facilities will be disposed of inshore because of the difficulty and expense in completely removing all of the hydrocarbons and their byproducts at the installation site rather than shore side. When disposal inshore is chosen, the structural component will be either totally or partially cut up for scrap.

- *Toppling in place, or emplacement.* This method is generally performed after the topsides have been removed. The legs are severed selectively so that the jacket can be toppled with two legs acting as hinges. The toppled structure must maintain 55 m of clear water column clearance as required by the International Maritime Organization (IMO) guidelines. Emplacement is much the same procedure as toppling except that the top section is completely cut from the lower section, lifted off and placed next to the lower section.

- *Disposal to reef site.* It has been estimated by the Gulf of Mexico Fishery Management Council that oil and gas structures account for 23 percent of the hard bottom habitat in that area (Day 2008). Therefore, prior to the emplacement of petroleum-related structures, suitable habitats in which new species could expand their range did not exist. Countries may establish a rigs-to-reef program to maintain the hard bottom habitats that these structures provide.

- *Deep-water dumping.* This method is particularly reserved for huge floating systems located in the North Sea. Essentially, the structure is disconnected from its moorings and towed to the deep ocean waters, where it is then flooded and sunk. Prior to any dumping operations, all components placed in the ocean waters must be free of hydrocarbons in harmful quantities to avoid pollution of the open sea.

The final phase of the abandonment process involves restoration of the site to its original predevelopment conditions by clearing the seafloor of debris and obstructions after platform removal, leaving the site trawlable (in shallow areas) and safe for fishing or other maritime uses. Deeper water sites may not require trawling simulations to clear the area.

The successful abandonment and disposal requires awareness of the social and political climate in the region, since public perception will play a key role in performing a successful disposal program. This necessitates that the environmental issues should be addressed by the operator up front, keeping in the loop all stakeholder groups and regulatory agencies with transparent disposal plans and environmental effects of the plan and alternative plans (Day 2008).

8 TRENDS OF ENVIRONMENTAL IMPACT IN THE E&P INDUSTRY

For more than 100 years, oilfield science and technology have been continually improving. The oil industry has evolved from one that was interested mainly in inventing tools and equipment to one that is not only economically, but also environmentally, conscious. In the 1980s, low oil prices forced oilfield technology to focus on economic efficiency and productivity. Simultaneously, environmental regulatory pressure added a new factor to petroleum engineering economics: the cost of working within the constraints of an environmental issue. Initial E&P industry activities for environmental protection were motivated by gradually tightening environmental regulations, being felt a burden to corporate profitability. There has been a paradigm shift, however, in that environmental performance is now considered by many oil operators as a potentially important contributor to the bottom line.

8.1 Progress in Environmental Compliance

In the 1990s, the industry has absorbed this cost and made a considerable progress in pollution control. In this short span of time, environmental management has also shifted gears from *end-of-pipe* pollution control toward pollution prevention. A great stride in technological explosions has caused this transformed in environmental management in the oil industry. Progress in oilfield technology, including modifications and improvements has been directed, of course, primarily toward productivity enhancement. Improved environmental performances have spun off in the process. This shift of focus can be judged from three facts, which should be taken in the backdrop of sevenfold increase in the average new discovery of oil and gas reserves compared to that in the late 1980s (Wojtanowicz 2008a,b,c; EIA 2004):

1. Since the early 1990s, emissions of air toxics decreased by almost 24 percent.
2. The rate of annual wetland losses decreased from almost 500,000 acres per year three decades ago to less than 100,000 acres per year, on average, since 1986.
3. Between 1991 and 1997, volumes of the 17 most toxic chemicals in hazardous waste fell 44 percent.

The technological advances have led to reduced environmental impacts by these three mechanisms (Wojtanowicz 2008a,b,c):

1. *Drilling fewer wells to add the same reserves.* Currently, the U.S. E&P industry adds two to four times as much oil and gas to the domestic reserve base per well than in the 1980s.
2. *Generating lower drilling waste volumes.* Currently, the same level of reserve additions is achieved with only 35 percent of the generated waste.
3. *Leaving smaller footprints.* Currently, the average well site footprint is 30 percent of the size it was in 1970, because through the use of *extended reach horizontal drilling*, an average well can now contact over 60 times more subsurface area, thereby requiring far less number of drilling sites.

Etkin (2001) examined the trends in U.S. oil spillage with respect to historical and current trends in the United States and worldwide and analyzed potential influences on spill frequencies. Contrary to popular perceptions, the number of oil spills, as well as the amount spilled, has decreased significantly over the last two decades, particularly in the last few years despite overall increases in oil transport. Decreases are pronounced for vessels. U.S. pipelines now spill considerably more than tankers. Overall, U.S. oil spillage decreased 228 percent since the 1970s and 154 percent since the 1980s. These trends are shown in Figures 18 and 19. This decrease mirrors international trends and is likely responsible for reduced accident rates due to preventive measures and increased concerns over escalating financial liabilities.

The decrease in oil spills worldwide, but in the United States in particular, may be attributed to a variety of influences. The impacts and repercussions of

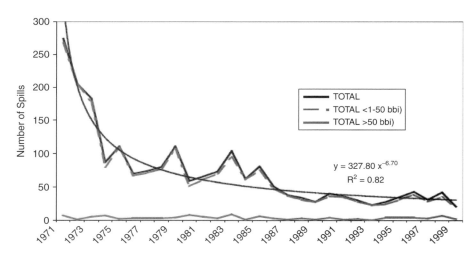

Figure 18   Annual number of oil spills from U.S. OCS exploration and production facilities: 1971–1999 (Etkin 2001)

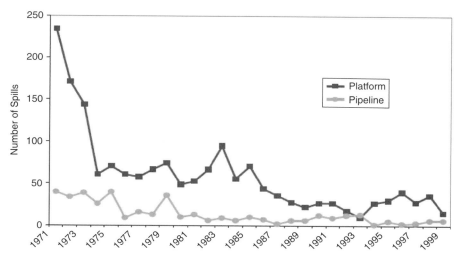

Figure 19   Annual number of U.S. OCS platform and pipeline spills: 1971–1999 (Etkin 2001)

the 1989 *Exxon Valdez* spill, largely the impetus for passage of the Oil Pollution of 1990 (OPA 90), were watched worldwide. With OPA 90, the United States has the strictest tanker regulations. Improved safety standards, contingency planning, exercise programs, and other measures have helped reduce U.S. spillage. International conventions and national legislation have reduced worldwide oil spillage.

Much of this reduction, however, was realized before the implementation of OPA 90 and other regulations and conventions. Another important influence on spillage rates has been realization by tanker owners and other potential spillers that spills in the United States could result in astronomical costs for which the spiller could have unlimited liability (a result of OPA 90). Cleanup, penalty, and damage costs associated with the *Exxon Valdez* spill and other significant events in its aftermath have seriously raised the stakes for potential oil spillers. Cleanup costs have risen dramatically in the last two decades even when the price tag is adjusted for inflation. Implementation of pending increases in environmental damage liability nationally or as part of international liability and compensation conventions, along with increases in oil spill damage liability and penalties for all spill sources, could change the oil spill picture on a worldwide basis.

Although the statistics show encouraging downward trends, there is no room for complacency. An ill-timed oil spill that occurs in a sensitive location, regardless of spill size, can cause devastating damage to natural environments, property, and business, and, occasionally, to human lives. Aging pipeline and facility infrastructures, as well as aging vessel fleets, may be "ticking time bombs," especially as they become subjected to increasing oil throughput and transport in future years (Etkin 2001).

There is no denying that environmental performance is an important factor impacting corporate image. Petroleum industry is particularly vulnerable to public image because, on one hand, it must seek public approval for accessing geographical areas and developing natural reserves, while on the other hand, its image can be easily damaged by highly visible accidents of oil spills or well blowouts. Moreover, the investment community now considers a company's environmental performance in corporate assessments. The petroleum industry is expected to perform concurrently in three areas: productivity, environment, and society. This *triple bottom line* concept operates on the principle that better performance of one of the three pillars—representing economic, environmental, and social considerations—cannot be considered substitutable for underperformance in another (Whitaker 1999). Therefore, a successful technological progress must address a technology that combines productivity advantage with environmental protection and, as such, makes the operator accountable to the public.

8.2   Integration of Environmental Impact Assessment and Environmental Management

Environmental impact assessment (EIA) for upstream oil and gas projects has a history that extends for at least 30 years. Initially, they were only required in a handful of highly regulated environments. Now, it is a rarity to find a major oil and gas project anywhere in the world that is not required, either by legislation or corporate standards, to undertake an EIA. They are now almost universally expected not as a part of doing business, but as an integral part of doing business the right way. EIA practices including methodologies, interpretation of data, and determination of levels of significance are well developed. Mitigation measures are standardized in many circumstances, and comparisons across companies, industries, and countries are relatively easy.

9   CALIFORNIA: A SHOWCASE FOR ENVIRONMENTLY CONSCIOUS PETROLEUM ENGINEERING

The state of California is a showcase of environmentally conscious petroleum engineering, where both public and government awareness and control vis-à-vis E&P industry initiatives have created many examples of minimizing environmental impacts to acceptable limits.

9.1   California Department of Conservation

The California Department of Conservation started as California Division of Oil and Gas in 1915 in the early days of oil for the purpose of supervising the drilling, operation, maintenance, and abandonment of wells to prevent waste or damage to oil and gas deposits. These duties were expanded in 1931 to include well spacing regulation and freshwater protection. But it took a long time before more serious attention to environment was given. In 1970, the Public Awareness Code was amended to include the responsibility of preventing "*damage to life, health, property.... By reason of the drilling operation, maintenance or abandonment of wells*." The Department issued written guidelines for oil operators regarding such environmental problems as oil field sump construction and placement, waste fluid transport, tank settings, waste disposal, production leakage, air pollution, hazardous operations, and scenic values (California Department of Conservation 2007).

The first instance of oil operator initiatives to protect the environment can be traced back to 1932, when several oil companies formed the *Valley Waste Disposal, Inc.* in Kern Country. The nonprofit corporation was responsible for disposing of a portion of oil field waste fluids. Many oil companies undertook such activities as fencing or filling sumps that were located in natural drainage channels or those that have shown wildlife loss. But a government report published in 1973 shows the gap between what have been done and what is expected (California Department of Conservation 2007):

> Over the last twenty years and primarily since 1969, the oil industry has often been the focus of considerable public criticism regarding its treatment of the environment, but little constructive advice has been offered to help alleviate and solve problems. .... After 80 years of indiscriminate drilling and production techniques, the oil industry needs to work more with its surroundings, to act as a habitat and wildlife preserver, and to properly manage all the resources over which it has purview.

9.2 Long Beach, California, Production Operations

As already shown, the 1970s is the decade for the surge of public consciousness for environment, heralded by William H. Curry (president, American Association for Petroleum Geologists) early in 1971: "Environment, Ecology, Pollution—these are called the bywords of today!" (Allen 1972).

Unlike other industries, the unique problem with the petroleum upstream industry is that petroleum is where you find it, and the drilling and producing operations cannot be moved to an area zoned for heavy industries. Yet the city of Long Beach, California, created an environmental showcase in cooperation with a consortium of oil companies (THUMS Long Beach Co., which stand for five oil companies, namely, Texaco, Humble, Union, Mobil, and Shell) in the 1970s in order to control a host of environmental problems related to oil drilling and production (Allen, 1972). The company operates the eastern offshore section of California's Wilmington oil field (which lies beneath Long Beach Harbor), having produced 930 million barrels of oil equivalent from the Wilmington field, with an estimated 100 million barrels still to be recovered (California Department of Conservation 2007; Hoovers 2009).

One of the earliest environmental problems tackled was *land subsidence* caused by oil withdrawal from subterranean geological structures. Land subsidence is a major problem because the center of the Wilmington oil field is located under the highly industrialized Port of Long Beach. Oil production since 1890 caused over four feet of subsidence by 1945, which caused inundation and damage to buildings in many city areas. Numerous studies associated land subsidence with subterranean voidance created by oil withdrawal, and therefore, water injection into the subterranean rock was recommended on a unitized basis by the operating oil companies. By 1960, large-scale water injection was underway and land subsidence was almost immediately halted in the areas of injection. With minor exceptions, subsidence has been halted throughout Wilmington field (Allen 1972).

Drilling and production facility development in Wilmington oil field were confined by city charter to the industrialized zones. In 1947, a portion of the undeveloped oil field was leased for development by a new drilling technology—*directional drilling*, from sites on the pier facing the civic center (see Figures 20 and 21). These wells were the first in the oilfield to be completed using hydraulic pumping units recessed in well cellars that left an unobstructed view of the city from the harbor area.

Air and water pollution have been reduced in response to county and state regulations. Open-pit brine skimming systems were developed as a result of regulations on the disposal of oil-contaminated water in the harbor. The open-pit skimmers were later replaced with closed systems in order to eliminate undesirable odors and vapor emissions. Initially, the oil-free produced water was released into the sea, which caused destruction of marine life due lack of dissolved oxygen. The produced water is now injected in shallow brine-bearing strata, and

Figure 20   Oil and gas drilling in the Los Angeles area. A camouflaged rig on Island Grissom, a manmade drilling island near Long Beach, CA. Photo by J. Jepson (California Department of Conservation 2007)

Figure 21 Four camouflaged rigs on Island Grissom, Long Beach, CA. Photo courtesy of the City of Long Beach (California Department of Conservation 2007)

Figure 22 *Signal Hill* oil field in Long Beach, California, circa 1930. Courtesy of the Long Beach Public Library Collection (California Department of Conservation 2007)

also in deep strata for waterflooding operations in the oilfield. The innovative produced water disposal system won the city industrywide awards (Allen 1972).

Until 1965, the city of Signal Hill, which is located in the center of the Long Beach field, had hundreds of derricks still standing (see Figure 22), many of

Figure 23   Reclamation of old oilfield property on Pacific Coast Highway in Huntington Beach, CA. New houses and roads where oil wells were first drilled here in 1922. Photo by J. Jepson (California Department of Conservation 2007)

them wooden and dating from early 1920s. An amended city oil code helped dismantle these derricks and reclaim the city landscape (see Figure 23).

After it was proved that a properly implemented water-injection program would stop and prevent land subsidence, drilling restrictions were lifted to help develop the remaining half of the Wilmington field. Detailed *environmental controls* were placed by the city on drilling and development operations from the standpoint of subsidence, beautification, and pollution, part of which are carried out by the THUMS Co. and others by the city directly (Allen 1972; Hoovers 2009).

Control of Land Subsidence

Accordingly, subsidence is controlled by waterflooding, started simultaneously with production of oil from a new well in order to ensure a minimum loss of subterranean reservoir pressure loss. To this end, special waterflood monitoring tests and pressure surveys are routinely made. Reservoir voidance maps are calculated and updated monthly and correlated with changes in surface elevation. Under this program, the elevation of more than 350 new benchmarks is established twice a year. Tide gauge stations are located on two of the islands to check on leveling from shore. Measurements of horizontal earth movement are also monitored by use of a geodimeter. Reports on surface stability are prepared and distributed twice a year.

Special subsidence and compaction monitoring instruments were developed, which include the following:

- Precision casing string joint-length measuring tool for detecting possible deep compaction
- Test wells for measuring shallow compaction by a cable extensometer.

All data are continually reviewed and operational measures are taken when necessary to ensure that no subsidence occurs (Allen 1972).

Control of Water Pollution

Since the THUMS artificial drilling islands were to be constructed directly offshore from the civic center and beaches, special care had to be exercised to avoid water pollution. The city adopted a policy of not allowing dumping or run-off of water over the side. This included storm water that might pick up oily wastes. All solid wastes, such as drill cuttings, are collected and barged away for dumping. Oil-stained wastewater is removed by vacuum truck or sent ashore by pipeline. All liquid production is piped to shore for separation, and after filtering sent back to the islands for reinjection into the oil zones. These pipelines are coated, as well as fitted with cathodic protection devices against corrosion and pressure-sensitive automatic shut-off valves.

Weekly aerial inspections are made from a helicopter to locate spills or leaks. In addition, emergency spill containment measures, such as floating boom equipment, is kept on each drilling island, ready to be use to corral any leak or spill that might occur despite all precautions. Tug boats and other marine equipment are always kept ready for placing the floating booms. A coordinated oil-spill disaster program has been established that, if needed, can deploy and integrate the services of all major parties—both public and private (Allen 1972).

Control of Air Pollution

Vapor-recovery systems are installed on all tank facilities; this not only keeps the air clean but also generates revenue from the sale of recovered hydrocarbon products. Traces of hydrogen sulfide gas (H_2S) are extracted from all produced gas before marketing in order to reduce or eliminate undesirable odors. To this end, the produced gas is processed in a plant at the industrial zone.

Control of Noise and Visual Pollution

The drilling rig enclosures on Island Grissom and Island White are all sound-proofed, electric-powered, and self-contained structures that are moved from drill site location to location on rails. Noise level is checked using a decibel meter.

Prevention of Oil Spills

The manmade drilling islands were constructed to contain spills from any leak around a well. Surface conductor pipe was set from ground surface to 800 to 1,000 feet on all wells, exceeding required standards. Multiple blowout preventer (BOP) stacks were mounted on all rigs. All producing wells are fitted with hydraulically operated automatic shut-off valves installed hundreds of feet down the well, to stop the well from flowing in case of a disaster.

THUMS employed stringent environment control measures that cost millions of dollars and yet sustained commercial oil production, thus becoming the first showcase of *environmentally conscious petroleum engineering* in history. The manmade drilling islands have won awards for engineering and esthetic design and excellence. The investment in beautification and environmental protection has paid out: While a drilling ban was in effect on other state-controlled tidelands, the 150 wells drilled in Long Beach netted about $25 million annually. It can be said that if the oil industry does not *lead* the way in establishing strong environmental standards and in complying with them, it will be a follower and eventually be forced to comply with standards set by persons outside the oil patch not likely to be either knowledgeable about or sympathetic to the E&P industry's unique problems (Allen 1972).

10   POTENTIAL SOURCES OF THREATS TO ENVIRONMENT

The academia and the E&P industry have overlooked certain aspects of environmental impact assessments that deserve attention. One is the fate of the abandoned oil and gas fields. The other is the recent drive for CO_2 sequestration in deep geological structures—namely brine aquifers, depleted oil and gas reservoirs, and deep coal mines. Some countries (e.g., Norway) have even embarked on pumping liquid CO_2 at deep ocean floors.

The deep Earth geological strata have reached equilibrium in terms of fluids movement and pressure over millions of years, which have been disturbed irreversibly by drilling hundreds of thousands of boreholes all over the world, ranging from a few hundred to tens of thousands feet deep.

10.1   Environmental Threats of Abandoned Wells and Fields

Abandoned or orphaned wells are oil or gas wells that were drilled since the mid-1800s and subsequently abandoned by unidentifiable owners and/or operators. As production declined and operating costs rose in the late 1970s and early 1980s, many wells and entire well fields have been abandoned by owners or operators that no longer exist or otherwise are not identifiable, and most of the abandoned wells have not been plugged or maintained according to state requirements. As a result, according a recent report by the U.S. Environmental Protection Agency (Worley et al. 2007) these wells pose a threat to surrounding soils, groundwater, and navigable waters of the United States. An unplugged well bore could potentially pose these threats:

- The well bore could leak oil reservoir fluids into groundwater.
- Oil reservoir fluids could flow to the surface and contaminate surface soils.

- Oil reservoir fluids could discharge into the navigable waters of the United States.
- The open well hole could present a public safety hazard.

The orphan well is a category of abandoned wells, which is in the larger context of *idle* wells. Some idle wells have identifiable owners who are responsible for the well until the end of its productive life and through its plugging.

There are thousands of orphan wells in the oil-producing states, many of which are creating environmental hazards. Additionally, orphan wells represent a small percentage of idle wells in the United States, and the majority of these wells do not pose a risk to human health and environment.

The environmental problem lies in the fact that these wells were not properly cleaned out and plugged before they were abandoned, and thus pose the threat of leaking fluids or gases from the reservoirs to which they lead. Oilfield theft of wellheads, valves, and pumping equipment has exacerbated the problem.

These idle wells can be idled with state approval, where they are deemed as having economic potential once the prices of the natural resource rises to a level that is profitable to resume production, or they can be without state approval, in which case they differ from orphan only in that they have an identifiable owner with responsibility over its proper handling.

It may be difficult to garner additional support from the E&P industry to shoulder the burden of producing revenues for the orphan well plugging funds due to issues of equity.

Most states have established abandoned well-plugging funds or orphan well-plugging funds using revenues from fines, penalties, taxes on the gas and petroleum industry, and so on. The current trend of extracting revenues for generating and sustaining orphaned well funds from the petroleum industry is not the most viable option for addressing such a large problem. The amount of funding required will take hundreds of years to collect. The cost of plugging an abandoned or orphan well varies widely. It can range from less than $1,000 to more than $250,000, depending on the depth of the well, the ease or difficulty of access to the well location, obstructions in the well bore, cave-ins in or around the well, stuck casing or tubing, and other factors. For example, by the end of 1997, Pennsylvania had spent $2,196,677.49 plugging 94 orphan wells. At that time, there were approximately 7,554 orphan wells still remaining to be plugged, 550 of which were known to be threatening the environment. However, in January 1998, there was only a balance of $119,101.56 in Pennsylvania's well-plugging funds, and the state only generates approximately $300,000 in revenues per year to add to the funds. Thus, the expense of plugging these wells exceeds Pennsylvania capacity to generate timely and adequate revenues for its abandoned or orphan well-plugging funds. Using of Pennsylvania as an example, the expense of plugging wells exceeds most states' financial capabilities as well (Woorley et al. 2001).

Cement-Plug Deterioration over Long Term

The integrity of groundwater aquifers must be maintained and protected to preserve our groundwater resources. This goal can be achieved by properly casing and cementing new production and injection wells, as well as cement-plugging abandoned or old wells based on knowledge of the well site's soils and geology. If these abandoned water wells are not properly sealed, they can provide a direct conduit for surface water carrying pollutants to groundwater. Proper plugging of wells restores barriers to contamination.

In theory, well construction requires that the subsurface isolation of aquifers and other strata to be restored with annular seals (casing pipe—backed by cement, grout, and resin mixtures). Failure of these seals would provide conduits for vertical transport of pollutants. The pollutants may originate from either well-bore fluids (drilling mud or injected wastewater) or formation fluids (oil, gas, or brine).

Pollution is caused by the loss of external integrity of injection or production wells resulting in upward migration of fluids outside cemented well bores. Pollution of air, surface waters, or groundwater aquifers may result from the migration of produced petroleum hydrocarbons, injected brines, or other toxic waste fluids. The migration takes place in the annular space between the well casing string and borehole walls. This phenomenon has long been known in petroleum terminology as *fluid cross-flow, channeling*, or *gas migration*. Most of these terms refer to the failure of well cements (Wojtanowicz 2008a,b,c).

In recent years, annular seal integrity has been achieved through improvements in well cementing technology in three main areas:

1. Steel–cement bonding techniques
2. Mud displacement practices
3. Cement slurry design to prevent fluids from migrating after placement

The most recent techniques have been developed to prevent the formation of channels due to gas migration in annuli after cementing. As the annular cement—still in liquid state—loses hydrostatic pressure, the well becomes under-balanced, and formation gas invades the slurry and finds its way upward, resulting in the loss of well's integrity. Cement slurry vibration using a low-frequency cyclic pulsation is used by the construction industry for improving quality of cement in terms of better compaction, compressive strength, and fill-up. In the oil industry, the idea of keeping cement slurry in motion after placement has been postulated a promising method for prolonging slurry fluidity in order to sustain hydrostatic pressure and prevent entry of gas into the well's annulus (Wojtanowicz 2008a,b,c).

Industrial use of the technology has been carried out by two companies in three oilfields of Eastern Alberta, Canada (Woorley et al. 2001). It is reported that the top pulsation method showed a 91 percent success rate in preventing gas flow after cementing.

Casing Corrosion

Underground casing strings, along with the cement sheathes behind them, serve as effective barriers to interstrata fluid migrations and leaks to the surface. The casing strings are regularly protected against corrosion, to which they are particularly subject because of the salt water, by cathodic protection method. The cathodic protection system is usually in place during the active life of a well, production, or injection. The system requires supply of electricity as a utility and subsurface wiring. Without the cathodic protection system, the buried casing strings are vulnerable to a high rate of corrosion, rendering them unfit for the purpose for which they were placed. The cathodic protection system can keep the casing strings during the expected life of the well, usually 20 years. Once the well is abandoned, including the cathodic protection system, the casing pipes deteriorate rapidly, and soon after, pressure leaks develop at locations undergoing maximum metal removal from corrosion.

Therefore, the fate of the wells abandoned for decades does not require a genius to imagine. As a matter of fact, thousands of abandoned wells have been found to have caused environmental havocs. These are briefly reported in the following case study.

Case Study of Abandoned Wells in the United States

Since the first commercial oil well was drilled in Pennsylvania in 1859, DEP estimates as many as 350,000 oil and gas wells have been drilled in the commonwealth, with many of those wells having been abandoned without proper plugging (Rathbun 2009).

It was not until the mid-1960s that the oil-producing states in the United States enacted regulations to protect freshwater supplies by requiring that hundreds of feet of cement be poured into the wells at different levels in the process of closing them properly. A majority of the wells listed as plugged in Texas records were plugged before the rules were in place. Brine pumped up with oil had been disposed in open pits from the 1920s to 1960, when that use of pits was prohibited. There were at least 44 abandoned, unplugged wells in the field, and since 1972 brine had been pumped through injection wells at high pressures.

Suro (1992) reported that from the Louisiana bayous to the arid plains of Texas and Oklahoma, thousands of oil and gas wells, abandoned at the end of their productive life, have become conduits for noxious liquids that bubble up from deep below the earth's surface to kill crops and taint drinking water. For state governments in America's oil patch, these abandoned wells have become an expensive legacy left by a fading industry.

The Federal Environmental Protection Agency estimates that there are about 1.2 million abandoned oil and gas wells nationwide and that some 200,000 of them may not be properly plugged. In Texas alone, officials calculate there are 40,000 to 50,000 abandoned wells that could pose pollution problems.

Often drilled to depths of a mile or more, oil wells typically tap into sandy formations permeated with a brine that is up to four times saltier than seawater and that is laced with radioactivity, heavy metals, and other toxins. Without extensive and costly plugging, that brine can flow up the well shaft and seep into freshwater aquifers or sometimes reach the surface.

Occasionally, the brine from abandoned wells hits the surface with explosive force. In the last few years it has erupted through a parking lot in San Angelo, Texas. Mixed with natural gas, it spewed into the backyard of a home in Bartlesville, Oklahoma, and oozed onto a freeway construction site in Tulsa.

But the damage is usually slower and more insidious. A single unplugged exploration hole in West Texas leaked brine for 22 years before being discovered, polluting the ground water beneath 400 to 600 acres of land, a 1990 study by the Bureau of Economic Geology at University of Texas found.

Many of the problem wells date to a freewheeling time before the oil business kept records or was regulated. In Texas, state records indicate that the sites of some 386,000 wells were never registered.

Oilfields have usually been overseen by state agencies like the Texas Railroad Commission or the Oklahoma Corporation Commission whose primary purpose is to promote the oil industry rather than protect the environment. The dangers posed by abandoned and improperly plugged wells began to attract attention from state officials only in the mid-1980s. The oil industry was a cherished source of revenue, and there was little research to distinguish contamination in oil fields from the naturally occurring salinity common in places like West Texas.

Since 1990, both Texas and Oklahoma have created funds specifically dedicated to plugging abandoned wells by imposing new taxes or fees on companies drilling or operating oil wells.

With exhausted fields and low prices driving down production, the number of wells abandoned by bankrupt operators has increased rapidly since the mid-1980s. In 1991, Oklahoma began plugging 260 wells that were orphaned when a single company went bankrupt. The average cost is $4,000 a well.

In the oil patch of Oklahoma and West Texas, brine formations lay at depths of 1,000 to 8,000 feet, where they are under tremendous pressure. Many strata of virtually impermeable rock lay between the brine and the freshwater aquifers, which are near the surface down to depths of 500 feet. Water moves laterally across the aquifers, slowly but across great distances, to streams and creeks.

The future health of these aquifers is a growing concern in Texas because they feed big river systems like the Colorado and the Brazos. Major cities, ranches, farms, and electric power plants all the way to the Gulf Coast depend on these rivers.

Approaches to Locate and Handle Abandoned Wells

Borton et al. (2007) reported a study to locate abandoned wells in northwestern Ohio; it was subjected to extensive oil production at the end of the 1800s. As a result, there is a high density of abandoned wells throughout the region, which can present a groundwater pollution hazard. Although some wells can be located by their casing protruding from the ground, many landowners have remove the casing, making the well impossible to locate by inspection. The focus of this study was to develop and refine a rapid and low-cost method for the location of abandoned oil wells in northwestern Ohio using resistivity, magnetics, and a hand-held spectrometer. The resistivity, measured with a capacitively coupled resistivity apparatus, was relatively low around the location of the well. Interpretation of resistivity contour maps generally yielded a likely location for the well within approximately 5 meters. Depth inversions show a relative low in resistivity almost directly under the pipe in all but 4 of the 10 wells studied. The magnetic intensity contour maps showed a high within 2 meters of the known wells. The spectra of soil samples taken from each well site show a small absorption band near 2.2 μm. Though this band was found at every site, it is so small that this study found it impractical for locating wells. Overall, the most effective approach involves collection of magnetic data first, followed by electrical resistivity and inversion modeling over sites indicated as probable by the magnetic data.

Southern Oil Company has identified an attractive and profitable opportunity to exploit the huge potential of orphaned wells in the northwest part of the state of Louisiana and participate in the recovery of 6.7 billion barrels of known reserves. With approximately 9,000 active wells and 16,000 nonproductive wells, the potential to activate as profitable producing units stemmed from the recent surge in oil prices.

Oil was first discovered in Louisiana in 1905. In 1908, the annual production reached 500,000 barrels (1,370 barrels per day). Since the first drillings, the Caddo Pine Island Field has produced approximately 400 million barrels over a period of 100 years. Production has been primarily from the Annona chalk formation. Publically available geological reports indicate a remaining reserve of approximately 6.7 billion barrels available for extraction.

In 1920, about 25,000 wells were in full production. Oil prices dropped to historical lows, and more wells were removed from production and became classified as abandoned or orphaned. Producing wells declined to 15,000 in the 1980s, and only 9,000 remain in production. Today 16,000 nonproducing wells have the potential to be activated as profitable producing wells, based on current oil prices.

To reactivate these wells to full production using local contractors, Southern Oil funds the reactivation and maintenance of oil well leaseholds and receives perpetual and very long-term revenue rights (see Figure 24).

The reactivation includes the following:

- Acid cleaning and replacing well casings
- Refurbishing or replacing holding tanks
- Ecologically correct containment barrier

Figure 24 (a) A typical orphaned well, (b) A well after reactivation (California Department of Conservation 2007)

- Refurbishing or replacing pipelines
- Supplying and refurbishing wellhead machinery and electrification

The primary field of operation is the Caddo Pine Island Field, located approximately 20 miles northwest of Shreveport, in Caddo Parish, Louisiana.

The Caddo Pine Island Field was discovered in 1905. Since then, there have been over 25,000 wells drilled by various operators. Today, the entire Caddo Pine Island Field covers over 80,000 surface acres (32,400 hectares) with the primary productive zone being the Annona Chalk formation encountered at approximately 1,500 feet. The initial discovery wells were completed in the Nacotosh Formation at approximately 1,100 feet.

10.2 Monitoring Long-Term Storage of CO_2 in Geologic Structures

International agreements on the need to reduce the emissions of greenhouse gases into the atmosphere may bring forward geo-storage of CO_2. Underground storage in petroleum reservoirs and aquifers has a large capacity and is considered environmentally safe. Underground storage of CO_2 both in petroleum reservoirs and aquifers offers new business opportunities for the petroleum industry. The petroleum industry has long experience with all the operations and aspects involved with transport and injection of CO_2 and other fluids into underground formations.

For underground storage, both oil and gas reservoirs and aquifers are potential storage sites. The main requirement for a storage site is that the retention time for CO_2 is long compared to the length of the fossil fuel era (i.e., the approximately 500 years time period that began in this century) (Holt et al. 2000). If some CO_2 leaks out from the deposition time over a period of 5,000 to 10,000 years, natural processes like absorption in the oceans and biofixation will avoid overload into the atmosphere.

Oil and gas reservoirs have proven their storage capacities for hydrocarbons over millions of years, and are thus considered as safe storage sites for CO_2. In recent years, millions of tonnes of CO_2 have been injected into oil reservoirs for profitable EOR, and no serious leaks to the atmosphere have been reported. CO_2 has for a long time been injected into oil reservoirs, and through a large number of projects CO_2-injection has proven its effectiveness as an injection fluid for enhanced oil recovery. In a review of the performance of 25 CO_2 injection projects, pilots and field scale (Holt et al. 2000), the average incremental oil recovery obtained was 13 percent of original oil in place.

Regarding aquifer storage, the feasibility of each potential aquifer must be considered. At depths of more than 700 meters, the CO_2 will typically be stored in dense phase (> 600 kg/Sm^3) and the formation should have a high porosity and a well-defined cap rock. If the integrity of the cap rock is less than perfect, possible escape of CO_2 from the aquifer can be predicted by simulation if the formation is well characterized. The result of such simulations is shown in Figure 25. In these examples, CO_2 is injected into a reservoir with 2,000 millidarcy horizontal permeability only 8,000 meters from a spill point (e.g., an open fracture or the boundary of the cap rock). Although CO_2 migrates toward the spill point, some of the CO_2 will gradually dissolve in the reservoir water, reducing CO_2 loss to the atmosphere. Even in these unfavorable situations, a considerable residence time can be expected. In a similar reservoir with 500 millidarcy horizontal permeability, the loss over a 5,000-year period would be negligible.

Possible leakage of CO_2 through existing wells appears to be especially important in mature sedimentary basins that have been intensively explored and

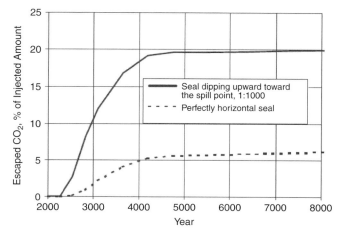

Figure 25   Simulated CO_2 escape profiles from a reservoir with a spill point 8000 m from the CO_2 injection well—6 million tonnes CO_2 was injected yearly for 25 years (Holt et al. 2000)

exploited for hydrocarbon production, such as those in North America. Leakage of CO_2 back into the atmosphere has been recognized in the context of the global carbon balance as being unavoidable in the long term but acceptable if it is small enough. Recent global analyses indicate maximum acceptable leakage rates between 0.01 percent and 1.0 percent leakage per year, where the percent fraction is defined as the volume leaked globally in that year, as compared to total cumulative volume stored. Leakage rates below these estimated rates are required to maintain atmospheric levels of CO_2 within the targets suggested by the International Panel on Climate Change (Nordbotten et al. 2005).

Nordbotten et al. (2005) recently provided mathematical modeling on the potential of CO_2 leakage in regions of high-density abandoned wells, which may serve as potential conduits through which sequestered CO_2 may leak. Although capture and subsequent injection of carbon dioxide into deep geological formations is being considered as a means to reduce anthropogenic emissions of CO_2 to the atmosphere, for such a strategy to be successful, the injected CO_2 must remain within the injection formation for long periods of time—at least several hundred years.

The problem is, mature continental sedimentary basins have a century-long history of oil and gas exploration and production, with a large numbers of oil and gas wells—both existing and abandoned. For example, more than 1 million such wells have been drilled in the state of Texas (United States) alone. These existing wells represent potential leakage pathways for injected CO_2. Figure 26 illustrates the potential pathways for leakage along an abandoned well, including flow along material interfaces and through well cements and casing (Nordbotten et al. 2005). A large number of wells are abandoned, approaching or even passing 50 percent of all wells, depending on basin maturity. Well completion and abandonment practices have evolved in time, from simply walking away in the nineteenth century to currently stringent procedures imposed by regulatory agencies. Considering the extreme variability in field practices, regulatory requirements, material quality, and subsurface conditions, and the fact that the injected CO_2 is supposed to remain underground for several centuries—during which even wells drilled and abandoned by today's standards are likely to degrade—the issue of CO_2 leakage from the injection formation requires special attention (Nordbotten et al. 2005).

To analyze leakage potential, modeling tools are needed that predict leakage rates and patterns in systems with injection and potentially leaky wells. The semianalytical modeling by Nordbotten et al. (2005) allows simple and efficient prediction of leakage rates for the case of injection of supercritical CO_2 into a brine-saturated deep aquifer. The solution predicts the extent of the injected CO_2 plume, provides leakage rates through an abandoned well located at an arbitrary distance from the injection well, and estimates the CO_2 plume extent in the overlying aquifer into which the fluid leaks. Comparison to results from a numerical

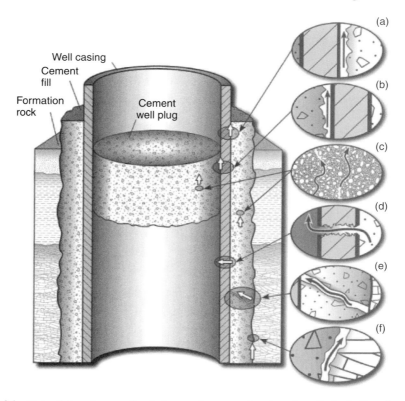

Figure 26 Potential pathways for leakage along an abandoned well, including flow along material interfaces (a, b, f) and through well cements and casing (c, d, e) (Nordbotten et al. 2005)

multiphase flow reservoir simulator show excellent agreement. Example calculations show the importance of outer boundary conditions (e.g., cap rock and other geological seals), the influence of both density and viscosity contrasts in the resulting solutions, and the potential importance of local upconing around the leaky well. Although several important limiting assumptions are required, the new semianalytical solution provides a simple and efficient procedure for estimation of CO_2 leakage for problems involving one injection well, one leaky well, and multiple aquifers separated by impermeable layers.

11 CLOSING REMARKS

The petroleum industry has emerged from a gloomy past to a credible environmental compliance in the past three decades. The E&P industry motto has transformed from concerns for personnel and materials safety to environmental regulatory conformance to outright monetization, as reflected by CO_2-injection for enhanced oil recovery.

Table 5 Petroleum Pollutions During Extraction and Remedies

Event/Activity Well Site	Environmental Damage	Remedy
(i) Blowouts	(a) Millions of gallons crude oil spill, and/or billions of cubic feet burnt/unburnt gas emission (b) Sour gas containing H_2S	Well-maintained BOP stack
(ii) Drilling fluids and cuttings disposal	Soil contamination, aquatic life endangered	Treatment of fluids and solid wastes before disposal, reinjection
(iii) Corrosion of casing pipes	(a) Contamination of subsurface freshwater aquifers (b) Gas or oil leakage to surface	Cathodic protection, placing cement plugs in wells at abandonment
Pipeline:	(a) Surface spill	Preventive maintenance
Refining:	(a) Spills (b) Atmospheric emissions (c) Low-pressure gas flaring	Preventive maintenance; containment; site remediation
Oil field:	Land subsidence or rebound Leakage through abandoned wells in CO_2 sequestration projects	Water injection ?

A summary of various sources of pollutions that may occur during extraction of petroleum and the preventive and remedial measures is given in Table 5. The environmental impacts of these measures are also indicated in Table 6. Indeed, the petroleum industry has come a long way from unbridled polluter from early years of the advent of the Age of Oil, known as *black gold,* to a proactive partner in respecting environmental sensitivity in its multifarious workflow—from extraction in oilfields to processing, refining, and bulk transportation of fuels and petrochemicals through a vast network of distribution system for consumption. Given the enormous quantity of petroleum that is handled daily all over the world, it is amazing the sheer reduction in environmental impacts compared to the first hundred years of petroleum usage.

The first credit in this environmental achievement goes to establishment of *petroleum engineering* as an academic discipline that started circa 1940s in the United States, when the engineering science of optimum drilling density (i.e., well spacing) was established. A direct consequence of this science was that the drilling footprint was drastically reduced, sending pictures depicting jungle of

Table 6 Environmental Impact of Petroleum, from Source to Sink

Activity	Old Practices (OP)	New Practices (NP)	Impact(+ve/−ve)*	Remarks
Petroleum Exploration	Shot-hole explosions for seismic vibration	Convoy of vibrosis trucks (Thumpers):	OP: −ve; NP: +ve	*Desirable:* Thumpers with smaller footprints
Drilling				
(i) Drilling fluids and chemicals	Water /polymer and oil-base muds	Nontoxic Ester ester based	OP: −ve; NP: +ve	*Desirable:* Foam and air drilling
(ii) Cuttings disposal	Untreated	Treated or cuttings reinjection	OP: −ve; NP: +ve	Satisfactory
(iii) Blowouts	More frequent	Rare	OP: −ve; NP: +ve	Satisfactory
(iv) Well testing	Wells flared	Smokeless flaring to containment	OP: −ve; NP: +ve	Satisfactory
(v) Hydraulic fracturing	Water to gelled water	Gelled water to foam	OP: −ve; NP: +ve	*Desirable:* Total control over fracture containment
Production				
(i) Wastewater disposal	Untreated	Treated or reinjected	OP: −ve; NP: +ve	Satisfactory
(ii) Chemical injection for enhanced oil recovery (EOR)	Untreated	Treated or reinjected	OP: −ve; NP: +ve	*Desirable:* EOR without chemical injection?
Abandoned wells/fields	Plugging with cement	Plugging with cement	?	*Desirable:* Major government awareness and involvement of global warming scale
(i) Groundwater contamination	Cased boreholes, with cement plugs inside	Improved cement, cathodic protection against casing corrosion	OP: −ve; NP: +ve	*Desirable:* Increased government awareness and involvement

(*continues*)

59

Table 6 (*continued*)

Activity	Old Practices (OP)	New Practices (NP)	Impact$(+ve/-ve)^*$	Remarks
(ii) Land subsidence CO_2 Sequestration	Water injection For EOR	Water injection For EOR and disposal in deep brine aquifers, abandoned gas fields, deep coal mines for coal bed methane extraction	Always +ve OP: −ve; NP: +ve	Satisfactory *Desirable:* Major awareness needed to address potential CO_2 leakage through abandoned wells

*+ve means more environmentally friendly; −ve means less environmentally friendly oil derricks (e.g., Figure 22) to the past forever. Added to this was the *theory of organic origin of oil*, which dawned on the stakeholders that the notion that petroleum is a *one-time crop*, not to be replenished in millions of years once an oilfield was exhausted. The oil industry fell in line to conserve petroleum in its entire gamut of operations, with minimal tolerance to losses that also created minimum environmental impact. As a matter of fact, Figure 27 shows the seriousness with which the petroleum industry takes environment, spending a whopping 25 percent of total capital expenditures on pollution abatement, second only to the paper manufacturing industry. It is hoped other countries, especially fast-developing countries such as China and India, will follow the U.S. example to arrest environmental impacts of petroleum.

The only avenue left open has been *greenhouse gas emissions* resulting from fossil fuels combustion for daily energy (approx. 94 percent of total petroleum is used for combustion as fuel), which is now a case with the entire global community, thanks to the adoption of the *1998 Kyoto Protocol* as an international law in 2004. As a matter of fact, the growing concern for global warming due to the anthropogenic emission of CO_2 gas has found a viable means to remove and sequester CO_2 "safely" for a long period of time. The world as a whole has trusted the safe technologies that the E&P industry has developed for containing and eliminating environmental pollutions during exploration, drilling, and production. *Petroleum engineering* has helped develop earth sciences to an extent

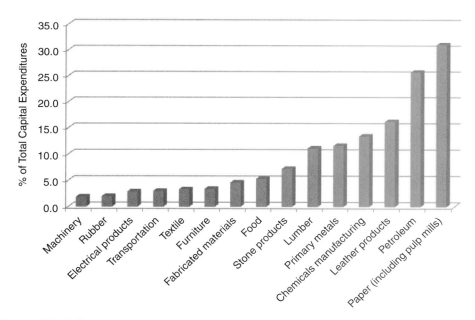

Figure 27 Pollution abatement expenditures by U.S. manufacturing industries (adapted from Allen and Shonnard 2002)

that the international community finds it safe to sequester CO_2 in deep geological structures used and identified by the E&P industry. The use of depleted gas fields, coalbed methane reservoirs, and oil fields all have provided a platform for continuing to benefit from fossil fuels for energy.

However, a word of caution must be sounded before concluding the story of petroleum vis-à-vis environmental safety. With many oil and gas fields aging to abandonment, another problem with potential threat to the environment is lurking on the horizon. The thousands of abandoned oil and gas wells pose two kinds of environmental problems: (1) uncontrolled flow of high salinity brines, with or without heavy metals, oil and gas seepage with and without hydrogen sulfide, and (2) channels for sequestered CO_2 to seep into the atmosphere.

Both problems can be tackled to a great extent with the current state of art in science and technology, but practical solutions and applications methodologies must be developed to assure environmental security to future generations. Klusman (2003), and Celia and Bachu (2003) heralded the timely call on this threat, and Nordbotten et al. (2005) have opened the science that is needed to write a new chapter in petroleum engineering. Here lies an opportunity, for both academia and the energy industries, to embark on cross-cutting research and development programs with active participation by all governments.

REFERENCES

Allen, D. R. (1972), "Environmental Aspects of Oil Producing Operations—Long Beach, California," *Journal of Petroleum Technology*, February.

Allen, D. T., and Shonnard, D. R. (2002), *Green Engineering: Environmentally Conscious Design of Chemical Processes*, Prentice Hall, Upper Saddle River, NJ.

American Petroleum Institute (2009), *API Environmental Guidance Document: Onshore Solid Waste Management in Exploration and Production Operations* Washington, D.C.

Baycroft, P. (2005), "Low-Toxicity Frac Fluid Cuts Costs Offshore," *Hart E&P Magazine*, (November 1).

Baycroft, P. D., McElfresh, P. M., Crews, J. B., Spies, B., Wilson, B., Conrad, M., and Blackman, M. (2005), "Frac-packing with Low Environmental Impact Hydraulic Fracturing Fluid Benefits the Operator," SPE/EPA/DOE Exploration and Production Environmental Conference, March 7–9, Galveston, TX.

Borton, T., Vincent, R. K., and Onasch, C. (2007), "A Method for Locating Abandoned Oil Wells in Wood County, Ohio Using Remote Sensing and Geophysical Techniques," The Geological Society of America (GSA), Joint South-Central and North-Central Sections 41st Annual Meeting, April 11–13, Lawrence, KS, http://gsa.confex.com/gsa/2007SC/finalprogram/index.html.

Burke, J. C., and Veil, A. J. (1995), "Synthetic-Based Drilling Fluids Have Many Environmental Pulses," *Oil and Gas Journal*, Vol. 93, pp. 59–64.

California Department of Conservation (2007), http://www.conservation.ca.gov/dog/photo_gallery/Pages/Index.aspx.

Celia, M. A., Bachu, S. (2003), "Geological Sequestration of CO2: Is Leakage Unavoidable and Acceptable?" *Proceedings of the Sixth International Greenhouse Gas Technologies Conference* Kyoto, Japan), October 1–5, 2002; Gale, J., Kaya, Y. (eds.), Pergamon, Vol. I, pp. 477–482.

Cone, A., and Johns, W. R., (1870), "Petrolia, A Brief History of the Pennsylvania Petroleum Region," d. Appleton and Company, New York, 652 p.

Day, M. D. (2008), "Decommissioning of Offshore Oil and Gas Installations," in Stefan T. Orszulik (ed.), *Environmental Technology in the Oil Industry,* Oxoid Ltd, Hampshire, U.K. Springer.

Dickey, P. A. (1941), "Oil Geology of the Titusville Quadrangle, Pennsylvania," Pennsylvania Topographic and Geologic Survey, Mineral Resource Report 22.

Dickey, P. A. (1959), "The First Oil Well, Oil Industry Centennial," *Journal of Petroleum Technology*, Vol. 59, pp. 14–25.

Dyrynda, P., and Symberlist, R. (2002), "Oil Pollution Following the Sea Empress Spill," http://www.swan.ac.uk/empress/oil/pollute.htm.

ECA Watch, "Photo Gallery of Shell's Oil Spill at Kholmsk, Sakhalin Island," http://www.eca-watch.org/problems/eu_russ/russia/sakhalin/ShellOilSpillPics.html.

Energy Industry Photos, www.energyindustryphotos.com.

Energy Information Administration (EIA) (2000), "Voluntary Reporting of Greenhouse Gases Program: Fuel and Energy Source Codes and Emission Coefficients," http://www.eia.doe.gov/oiaf/1605/coefficients.html.

Energy Information Administration (2004), U.S. Crude Oil, Natural Gas, and Natural Gas Liquids Reserves, 2003 Annual Report, DOE/EIA-0216 (2003) Advanced Summary, September 2004.

Energy Information Administration (EIA) (2005), "*Annual Energy Outlook 2005*," http://tonto.eia.doe.gov/ftproot/forecasting/0383(2005.pdf.

Etkin, D. S. (2001), "Analysis of Oil Spill Trends in the United States and Worldwide," The International Oil Spill Conference, Tampa, FL, March 26–29.

Giddens, P. H. (1938), *The Birth of the Oil Industry*. The Macmillan Company, New York, 216 p.

Guo, B., Lyons, W. C., and Gholambor, A. (2007), *Petroleum Production Engineering—A Computer-Assisted Approach*. Gulf Professional Publishing, Houston, TX.

Hanighen, F. C. (1934), *The Secret War*, a volume in the Hyperion reprint series: The History and Politics of Oil, Hyperion Press, Inc., Westport, CT (Original publisher information at: http://catalogue.nla.gov.au/Record/139578.)

Holt, T., Lindeberg, E. G. B., and Taber, J. J. (2000), "Technologies and Possibilities for Larger-Scale CO_2 Separation and Underground Storage," SPE 63103, SPE Annual Technical Conference and Exhibition held in Dallas, TX, October 1–4.

Hoovers (2009), "THUMS Long Beach Company," http://www.hoovers.com/thums-long-beach-co./--ID__47660--/free-co-profile.xhtml.

Klusman, R. W. (2003), "Evaluation of Leakage Potential from a Carbon Dioxide EOR/Sequestration Project," *Energy Convers. Manage.*, Vol. 44, pp. 1921–1940.

Kvonvolden, K. (2004), "Oil Pollution in Prince William Sound, Alaska," http://images. google.com/imgres?imgurl=http://menlocampus.wr.usgs.gov/50years/accomplishments/ images/PWS_tanker_oil_spill.jpg&imgrefurl=http://menlocampus.wr.usgs.gov/50years/

accomplishments/oil.html&usg=__wK08rZ_0uRHrdXFdDUwouryqQRo=&h=371&w=580&sz=56&hl=en&start=1&um=1&tbnid=FfGGuf1kP59QsM:&tbnh=86&tbnw=134&prev=/images%3Fq%3Doil%2Bpollution%26hl%3Den%26rlz%3D1T4SUNA_enUS266US284%26um%3D1.

Malachosky, E., Shannon, B. E., and Jackson, J. E. (1993), "Offshore Disposal of Oil-based Drilling Fluid Waste: An Environmentally Acceptable Solution," SPE 23373, Proc. First International Conference on Health Safety and Environment, The Hague, The Netherlands, November 10–14, pp. 465–473; also SPE Drilling and Completion, December, 1992, pp. 283–287, 27.

McMordie, W. C. (1980), "Oil Base Drilling Fluids," Symposium on Research on Environmental Fate and Effect of Drilling Fluids and Cuttings. Lake Buena Vista, FL.

Merriam-Webster, "Crude Oil Pipeline," http://visual.merriam-webster.com/energy/geothermal-fossil-energy/oil/crude-oil-pipeline.php; and"Drilling Rig,"http://visual.merriam-webster.com/energy/geothermal-fossil-energy/oil/drilling-rig.php.

Melton, H. R., Smith, J. P., Martin, C. R., Nedwed, T. J., Mairs, H. L., and Raught, D. L. (2000), "Offshore Discharge of Drilling Fluids and Cuttings; A Scientific Perspective on Public Policy," Rio Oil and Gas Conference. Rio de Janeiro, Brazil.

National Geographic (2007), "Global Warming Fast Facts," National Geographic News, http://news.nationalgeographic.com/news/2004/12/1206_041206_global_warming.html.

Natural Gas at: http://www.naturalgas.org/images/offshore_drill_platform.gif.

Nordbotten, J. M., Celia, M. A., Bachu, S., and Dahle, H. K. (2005), "Semianalytical Solution for CO_2 Leakage through an Abandoned Well," *Environmental Science and Technology*, Vol. 39, No. 2, pp. 602–611.

Owen, J. W. (1975), "Trek of the Oil Finders," American Association of Petroleum Geologists, Tulsa, Oklahoma, Memoir 6, Semi-Centennial Convention Volume, http://www.petroleumhistory.org/OilHistory/pages/drake/beginning.html.

Oil Watch, http://www.oilwatch.org/doc/boletin/bole56en.pdf.

Pees, S. T. (1989a), *Oil History*, in http://www.petroleumhistory.org/OilHistory/pages/Whale/whale.html.

Pees, S. T., ed. (1989b), *History of the Petroleum Industry Symposium,* American Association of Petroleum Geologists, Guidebook, 84 p.

Pees, S. T., and Stewart, A.W. (1995), "The Setting in Oil Creek Valley, Pa., and the Chronological Progress of the Drake Well Museum, an Important Repository of Oil Industry History." *Northeastern Geology and Environmental Sciences,* Vol. 17, No. 3, p. 282–294.

Rathbun, T. (2009), "Pennsylvania Dept. of Environmental Protection Completes 12 Abandoned Oil and Gas Well Plugging Projects in 2008," in PR Newswire, March 4, 2009, available at http://news.prnewswire.com/ViewContent.aspx?ACCT=109&STORY=/www/story/03-04-2009/0004982825&EDATE=.

RCEP (2000), "Energy—The Changing Climate," The Royal Commission on Environmental Pollution, June, HMSO, London.

Records of the Petroleum Administrative Board, on *Federal Oil Conservation Board. quoted from*: http://www.archives.gov/research/guide-fed-records/groups/232.html.

Reis, J. C. (1996), *Environmental Control in Petroleum Engineering*, Gulf Publishing Company, Houston, TX.

Soegianto, A., Irawan, B., and Affandi, M. (2008), "Toxicity of Drilling Waste and Its Impact on Gill Structure of Post Larvae of Tiger Prawn (*Penaeus monodon*)," *Global Journal of Environmental Research,* Vol. 2, No. 1, pp. 36–41.

Standard Oil Company (2009), http://www.ohiohistorycentral.org/image.php?rec=1573& img=2144; Also available at: "JDR Photo Album," http://www.pocanticohills.org/ rockefeller/refinery1.htm.

Suro, R. (1992), "Abandoned Oil and Gas Wells Become Pollution Portals," *The New York Times,* May 3, 1992, http://query.nytimes.com/gst/fullpage.html?res=9E0CE7DF133CF930 A35756C0A964958260&sec=&spon=&pagewanted=all.

Southern Oil Company, Louisiana, http://southern-oil.com/aboutSO/missionstatement.html.

Therkelsen, R. L. (1973), "Wildlife Habitat Enhancement of Oil Fields in Kern County, California," California Division of Oil and Gas, Sacramento, CA, pp. iii and 14.

Whitaker, Marvin (1999), "Emerging 'Triple Bottom Line' Model for Industry Weighs Environmental, Economic, and Social Considerations," *Oil and Gas Journal*, (December 20).

Wojtanowicz, A. K. (2008a), "Environmental Control of Drilling Fuilds and Produced Water," in Stefan T. Orszulik (ed.), *Environmental Technology in the Oil Industry,* 2nd ed., Oxoid Ltd, Hampshire, U.K. Springer.

Wojtanowicz, A. K. (2008b), "Environmental Control of Well Integrity," in Stefan T. Orszulik (ed.), *Environmental Technology in the Oil Industry,* 2nd ed., Oxoid Ltd, Hampshire, U.K. Springer.

Wojtanowicz, A. K. (2008c), "Environmental Control Technology for Oilfield Processes," in Stefan T. Orszulik (ed.), *Environmental Technology in the Oil Industry,* 2nd ed., Oxoid Ltd, Hampshire, U.K. Springer.

Worley, R., Franklin, R., and Zenone, V. E. (2001), "The National Problem of Oil Well Fields," The International Oil Spill Conference, Tampa, FL, March 26–29.

Wyoming Dept. of Environmental Quality (1998), "Plugging Abandoned Wells," http://deq.state.wy.us/wqd/groundwater/downloads/Private%20Wells/wellheadplug.asp.

APPENDIX A[*]

Clean Water Act
"Spill Prevention Control and Countermeasure Plans"

[33 United States Code § 1251 et seq.]

[40 Code of Federal Regulations § 112]

The U.S. Federal Clean Water Act prohibits the discharge of oil in harmful quantities into or upon the navigable waters of the United States. In order to prevent pollution damage, the Act requires owners of certain petroleum facilities to develop and implement a Spill Prevention Control and Countermeasure (SPCC) Plan. SPCC Plans must contain information regarding prevention and containment equipment, training procedures, and contingency plans to deal with oil spills from covered facilities.

Facilities exempt from the SPCC requirements include those having oil storage capacities of less than 1320 gallons (approx. 31 barrels)(with no single tank holding more than 660 gallons [approx. 15 barrels]) in aboveground tanks and less than 42,000 gallons (1000 barrels) in underground tanks or facilities which, due to their location, could not reasonably be expected to discharge oil into navigable waters.

*Note: "Navigable waters" include, but are not limited to, lakes, rivers, streams (in-*cluding *intermittent streams), mudflats, sandflats, and wetlands. Wetlands can include playa lakes, swamps, marshes, bogs, sloughs, prairie potholes, wet meadows and natural ponds.*

A "harmful" discharge of oil includes a discharge that causes a film or sheen upon or discoloration of the surface water.

<u>Summary</u>

A Spill Prevention Control and Countermeasure Plan is required at all non-transportation related onshore and offshore facilities engaged in drilling, producing, gathering, storing, processing, refining, transferring, distributing or consuming oil and oil products unless: (1) the facility, due to the location, could not reasonably be expected to discharge oil into or upon the navigable waters of the United States; or, (2) the oil storage capacity is less than 42,000 gallons (1000 bbls) underground and above ground oil storage capacity does not exceed 1320 gallons (approx. 31 bbls) with no single container holding more than 660 gallons (approx. 15 bbls).

Since the effective date of July 10, 1974, SPCC Plans must be prepared by the owner or operator of each existing facility covered by this act and within six months after the date of initial operation of any new facility. Copies of the Plan must be maintained at the facility if someone is present at least eight hours per day. For facilities with employees present less than eight hours a day, the SPCC Plan may be kept at the nearest field office. Each individual Plan must be reviewed by a Registered Professional Engineer, who must certify that the Plan has been prepared in accordance with good engineering practices. Any and all amendments to the Plan must also be certified by a Professional Engineer. SPCC Plans must be reviewed and amended when there is a change in facility design, construction, operation or maintenance which could materially affect the potential for oily discharge into or upon the navigable waters. Regardless of whether changes have been made at a facility, the SPCC Plan must be reviewed and recertified every three years.

It is not necessary for operators to submit completed SPCC Plans to the Environmental Protection Agency (EPA) unless a facility has discharged more than 1000 gallons of oil in a single spill or has had two reportable spills within a twelve-month period. However, the Plan must be available at the facility or field office for review by EPA during normal business hours. Owners or operators of onshore and

---

[*]*Source*: Amy L. Gilliand, *Environmental Reference Manual*, 2nd ed., Texas Independent Producers & Royalty Owners Association, Austin, TX, 1993–1994, pp. 96–98.

offshore mobile or portable facilities, such as onshore drilling or workover rigs, barge mounted offshore drilling or workover rigs, and portable refueling facilities need not prepare a new SPCC Plan each time the facility is relocated. Instead, operators may prepare a general SPCC Plan. This general Plan, prepared according to the SPCC guidelines, will apply only while the facility is in a fixed (non transportational) operating mode.

According to EPA, issues which should be addressed in the SPCC Plan include:

> written descriptions of certain spills at the facility, corrective action taken and plans for preventing recurrence.

> predictions of the direction, rate of flow, and total quantity of oil which could be discharged from the facility as a result of equipment failure.

> appropriate containment and/or diversionary structures or equipment to prevent discharged oil from reaching a navigable watercourse (including wetlands).

> secondary means of containment for tank batteries.

Inspections of equipment and facilities required by section 40 CFR Part 112 should be in accordance with written procedures developed for the facility by the owner or operator. These written procedures and a record of the inspections should be made part of the SPCC Plan and maintained for a period of three years.

Owners or operators are responsible for properly training facility personnel in the operation and maintenance of equipment to prevent the discharge of oil and oil products.

Note: In October of 1991, EPA proposed significant

revisions to the SPCC Plan requirements. The proposed revisions include: making certain guidelines in the regulations mandatory rather than discretionary; adding a facility notification provision; requiring equipment upgrades in certain situations; and mandating immediate removal of oil or soil contaminated with oil from small leaks. These proposed changes have not been adopted as of December 1, 1992. Changes adopted after publication of the *Environmental Reference Manual* will be addressed in the TIPRO *Target* or may be found in the *Federal Register.*

Oil Spills

If a facility subject to this act discharges more than 1000 gallons (approx. 23 bbls) of oil into or upon the navigable waters of the U.S. in a single spill event, or discharges oil in harmful quantities in two spill events occurring within a 12 month period, the owner or operator of such facility shall submit a report to the EPA Regional Administrator within 60 days. This report is to contain information such as: the name, location and owners of the facility; a copy of the SPCC Plan and any amendments made to the Plan; the cause of the spill; corrective actions taken; and preventative measures taken to minimize the possibility of a recurrence.

A copy of the report must also be provided to the state agency in charge of water pollution control activities. According to a 1987 Memorandum of Understanding between the Texas Railroad Commission, the Texas

Water Commission and the Texas Department of Health, jurisdiction over spills resulting from the oil and gas industry is explained as follows: "Generally, the Railroad Commission of Texas has spill response authority over spills from activities associated with the exploration, development, and production of oil, gas, or geothermal resources and the Texas Water Commission has spill response authority over other spills, including those that occur at oil refineries. Incidents involving radioactive materials are handled by the Texas Department of Health. Both the Railroad Commission of Texas and the Texas Water Commission have spill response authority over spills of harmful quantities of crude oil that occur during transportation or in coastal waters." In addition, the Texas Oil Spill Prevention and Response Act of 1991 (See Chapter VII. "OSPRA") gives the General Land Office jurisdiction over large spills and discharges into or threatening coastal waters.

The Clean Water Act requires operators to immediately notify the National Response Center (NRC) in the event of a spill of a harmful quantity of crude oil into or upon the navigable waters of the United States. In addition to notifying the NRC, operators must also notify the appropriate state agencies. In most cases, crude oil spills at the facility will need to be reported to the Railroad Commission District Office with jurisdiction over the facility. (Please refer to Chapters I.A. "CERCLA", I.B. "SARA Title III", I.G. "Oil Pollution Act of 1990", III.B. "HAZWOPER", V.A. "RRC Rule 8", V.D. "RRC Rule 20", and VII. "OSPRA" for a more complete description of the notification requirements in the event of an oil spill at onshore and offshore facilities.)

Note: The Federal Oil Spill

APPENDIX B[*]

Chapter I.D.   **Environmental Protection Agency**

Safe Drinking Water Act Part C
"Underground Injection Control"

[42 United States Code § 300h]

[40 Code of Federal Regulations § 147.2201]

Part C of the Safe Drinking Water Act provides for the protection of underground sources of drinking water through the regulation of underground injection by the Environmental Protection Agency or by an EPA-approved state program. Texas has an EPA-approved underground injection control program.

Summary

Provisions of the Safe Drinking Water Act (SDWA) Part C require the Environmental Protection Agency (EPA) to ensure the protection of underground drinking water sources through the regulation of underground injection.

Under the SDWA, EPA was required to develop minimum guidelines for underground injection regulations to assist states in establishing their own underground injection control (UIC) program. States may apply for primary enforcement authority once a program consistent with federal statutory requirements is adopted. In 1982, Texas became one of the first states to receive primary enforcement responsibility.

Texas' underground injection control program is jointly enforced by the Texas Water Commission and the Texas Railroad Commission. The Railroad Commission (RRC) has jurisdiction over Class II injection wells, with the exception of any wells on Indian lands. (EPA administers the UIC program on Indian lands.) A Class II well is defined by the SDWA as a well designed for the injection of oil and natural gas fluid wastes (generally the disposal of salt water and other production fluids), injection arising out of and incidental to the operation of gasoline plants, natural gas processing plants and pressure maintenance or repressuring plants, a well used for enhanced recovery of oil or gas, or a well used for underground storage of liquid (not gas) hydrocarbons. Specific requirements of the RRC Class II UIC program are addressed in Chapter V.J.

Penalties

See Chapter V.J. "Underground Injection Control Program."

Informational Sources

U.S. Environmental Protection Agency
Region VI
1445 Ross Avenue

Dallas, Texas 75202-2733
(214) 655-6444

Railroad Commission of Texas
Environmental Services
P.O. Drawer 12967
Austin, Texas 78711-2967
(512) 463-6790

Texas Water Commission
P.O. Box 13087
Capitol Station
Austin, Texas 78711-3087
(512) 463-7830

Additional Reference Sources

See Chapter V.J. "Underground Injection Control Program."

---

[*]*Source*: Amy L. Gilliand, *Environmental Reference Manual*, 2nd ed., Texas Independent Producers & Royalty Owners Association, Austin, TX, 1993–1994, p. 202.

APPENDIX C[*]

Chapter I. G.   Environmental Protection Agency

The Oil Pollution Act of 1990

(P.L. 101-380)

The Oil Pollution Act of 1990 was enacted in response to several major oil spills into navigable waters in the past few years, including the Exxon Valdez spill into Alaska's Prince William Sound and the Mega Borg spill in the Gulf of Mexico. The Act is intended to reduce the number of oil spills and to improve the nation's preparedness and ability to respond to spills. The Act creates a comprehensive prevention, response, and compensation program for oil spills into navigable waters from vessels or from onshore or offshore facilities.

Summary

The Oil Pollution Act (OPA) was signed into law by President Bush on August 18, 1990. The OPA provides for a number of oil spill prevention measures, significantly increases both civil and criminal liability for responsible parties, and mandates augmentation of oil spill response capabilities by owners and operators of covered facilities and by local, state and federal authoritites.

The United States Environmental Protection Agency (EPA) has been delegated jurisdiction over certain onshore facilities under OPA. Non-transportation related offshore facilities and transportation related pipelines that link offshore oil production platforms to onshore facilities are under the jurisdiction of the Minerals Management Service. The United States Coast Guard has jurisdiction over vessels and marine transportation-related facilities under OPA. This overview addresses the requirements for onshore facilities subject to EPA jurisdiction.

Facility Response Plans

The Act requires owners or operators of certain onshore facilities to prepare and submit a facility response plan by February 18, 1993. Currently, regulations have not been adopted by EPA to specifically identify which onshore facilities will be required to submit a plan or what the plan should contain. The Act generally identifies affected facilities as ones that could cause "substantial harm" to the environment by discharging into navigable waters or on the adjoining shorelines.

> **Note:** The facility response plans required under OPA are in addition to similar plans required for certain affected facilities by the Texas Oil Spill Prevention and Response Act (See Chapter VII. "OSPRA") and by the Clean Water Act (See Chapter I.C.1. "Spill Prevention Control and Counter-

measure Plans").

According to EPA, a two year extension from this requirement may be granted if an owner or operator of a facility can prove that they are capable of responding to a worst-case discharge from the facility. A worst-case discharge is defined by the Act as "the largest forseeable discharge in adverse weather conditions."

The OPA prohibits facilities required to have a response plan from handling, storing, or transporting oil after February 18, 1993 unless a plan has been submitted. EPA intends to publish in the *Federal Register* regulations describing the facilities required to prepare and submit a plan and outlining the requirements of the response plan prior to the February deadline.

Oil Spill Liability and Reporting

The Federal Water Pollution Control Act (also known as the Clean Water Act) requires the responsible party of a facility to notify the National Response Center upon discovering an oil spill from the facility into or along navigable waters. The OPA applies to such incidents occurring on or after August 18, 1990. The Act substantially increases the liability exposure

---

[*]*Source*: Amy L. Gilliand, *Environmental Reference Manual*, 2nd ed., Texas Independent Producers & Royalty Owners Association, Austin, TX, 1993–1994, p. 408–409.

of owners and operators of onshore and offshore facilities from that imposed previously. In the event of an oil spill into navigable waters or onto the adjoining shorelines, the scope of damages for which the responsible party is liable is expanded to include: damages for injury to, or loss of the use of natural resources; damages for injury to property; loss of revenues, profits, or earning capacity; and costs of public services during or after oil removal activities.

The responsible party for an onshore facility is any person who owns or operates the facility, except the federal government, or a state or municipal government.

The only oil discharges that are excluded under the Act are those that are allowed by a permit under federal, state or local law. In addition, an owner or operator of a facility is not liable for costs if: (1) the discharge was caused solely by an act of God; (2) the discharge was caused solely by an act of war: or (3) the discharge was caused solely by the act or omission of a third party. The third party defense does not apply if the act or omission is done by an employee or an agent of the responsible party or by a person under any contractual relationship with the responsible party.

The Act establishes a limit on liability of the total of all removal costs plus $75 million for offshore facilities, and $350 million for onshore facilities and deepwater ports. These liability limits do not apply to oil spills if: the spill is caused by willful misconduct or gross negligence; the responsible party violates any applicable federal safety, construction or operating regulation; the responsible party fails to report an oil spill; or a responsible party fails to cooperate with the government.

The OPA establishes that the Oil Spill Liability Trust Fund (a $1 billion fund funded by a 5 cent per barrel tax on imported and domestic crude oil and crude oil products) is to be used to pay for removal costs and damages not recovered from responsible parties.

Penalties

The Act increases substantially the amount of civil penalties, criminal fines, and terms of imprisonment which may be imposed upon facility owners and operators. For instance, a failure to notify the federal government of a discharge of oil may result in a criminal fine of up to $250,000 for an individual and up to $500,000 for an organization, or not more than five years imprisonment or both.

Previously, under the Federal Water Pollution Control Act (FWPCA or Clean Water Act), a person discharging oil into navigable waters of the United States was not subject to criminal punishment. Under OPA, however, a person who negligently discharges oil into navigable waters is subject to a criminal fine of not less than $2500 nor more than $25,000 per day of violation, or one year imprisonment, or both. For a second conviction, the fine may increase to $50,000 per day of violation, or two years imprisonment, or both.

In addition, the responsible party of an onshore or offshore facility from which oil is discharged into navigable waters or the adjoining shorelines is subject to a civil penalty of up to $25,000 per day of violation or an amount up to $1,000 per barrel of oil discharged. This penalty is mandatory. If the discharge is the result of gross negligence or willful misconduct, the person shall be be subject to

a penalty of not less than $100,000 plus not more than $3,000 per barrel of oil discharged.

Informational Sources

U.S. Environmental Protection Agency
Region VI
1445 Ross Avenue
Dallas, Texas 75202-2733
(214) 655-6444

National Response Center
(800) 424-8802

APPENDIX D § 3.13. RULE 13. CASING, CEMENTING, DRILLING, AND COMPLETION REQUIREMENTS (AMENDED EFFECTIVE AUGUST 13, 1991)*

(a) General.

(1) The operator is responsible for compliance with this section during all operations at the well. It is the intent of all provisions of this section that casing be securely anchored in the hole in order to effectively control the well at all times, all usable-quality water zones be isolated and sealed off to effectively prevent contamination or harm, and all potentially productive zones be isolated and sealed off to prevent vertical migration of fluids or gases behind the casing. When the section does not detail specific methods to achieve these objectives, the responsible party shall make every effort to follow the intent of the section, using good engineering practices and the best currently available technology.

(2) Definitions. The following words and terms, when used in this chapter, shall have the following meanings, unless the context clearly indicates otherwise.

(A) Stand under pressure--To leave the hydrostatic column pressure in the well acting as the natural force without adding any external pump pressure. The provisions are complied with if a float collar is used and found to be holding at the completion of the cement job.

(B) Zone of critical cement--For surface casing strings shall be the bottom 20% of the casing string, but shall be no more than 1,000 feet nor less than 300 feet. The zone of critical cement extends to the land surface for surface casing strings of 300 feet or less.

(C) Protection depth--Depth to which usable-quality water must be protected, as determined by the Texas Department of Water Resources, which may include zones that contain brackish or saltwater if such zones are correlative and/or hydrologically connected to zones that contain usable-quality water.

(D) Productive horizon--Any stratum known to contain oil, gas, or geothermal resources in commercial quantities in the area.

(b) Onshore and inland waters.

(1) General.

(A) All casing cemented in any well shall be steel casing that has been hydrostatically pressure tested with an applied pressure at least equal to the maximum pressure to which the pipe will be subjected in the well. For new pipe, the mill test pressure may be used to fulfill this

---

* *Source*: Amy L. Gilliand, *Environmental Reference Manual*, 2nd ed., Texas Independent Producers & Royalty Owners Association, Austin, TX, 1993–1994, p. 629–637.

requirement. As an alternative to hydrostatic testing, a full length electromagnet, ultrasonic, radiation thickness gauging, or magnetic particle inspection may be employed.

(B) Wellhead assemblies shall be used on w~ to maintain surface control of the well. Each component of the wellhead shall have a pressure rating equal to or greater than the anticipated pressure to which that particular component might be exposed during the course of drilling, testing, or producing the well.

(C) A blowout preventer or control head and other connections to keep the well under control at all times shall be installed as soon as surface casing is set. This equipment shall be of such construction and capable of such operation as to satisfy any reasonable test which may be required by the Commission or its duly accredited agent.

(D) When cementing any string of casing more than 200 feet long, before drilling the cement plug the operator shall test the casing at a pump pressure in pounds per square inch (psi) calculated by multiplying the length of the casing string by *0.2*. The maximum test pressure required, however, unless otherwise ordered by the Commission, need not exceed 1,500 psi. If, at the end of 30 minutes, the pressure shows a drop of 10% or more from the original test pressure, the casing shall be condemned until the leak is corrected. A pressure test demonstrating less than a 10% pressure drop after 30 minutes is proof that the condition has been corrected.

(E) Wells drilling to formations where the expected reservoir pressure exceeds the weight of the drilling fluid column shall be equipped to divert any wellbore fluids away from the rig floor. All diverter systems shall be maintained in an effective working condition. No well shall continue drilling operations if a test or other information indicates the diverter system is unable to function or operate as designed.

(2) Surface casing.
   (A) Amount required
      (i) An operator shall set and cement sufficient surface casing to protect all usable-quality water strata, as defined by the Texas Department of Water Resources. Before drilling any well in any field or area in which no field rules are in effect or in (~ which surface casing requirements are not specified in the applicable field rules, an operator shall obtain a letter from the Texas Department of Water Resources stating! the protection depth. In no case, however, is surface casing to be set deeper than 200 ~;!I feet below the specified depth without prior approval from the Commission.

(ii) Any well drilled to a total depth of 1,000 feet or less below the ground surface may be drilled without setting surface casing provided no shallow gas sands or ! abnormally high pressures are known to exist at depths shallower than 1,000 feet below the ground surface; and further, provided that production casing is cemented from the shoe to the ground surface by the pump and plug method.

(B) Cementing. Cementing shall be by the pump and plug method. Sufficient cement shall be used to fill the annular space outside the casing from the shoe to the ground surface or to the bottom of the cellar. If cement does not circulate to ground surface or the bottom of the cellar, the operator or his representative shall obtain the approval of the District Director for the procedures to be used to perform additional cementing operations, if needed, to cement surface casing from the top of the cement to the ground surface.

(C) Cement quality.

(i) Surface casing strings must be allowed to stand under pressure until the cement has reached a compressive strength of at least 500 psi in the zone of critical cement before drilling plug or initiating a test. The cement mixture in the zone of critical cement shall have a 72-hour compressive strength of at least 1,200 psi.

(ii) An operator may use cement with volume extenders above the zone of critical cement to cement the casing from that point to the ground surface, but in no case shall the cement have a compressive strength of less than 100 psi at the time of drill out nor less than 250 psi 24 hours after being placed.

(iii) In addition to the minimum compressive strength of the cement, the API free water separation shall average no more than six milliliters per 250 milliliters of cement tested in accordance with the current API RP 10B.

(iv) The Commission may require a better quality of cement mixture to be used in any well or any area if evidence of local conditions indicates a better quality of cement is necessary to prevent pollution or to provide safer conditions in the well or area.

(D) Compressive strength tests. Cement mixtures for which published performance data are not available must be tested by the operator or service company. Tests shall be made on representative samples of the basic mixture of cement and additives used, using distilled water or potable tap water for preparing the slurry. The tests must be conducted using the equipment and procedures adopted by the American Petroleum Institute, as published in the current API RP 10B. Test data showing competency of a proposed cement mixture to meet the above requirements must be furnished to the Commission prior to the cementing operation. To determine that the minimum compressive strength has been obtained, operators shall use the typical performance data for the particular cement used in the well (containing all the additives, including any accelerators used in the slurry) at the following temperatures and at atmospheric pressure.

(i) For the cement in the zone of critical cement, the test temperature shall be within 10 degrees Fahrenheit of the formation equilibrium temperature at the top of the zone of critical cement.

(ii) For the filler cement, the test temperature shall be the temperature found 100 feet below the ground surface level, or 60 degrees Fahrenheit, whichever is greater.

(E) Cementing report. Upon completion of the well, a cementing report must be filed with the Commission furnishing complete data concerning the cementing of surface casing in the well as specified on a form furnished by the Commission. The operator of the well or his duly authorized agent having personal knowledge of the facts, and representatives or the cementing company performing the cementing job, must sign the form attesting to compliance with the cementing requirements of the Commission.

(F) Centralizers. Surface casing shall be centralized at the shoe, above and below a stage collar or diverting tool, if run, and through usable-quality water zones. In nondeviated holes, pipe centralization as follows is required: a centralizer shall be placed every fourth joint from the cement shoe or to the bottom of the cellar. All centralizers shall meet API spec 10D specifications. In deviated holes, the operator shall provide additional centralization.

(G) Alternative surface casing programs.

(i) An alternative method of freshwater protection may be approved upon written application to the appropriate District Director. The operator shall state the reason (economics, well control, etc.) for the alternative freshwater protection method and outline the alternate program for casing and cementing through the protection depth for strata containing usable-quality water. Alternative programs for setting more than specified amounts of surface casing for well control purposes may be requested on a field or area basis. Alternative programs for setting less than specified amounts of surface casing will be authorized on an individual well basis only. The District Director may approve, modify, or reject the proposed program. If the proposal is modified or rejected, the operator may request a review by the Director of

Field Operations. If the proposal is not approved administratively, the operator may request a public hearing. An operator shall obtain approval of any alternative program before commencing operations.

(ii) Any alternate casing program shall require the first string of casing set through the protection depth to be cemented in a manner that will effectively prevent the migration of any fluid to or from any stratum exposed to the wellbore outside this string of casing. The casing shall be cemented from the shoe to ground surface in a single stage, if feasible, or by a multi-stage process with the stage tool set at least 50 feet below the protection depth.

(iii) Any alternate casing program shall include pumping sufficient cement to fill the annular space from the shoe or multi-stage tool to the ground surface. If cement is not circulated to the ground surface or the bottom of the cellar, the operator shall run a temperature surveyor cement bond log. The appropriate District Office shall be notified prior to running the required temperature surveyor bond log. After the top of cement outside the casing is determined, the operator or his representative shall contact the appropriate District Director and obtain approval for the procedures to be used to perform any required additional cementing operations. Upon completion of the well, a cementing report shall be filed with the Commission on the prescribed form.

(iv) Before parallel (nonconcentric) strings of pipe are cemented in a well, surface or intermediate casing must be set and cemented through the protection depth.

(3) Intermediate casing.

(A) Cementing method. Each intermediate string of casing shall be cemented from the shoe to a point at least 600 feet above the shoe. If any productive horizon is open to the wellbore above the casing shoe, the casing shall be cemented from the shoe up to a point at least 600 feet above the top of the shallowest productive horizon or to a point at least 200 feet above the shoe of the next shallower casing string that was set and cemented in the well.

(B) Alternate method. In the event the dIstance from the casIng shoe to the top of the shallowest productive horizon make cementing, as specified above, impossible or impractical, the multi-stage process may be used to cement the casing in a manner that will effectively seal off all such possible productive horizons and prevent fluid migration, 1 to or from such strata within the wellbore.

(4) Production casing.

    (A) Cementing method. The producing string of casing shall be cemented by the pump and plug method, or another method approved by the Commission, with sufficient cement to fill the annular space back of the casing to the surface or to a point at least 600 feet above the shoe. If any productive horizon is open to the wellbore above the casing shoe, the casing shall be cemented in a manner that effectively seals off all such possibly productive horizons by one of the methods specified for intermediate casing in paragraph (3) of this subsection.

    (B) Isolation of associated gas zones. The position of the gas-oil contact shall be determined by coring, electric log, or testing. The producing string shall be landed and cemented below the gas-oil contact, or set completely through and perforated in the oil-saturated portion of the reservoir below the gas-oil contact.

(5) Tubing and storm choke requirements.

    (A) Tubing requirements for oil wells. All flowing oil wells shall be equipped with and produced through tubing. When tubing is run inside casing in any flowing oil well, the bottom of the tubing shall be at a point not higher than 100 feet above the top of the producing interval nor more than 50 feet above the top of a line, if one is used. In a multiple zone structure, however, when an operator elects to equip a well in such a manner that small through-the-tubing type tools may be used to perforate, complete, plug back, or recomplete without the necessity of removing the installed tubing, the bottom of the tubing may be set at a distance up to, but not exceeding, 1,000 feet above the top of the perforated or open-hole interval actually open for production into the wellbore. In no case shall tubing be set at a depth of less than 70% of the distance from the surface of the ground to the top of the interval actually open to production.

    (B) Storm choke. All flowing oil, gas, and geothermal resource wells located in bays, estuaries, lakes, rivers, or streams must be equipped with a storm choke or similar safety device installed in the tubing a minimum of 100 feet below the mud line.

(c) Texas offshore casing, cementing, drilling, and completion requirements.

(1) Casing. The casing program shall include at least three strings of pipe, in addition to such drive pipe as the operator may desire, which shall be set in accordance with the following program.

    (A) Conductor casing. A string of new pipe, or reconditioned pipe with substantially the same characteristics as new pipe, shall be set and cemented at a depth of not less than 300 feet TVD (true vertical depth) nor more than 800 feet TVD below the mud line. Sufficient cement shall be used to fill the annular space back of the pipe to the mud line; however, cement may be washed out or displaced to a maximum depth of 50 feet below the mud line to facilitate pipe removal on abandonment. Casing shall be set and cemented in all cases prior to penetration of known shallow oil and gas formations, or upon encountering such formations.

(B) (i) Surface casing. All surface casing shall be a string of new pipe with a mill test of at least 1,100 pounds per square inch (psi) or reconditioned pipe that has been tested to an equal pressure. Sufficient cement shall be used to fill the annular space behind the pipe to the mud line; however, cement may be washed out or displaced to a maximum depth of 50 feet below the mud line to facilitate pipe removal on abandonment. Surface casing shall be set and cemented in all cases prior to penetration of known shallow oil and gas formations, or upon encountering such formations. In all cases, surface casing shall be set prior to drilling below 3,500 feet TVD. Minimum depths for surface casing are as follows.

Surface Casing Depth Table

Proposed Total Vertical Depth of Well to 7,000 feet	Surface 25% of proposed Total depth of well
7,000–10,000 feet	2,000 feet
10,000 and be	2,500 feet

(ii) Casing test. Cement shall be allowed to stand under pressure for a minimum of eight hours before drilling plug or initiating tests. Casing shall be tested by pump pressure to at least 1,000 psi. If, at the end of 30 minutes, the pressure shows a drop, of 100 psi or more, the casing shall be condemned until the leak is corrected. A pressure test demonstrating a drop of less than 100 psi after 30 minutes is proof that the condition has been corrected.

(C) Production casing or oil string. The production casing or oil string shall be new or reconditioned pipe with a mill test of at least 2,000 psi that has been tested to an equal pressure and after cementing shall be tested by pump pressure to at least 1,500 psi. If, at the end of 30 minutes, the pressure shows a drop of 150 psi or more, the casing shall be condemned. After corrective operations, the casing shall again be tested in the same manner. Cementing shall be by the pump and plug method. Sufficient cement shall be used to fill the calculated annular space above the shoe to protect any prospective producing horizons and to a depth that isolates abnormal pressure from normal pressure (0.465 gradient). A float collar or other means to stop the cement plug shall be inserted in the casing string above the shoe. Cement shall be allowed to stand under pressure for a minimum of eight hours before drilling the plug or initiating tests.

(2) Blowout preventers.

(A) Before drilling below the conductor casing, the operator shall install at least one remotely controlled blowout preventer with a mechanism for automatically diverting the drilling fluid to the mud system when the blowout preventer is activated.

(B) After setting and cementing the surface casing, a minimum of two remotely controlled hydraulic ram-type blowout preventers (one equipped with blind rams and one with pipe rams), valves, and manifolds for circulating drilling fluid shall be installed for the purpose of controlling the well at all times. The ram-type blowout preventers, valves, and manifolds shall be tested to 100% of rated working pressure, and the annular-type blowout preventer shall be tested to 1,000 psi at the time *of* installation. During drilling and completion operations, the ram-type blowout preventers shall be tested by closing at least once each trip, and the annular-type preventer shall be tested by closing on drill pipe once each week.

(3) Kelly cock. During drilling, the well shall be fitted with an upper kelly cock in proper working order to close in the drill string below hose and swivel, when necessary for well control. A lower kelly safety valve shall be installed so that it can be run through the blowout preventer. When needed for well control, the operator shall maintain at all times on the rig floor safety valves to include:

(A) full-opening valve of similar design as the lower kelly safety valves; and

(B) inside blowout preventer valve with wrenches, handling tools, and necessary subs for all drilling pipe sizes in use.

(4) Mud program. The characteristics, use, and testing of drilling mud and conduct of related drilling procedures shall be designed to prevent the blowout *of* any well. Adequate supplies of mud of sufficient weight and other acceptable characteristics shall be maintained. Mud tests shall be made frequently. Adequate mud testing equipment shall be kept on the drilling platform at all times. The hole shall be kept full of mud at all times. When pulling drill pipe, the mud volume required to fill the hole each time shall be measured to assure that it corresponds with the displacement of pipe pulled. A derrick floor recording mud pit level indicator shall be installed and operative at all times. A careful watch for swabbing action shall be maintained when pulling out *of* hole. Mud-gas separation equipment shall be installed and operated.

(5) Casinghead.

    (A) Requirement. All wells shall be equipped with casingheads of sufficient rated working pressure, with adequate connections and valves available, to permit pumping mud-laden fluid between any two strings of casing at the surface.

    (B) Casinghead test procedure. Any well showing sustained pressure on the casinghead, or leaking gas or oil between the surface casing and the oil string, shall be tested in the following manner. The well shall be killed with water or mud and pump pressure applied. Should the pressure gauge on the casinghead reflect the applied pressure, the casing shall be condemned. After corrective measures have been taken, the casing shall be tested in the same manner. This method shall be used when the origin *of* the pressure cannot be determined otherwise.

(6) Christmas tree. All completed wells shall be equipped with Christmas tree fittings and wellhead connections with a rated working pressure equal to, or greater than, the surface shut-in pressure of the well. The tubing shall be equipped with a master valve, but two master valves shall be used on all wells with surface pressures in excess of 5,000 psi. All wellhead connections shall be assembled and tested prior to installation by a fluid pressure equal to the test pressure of the fitting employed.

(7) Storm choke and safety valve. A storm choke or similar safety device shall be installed in the tubing of all completed flowing wells to a minimum of 100 feet below the mud line. Such wells shall have the tubing-casing annulus sealed below the mud line. A safety valve shall be installed at the wellhead downstream of the wing valve. All oil, gas, and geothermal resource gathering lines shall have check valves at their connections to the wellhead.

(8) Pipeline shut-otI valve. All gathering pipelines designed to transport oil, gas, condensate, or other oil or geothermal resource field fluids from a well or platform shall be equipped with automatically controlled shut-off valves at critical points in the pipeline system. Other safety equipment must be in full working order as a safeguard against spillage from pipeline ruptures.

(9) Training. Effective January 1, 1981, all tool pushers, drilling superintendents, and operators' representatives (when the operator is in control of the drilling) shall be required to furnish certification of satisfactory completion of a USGS-approved school on well control equipment and techniques. The certification shall be renewed every two years by attending a USGS-approved refresher course. These training requirements apply to all drilling operations on lands which underlie fresh or marine waters in Texas.

APPENDIX E § 3.14. RULE 14. PLUGGING (AMENDED EFFECTIVE MARCH 1, 1992)[†]

(a) Application to plug.

(1) Notification of intention to plug any well or wells drilled for oil, gas, or geothermal resources or for any other purpose over which the commission has 1urisdiction, except those specifically addressed in § 3.100(f)(1) of this title (relating to Seismic Holes and Core Holes) (Statewide Rule 100), shall be given to the commission prior to plugging. Notification shall be made, in writing, to the district office on the appropriate form.

(2) This written notification must be received by the District Office at least five days prior to the beginning of plugging operations and shall show the proposed procedure as well as the complete casing record. The work of plugging the well or wells shall not commence before the date set out in the notification for the beginning of plugging operations unless authorized by the District Director. The operator shall call the District Office at least four hours before commencing plugging operations and proceed with the work as outlined, unless the proposed plugging procedure is not approved by the District Director. Exceptions may be granted at the discretion of the District Director when either a workover or a drilling rig is already at work on location, ready to commence plugging operations. Operations shall not be suspended prior to plugging the well unless the hole is cased and casing is cemented in place in compliance with Commission rules.

(3) The landowner and the operator may file an application to condition an abandoned well located on the landowner's tract for usable quality water production operations, provided the landowner assumes responsibility for plugging the well and obligates himself, his heirs, successors, and assignees as a condition to the Commission's approval of such application to complete the plugging operations. The application shall be made on the form prescribed by the Commission. In all cases, the operator responsible for plugging the well must place all cement plugs required by this rule up to the base of the usable quality water strata, as determined by the Texas Department of Water Resources.

(4) Before plugging any well, notice shall be given to the surface owner of the well site tract, or the resident if the owner is absent. If they so desire, a representative of the surface owner, in addition to the Commission representative, may be present to witness the plugging of the well. Plugging shall not be delayed because of the inability to deliver notice to the surface owner or resident.

[†] *Source*: Amy L. Gilliand, *Environmental Reference Manual*, 2nd ed., Texas Independent Producers & Royalty Owners Association, Austin, TX, 1993–1994, p. 641–647.

(b) Plugging: Commencement of Operations, Extensions, and Responsibility.

(1) A plugging record shall be completed, duly verified, and filed, in duplicate, on the appropriate form in the District Office within 30 days after plugging operations are completed. A cementing report made by the party cementing the well shall be attached to, or made a part of, the plugging report.

(2) Plugging operations on each dry or inactive well must be commenced within a period of one year after drilling or operations cease and shall proceed with due diligence until completed. For good cause, a reasonable extension of time in which to start the plugging operations may be granted pursuant to the following procedures.

(A) The Commission or its delegate may administratively grant an extension of time of one year if the well is in compliance with all other laws and Commission rules relating to conservation and safety or the prevention or control of pollution, is not a pollution hazard; and

(i) provided that the operator pays the proper fee as provided in § 3.76 of this title (relating to Fees, Performance Bonds and Alternate Forms of Financial Security Required to be Filed) (Statewide Rule 78), obtains a permit for this extension, and no more than three extensions have been granted after June 1, 1992 for the well under the provisions of this clause; or

(ii) the operator files an individual or blanket performance bond as provided in § 3.76 or a letter of credit.

(B) Any administratively granted extension of time is subject to review by the Commission or its delegate at any time.

(C) If the Commission or its delegate declines administratively to grant or to continue an extension of time, the operator shall plug the well or request a hearing on the matter.

(D) The Commission or its delegate may allow a well to be the subject of more than four extensions granted after June 1, 1992 under the provisions of subparagraph (A)(i) of this paragraph, upon written application and a showing that no pollution of surface or subsurface water could occur as a result of granting the extension. If such application is administratively denied, the Commission may subsequently grant the extension.

(E) All wells more than 25 years old that become inactive and subject to the provisions of this paragraph shall be plugged or tested annually to determine whether the well poses a potential threat of harm to natural resources, including surface and subsurface water, oil and gas. In general, a fluid level test is a sufficient test for purposes of this subparagraph. However, the Commission or its delegate may require alternate methods of testing, and more frequent tests, if it is necessary to ensure the well does not pose a potential threat of harm to natural resources. Wells that are returned to continuous production, as evidenced by three consecutive months of production, within a year after the well becomes inactive need not be tested. Alternate methods of testing may be approved by the Commission or its delegate by written application and upon a showing that such a test will provide information sufficient to determine that the well does not pose a threat to natural resources. No test shall be conducted without prior approval from the District Office. The District Office shall be notified at least three days before the test is conducted. The test results shall be filed with the District Office, on a Commission-approved form, within 30 days of the completion of the test. Fluid level tests shall be conducted on an annual basis according to the following schedule.

(i) Wells that become both inactive and more than 25 years old after June 1, 1992 shall be tested within one year after the well becomes both inactive and more than 25 years old, unless the well has undergone a hydraulic pressure test within five years of the date a test is required under this clause and the results of such test are on file with the Commission.

(ii) Wells that are both inactive and more than 45 years old as of June 1, 1992 shall be tested before June 1, 1993.

(iii) Wells that are both inactive and 35 to 45 years old as of June 1, 1992 shall be tested before June 1, 1994.

(iv) Wells that are both inactive and 25 to 35 years old as of June 1, 1992 shall be tested before June 1, 1995.

(F) As of January 1, 1997, all wells that are, or become, both more than 25 years old and inactive for more than 10 years, shall be tested for mechanical integrity so long as the well remains inactive. Such tests shall be conducted every five years. However, the Commission or its delegate may require that a well undergo a mechanical integrity test more frequently than every five years if conditions indicate that more frequent testing is necessary to prevent the threat of harm to natural resources. A hydraulic pressure test is a sufficient mechanical integrity test for purposes of this section. Alternate methods of testing for mechanical integrity may be approved by the Commission or its delegate upon written application and a showing that such alternate method of testing will provide information sufficient to determine that the casing is sound and not subject to leakage. No test shall be conducted without prior approval from the District Office. The District Office shall be notified at least three days before the test is conducted. The test results shall be filed with the District Office, on a Commission-approved form, within 30 days of the completion of the test. Wells for which mechanical integrity tests are conducted pursuant to this subparagraph are not subject to the annual test requirements of subparagraph (E) of this paragraph.

(3) Proper plugging is the responsibility of the operator of the well. For purposes of plugging responsibility, the Commission will presume that the operator designated on the most recent Commission-approved Producer's Transportation Authority and Certificate of Compliance was the person responsible for the physical operation and control of the well at the time the well was abandoned or ceased operation. This presumption may be refuted at a hearing called for the purpose of determining plugging responsibility.

(c) General plugging requirements.

(1) In plugging wells, it is essential that all formations bearing usable quality water, oil, gas, or geothermal resources be protected. All cementing operations during plugging must be performed under the direct supervision of the operator or his authorized representative, who shall not be an employee of the service or cementing company hired to plug the well. Direct supervision means supervision on location at the well site.

(2) Cement plugs shall be set to isolate each productive horizon and usable quality water strata. A "productive horizon," as used in this rule, is defined as any stratum known to contain oil, gas, or geothermal resources in commercial quantities in the area.

(3) Cement plugs must be placed by the circulation or squeeze method through tubing or drill pipe.

(4) All cement for plugging shall be an appropriate API oil well cement without volume extenders and mixed in accordance with API standards. Slurry weights shall be reported on the cementing report. The District Director may require specified cementing compositions to be used in special situations; for example, when high temperature, salt section, or highly corrosive sections are present.

(5) Operators shall use only cementers approved by the Director of Field Operations. Cementing companies, service companies, or operators can qualify as approved cementers by demonstrating that they are able and qualified to mix and pump cement in compliance with this rule. If the Director of Field Operations refuses to administratively approve a cementing company, the company may request a hearing on the matter. After hearing, the examiner shall recommend final action by the Commission.

(6) The District Director may require additional cement plugs to cover and contain any productive horizon or to separate any water stratum from any other water stratum if the water qualities or hydrostatic pressures differ sufficiently to justify separation. The tagging of any such plugs and respotting may be required.

(7) For onshore or inland wells, a 10-foot cement plug shall be placed in the top of the well, and casing shall be cut off three feet below the ground surface.

(8) Mud-laden fluid of at least 9 1/2 pounds per gallon shall be placed in all portions of the well not filled with cement. The hole must be in static condition at the time the plugs are placed.

(9) Non-drillable material that would hamper or prevent re-entry of a well shall not be placed in any wellbore during plugging operations, except in the case of a well plugged and abandoned under the provision of 16 TAC §3.35. Pipe and unretrievable junk shall not be cemented in the hole during plugging operations without prior approval by the District Director.

(10) All cement plugs, except the top plug, shall have sufficient slurry volume to fill 100 feet of hole, plus 10% for each 1,000 feet of depth from the ground surface to the bottom of the plug.

(11) After plugging work is completed, the operator must fill the rat hole, mouse hole, and cellar, and must remove all loose junk and trash from the location. All pits must be backfilled within a reasonable period of time.

(d) Plugging requirements for wells with surface casing.

(1) When insufficient surface casing is set to protect all usable quality water strata and such usable quality water strata are exposed to the wellbore when production or intermediate casing is pulled from the well or as a result of such casing not being run, a cement plug shall be placed from 50 feet below the base of the deepest usable quality water stratum to 50 feet above the top of the stratum. This plug shall be evidenced by tagging with tubing or drill pipe. The plug must be res potted if it has not been properly placed. In addition, a cement plug must be set across the shoe of the surface casing. This plug be a minimum of 100 feet in length and shall extend at least 50 feet above and below the shoe.

(2) When sufficient surface casing has been set to protect all usable quality water strata cement plug shall be placed across the shoe of the surface casing. This plug shall be a minimum of 100 feet in length and shall extend at least 50 feet above the shoe and at least 50 feet below the shoe.

(3) If surface casing has been set deeper than 200 feet below the base of the deepest usable quality water stratum, an additional cement plug shall be placed inside the surface casing across the base of the deepest usable quality water stratum. This plug shall be a minimum of 100 feet in length and shall extend from 50 feet below the base of the deepest usable quality water stratum to 50 feet above the top of the stratum.

(e) Plugging requirements for wells with intermediate casing.

(1) For wells in which the intermediate casing has been cemented through all usable quality water strata and all productive horizons, a cement plug meeting the requirements of subsection (c)(10) of this section shall be placed inside the casing and centered opposite the base of the deepest usable quality water stratum, but extend no less than 50 feet above and below the stratum.

(2) For wells in which intermediate casing is not cemented through all usable quality water strata and all productive horizons, and if the casing will not be pulled, the intermediate casing shall be perforated at the required depths to place cement outside of the casing by squeeze cementing through casing perforations.

(f) Plugging requirements for wells with production casing.

(1) For wells in which the production casing has been cemented through all usable quality water strata and all productive horizons, a cement plug meeting the requirements of subsection (c)(10) of this section shall be placed inside the casing and centered opposite the base of the deepest usable quality water stratum and across any multi-stage cementing tool.

(2) For wells in which the production casing has not been cemented through all usable quality water strata and all productive horizons and if the casing will not be pulled, the production casing shall be perforated at the required depths to place cement outside of the casing by squeeze cementing through casing perforations.

(3) The District Director may approve a cast iron bridge plug to be placed immediately above each perforated interval, provided at least 20 feet of cement is placed on top of each bridge plug. A bridge plug shall not be set in any well at a depth where the pressure or temperature exceeds the ratings recommended by the bridge plug manufacturer.

(g) Plugging requirements for well with screen or liner.

(1) If practical, the screen or liner shall be removed from the well.

(2) If the screen or liner is not removed, a cement plug in accordance with subsection (c)(10) of this section shall be placed at the top of the liner.

(h) Plugging requirements for wells without production casing and open-hole completions.

(1) Any productive horizon or any formation in which a pressure or formation water problem is known to exist shall be isolated by cement plugs centered at the top and bottom of the formation. Each cement plug shall have sufficient slurry volume to fill a calculated height as specified in subsection (c)(10) of this section.

(2) If the gross thickness, of any such formation is less than 100 feet, the tubing or drill pipe shall be suspended 50 feet below the base of the

formation. Sufficient slurry volume shall be pumped to fill the calculated height from the bottom of the tubing or drill pipe up to a point at least 50 feet above the top of the formation, plus 10% for each 1,000 feet of depth from the ground surface to the bottom of the plug.

(i) The District Director shall review and approve the "Notification of Intention to Plug" in a manner so as to accomplish the purposes of this section. The District Director may approve, modify, or reject the operator's "Notification of Intention to Plug". If the proposal is modified or rejected, the operator may request a review by the Director of Field Operations. If the proposal is not administratively approved, the operator may request a hearing on the matter. After hearing, the examiner shall recommend final action by the Commission.

(ii) Plugging Horizontal Drainhole Wells. All plugs in horizontal drainhole wells shall be set in accordance with subsection (c)(10) of this section. The productive horizon isolation plug shall be set from a depth 50 feet below the top of the productive horizon to a depth either 50 feet above the top of the productive horizon, or 50 feet above the production casing shoe if the production casing is set above the top of the productive horizon. If the production casing shoe is set below the top of the productive horizon, then the productive horizon isolation plug shall be set from a depth 50 feet below the production casing shoe to a depth that is 50 feet above the top of the productive horizon. In accordance with subsection (c)(6) of this section, the Commission or its delegate may require additional plugs.

CHAPTER 2

CARBON MANAGEMENT AND HYDROGEN REQUIREMENTS IN OIL SANDS OPERATIONS

Ali Elkamel, J. Guillermo Ordorica-Garcia, Peter Douglas, and Eric Croiset
Department of Chemical Engineering
University of Waterloo
Waterloo, Ontario, Canada

1 THE OIL SANDS INDUSTRY

1.1 Introduction

Current global market and political environments have caused oil prices to soar in the past few years. Concurrently, oil supply concerns in North America have contributed to an explosion of new oil sands developments. Beyond Canada, markets in the United States and Asia have unequivocally expressed their interest in Canada's oil resources. Within this framework, the importance of the oil sands industry is evident both now and in the foreseeable future. The challenges to develop and exploit the oil sands resource, however, are substantial. In simple terms, these challenges are recoverability and economics. As with most underground oil deposits, only a fraction of the reservoir is either technically or economically recoverable. Current technology allows bitumen recovery levels between one third and one half of the total reserves in the Athabasca region (Lee 1996). Also, the recovered bitumen must be upgraded to synthetic crude oil (SCO) before it can be marketed. The combined capital investments required to extract and upgrade bitumen are elevated, which makes SCO production more expensive than conventional oil production.

Oil companies extract bitumen from the oil sands and upgrade a good portion of it to SCO. In the extraction and upgrading processes, vast quantities of energy in the form of electricity, hydrogen, steam, hot water, diesel fuel, and natural gas are consumed. This energy is almost entirely produced using fossil feedstocks/fuels (whether directly or indirectly), which inevitably results in significant greenhouse gas (GHG) atmospheric emissions. The combined CO_2 emissions from energy production for bitumen extraction and upgrading make the oil sands industry the single largest contributor to GHG emissions growth (McCulloch et al. 2006).

A key component in the development of effective CCS-based GHG mitigation strategies for the oil sands industry is an inventory of their energy demands and associated GHG emissions. It is vital for the decision-making processes of policy makers, as well as industries and investors, to forecast how much energy is required to realize any given future SCO and/or bitumen production level.

The energy demands for oil sands operations, being as substantial as they are, are intrinsically tied to the anticipated growth in bitumen extraction and upgrading. Energy commodities such as power, hydrogen, and steam are produced in power plants, hydrogen plants, boilers, and other units. Thus, if production levels rise, so will the energy demands of the oil sands industry, which will, in turn, require that more energy-producing units be built and commissioned to uninterruptedly sustain operations.

In addition to energy demands, growth strategies for the oil sands industry in an increasingly CO_2-constrained world must be based on a sound knowledge of the magnitude of its GHG emissions and their sources. To implement CCS as a GHG abatement strategy in an economic fashion, it is imperative to first develop a CO_2 inventory that quantifies emissions associated with fossil fuel use in energy-producing units within the oil sands industry. The resulting benefits for the industry would be significant reductions in CO_2 emissions, while ensuring affordable energy availability for sustained SCO and bitumen production growth.

The potential economic advantages derived from implementing CCS in oil sands operations include the potential revenue from CO_2 credits or commercially supplying CO_2 for EOR or ECBM to nearby users, once a suitable CO_2 distribution infrastructure is in place. Furthermore, future environmental legislation limiting GHG intensity and emissions in oil sands operations could well drive oil sands operators to incorporate CCS schemes in their operations.

As the oil sands industry continues its accelerated expansion, serious economic and environmental impacts are expected. These impacts will be largely shaped by the magnitude of the bitumen and SCO production levels and by the technologies used. These two variables combined will also dictate the composition and scale of the energy demands for oil sands operations. Ultimately, energy production for bitumen extraction and upgrading is responsible for all nonfugitive GHG emissions of the oil sands industry. Likewise, the cost associated with meeting these demands has a large influence on the production costs in oil sands operations.

The energy balance in oil sands operations consists of the energy demand side and the energy supply side. The former is related to the processes used to extract and upgrade bitumen and is a function of oil production. The latter is a combination of energy producing plants, typically employing different technologies, in numbers sufficient to match the energy demands—the *energy infrastructure*. This term shall be used extensively in this chapter, and is also the focus of both modeling and optimization efforts.

In the next few sections, we will review the bitumen-extraction and upgrading processes found in the oil sands industry as well as other relevant energy production technologies.

1.2 Bitumen Extraction

The only commercial technologies currently used for extracting bitumen from oil sands are in situ or surface mining. According to some estimates (Malherbe et al. 1983), less than one-tenth of the total in-place bitumen reserves can be extracted by mining. The remainder must be recovered by using in situ technologies.

Surface mining is a well-established technology that has seen its share of improvements since its inception in the 1960s. As such, the bitumen recovery rates are high, in excess of 95 percent. In situ technology will account for the majority of the growth in oil sands operations in the next 20 years (Alberta Chamber of Resources 2004). The technology used in in situ projects was developed in the 1970s, and to date, the research in this area is squarely aimed at improving recoveries and reducing steam consumption. Interest in this technology is high among oil companies. A recent example is the Long Lake project—operated by Opti Canada Inc. and Nexen Inc—which began operations in late 2007 (Long Lake 2007).

Surface mining involves three stages: overburden removal, oil sands mining, and bitumen extraction. The layer covering the oil sands must be removed prior to mining. This layer, commonly known as *overburden,* consists of sublayers of decaying vegetation, stagnant water, wet sands, and clay. Trucks and shovels strip off the rocky, claylike overburden and place it in mined-out pits. Once the overburden is removed, the thick deposit of oil sand is exposed.

Current oil sands mining technologies can be classified as conventional or hydrotransport. Conventional mining techniques were developed in the early 1950s, while the hydrotransport method was developed in the late 1980s by Syncrude. In conventional mining processes, walking dragline/reclaimers or shovel/trucks transport the sand to the bitumen-extraction plant. Conventional mining has been almost completely phased out in oil sands operations and has been replaced by oil sands hydrotransport. In the hydrotransport process, hydraulic shovels dig the oil sand and feed into the trucks, which deliver the material into a crusher. A mixer combines the oil sand from the crusher with hot water (35–50°C) to create slurry that is pumped via pipeline to the extraction plant. By the time the slurry reaches the plant, it is already conditioned, so the first step in the hot water process (see Figure 1) can be omitted. Hydrotransport technology improves the energy efficiency and environmental performance of mining operations, as less hot water is required to process the mined sand.

The only commercially proven process to extract bitumen from mined oil sands is known as the *hot water process* (HWP). It was developed between 1940 and 1960. A schematic of the HWP is shown in Figure 1. The process consists of three main steps: (1) conditioning, (2) separation, and (3) froth treatment.

In the conditioning stage, hot water (35–50°C) and caustic soda are added to the mined oil sand. The resulting slurry is agitated in rotary drums, known as *tumblers*. The temperature in the tumblers is maintained by steam injection. Bitumen is stripped from the individual sand grains in this step. The resulting slurry is a mixture containing water, sand grains, and bitumen globules of nearly identical size.

The conditioned slurry passes through a vibrating screen before entering the primary separation vessel (PSV). Oversized rocks, clay lumps, and metal pieces from excavation equipment are screened out and sent back to the mine as over-sized overburden.

Once screened, the slurry is diluted with water and a sand-rich mixture called *middlings*, which is recycled from a downstream unit. In the PSV, mineral

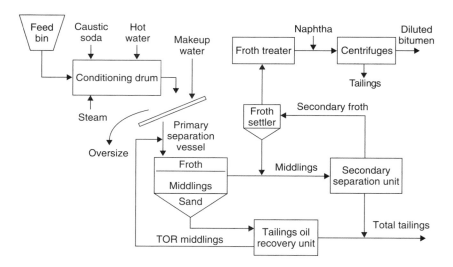

Figure 1 Hot water process flowchart

particles readily settle. At the same time, bitumen globules float to the top, forming a bitumen-rich froth. The bitumen froth layer is skimmed off and sent to treatment, which is the last stage in the bitumen-extraction process.

The bottom stream leaving the PSV contains mostly water-saturated sand, clay, and fines. However, some bitumen is entrained in this sludge, called *tailings*. The tailings are processed in the tailings oil recovery (TOR) unit, where additional bitumen is recovered and sent back to the PSV.

The intermediate layer in the PSV, the *middlings*, mainly consist of water and solids, but they also contain suspended silt and clay fines. A middlings stream is continuously withdrawn from the PSV and sent to the secondary separation unit (SSU) for further bitumen extraction. A bituminous froth is formed in the SSU, which is later sent to a settler, to improve its quality. The treated SSU froth is mixed with the PSV froth, heated, and deareated in the froth treater. The treated froth is later diluted with naphtha to reduce the bitumen's specific gravity/viscosity. In the final step, the entrained solids and water are removed in a two-stage centrifugation process. The resulting bitumen product contains less than 0.5 percent solids and 4–7 percent water (mass).

In addition to water-based bitumen-extraction technology, some development work has been conducted on solvent-based bitumen-extraction methods. A key problem with this approach is the solvent recovery from the sand after bitumen extraction. Most of the time, large quantities of solvent are lost, through intersitial transport in the sand.

No solvent-based method is commercially used for bitumen extraction from oil sands. Most of the methods have only been developed at a conceptual stage, or at laboratory scale. More details about solvent-based processes can be found in (Malherbe et al. 1983),

In situ technology was originally developed to recover heavy oil from deep underground reservoirs. However, the shared characteristics of bitumen and heavy oil favored the application of this technology in oil sand operations. Two main in situ recovery technologies for oil sands exist: thermal and emulsification processes. The former involve the injection of one or combinations of the following: steam, air, or water. The latter technology involves the use of steam plus chemicals. These chemicals promote emulsification so that the bitumen may be transported to the surface. There is almost no literature available on in situ emulsification techniques. The bulk of the in situ research is focused on thermal recovery techniques, as it is generally acknowledged as the best in situ recovery technology.

The focus of this chapter is exclusively on thermal in situ technologies, as they are the only ones that have been successful in commercial applications. Specifically, bitumen recovery by steam injection using steam-assisted gravity drainage (SAGD) is the technology of choice in this study. SAGD extraction involves drilling horizontal well pairs in the reservoir. Steam injected via the upper well rises through the deposit and heats the bitumen. The hot bitumen

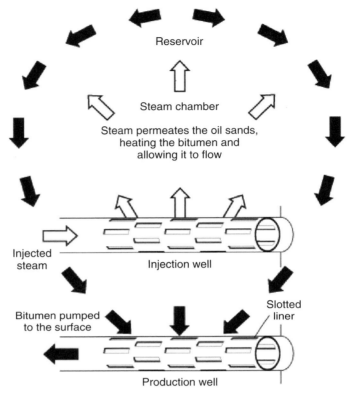

Figure 2 Steam-assisted gravity drainage (SAGD)

separates from the sand and is collected along with condensed steam (water) into the lower well, and is then pumped to the surface. A schematic of this process is shown in Figure 2.

1.3 Bitumen Upgrading

The raw bitumen recovered from oil sand resembles a thick black tar with an extremely high specific gravity of about 9 API. This makes it impossible in practice to pipeline the bitumen to refineries. One alternative is to dilute the bitumen with naphtha so that it can be transported. Alternatively, the bitumen can be processed at the oil sand site to produce a higher-quality product suitable to be transported by pipeline. Such a process is usually referred to as *upgrading,* and its product is known as synthetic crude oil (SCO) or *synthetic crude*.

Another reason to upgrade bitumen is to remove undesirable species —sulfur, nitrogen, carbon, aromatics, vanadium, and nickel—to a degree required for refinery processing. Many refineries, especially in the United States, are designed to handle conventional light and sweet crudes and are ill equipped to process heavy oils or bitumen. Thus, bitumen upgrading to SCO is required to reach these markets.

Meyers (1984) provides the following summary of six key properties of bitumen that have prompted the development of upgrading technologies:

1. Extremely high viscosity at ambient temperatures, which renders pipeline transportation virtually impossible without the addition of substantial quantities of diluent (natural gas condensate or naphtha)

2. Hydrogen deficiency relative to conventional light- and medium-gravity crude oils

3. Large percentage of high-boiling-point material, which limits the volume of virgin transportation fuels that may be recovered by simple separation processes

4. Substantial quantities of resins and asphaltenes, which act as undesirable coke precursors in high-temperature refining operations

5. High sulfur and/or nitrogen content, which necessitates severe hydroprocessing of the distillate fractions to produce fuels or intermediate products for refineries

6. High metals content, particularly vanadium and nickel, which causes deactivation of downstream cracking catalysts

Generally speaking, bitumen-upgrading technologies can be classified as: (1) coking processes and (2) hydrotreating processes. These processes convert raw bitumen into synthetic crude oil by using heat and hydrogen as cracking agents, respectively. Traditional oil sands upgrading plants relied heavily on thermal-based coking in the past. Currently, coking technologies are supplemented by hydrocracking processes in various process configurations. These combinations result in increased liquid fraction yields, lower-sulfur products, and higher bitumen conversions to SCO.

All of the upgrading plants in existing oil sands projects follow a similar process steps sequence for SCO production. A generalized sequence is shown in Figure 3. In each case, bitumen is fed to a primary upgrading process in which conversion of the high-boiling range components in the bitumen occurs. Overall, the products of upgrading are:

- Hydrocarbon off-gases—single, double, and triple-bonded with three to six C atoms
- Cracked liquid distillates—naphtha and light and heavy gas oils
- Residue fraction—petcoke or pitch

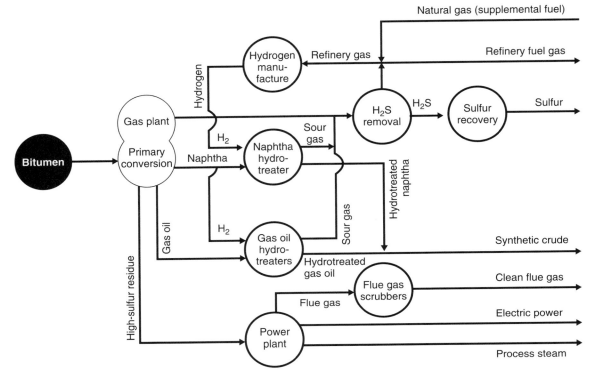

Figure 3 Generalized conventional bitumen upgrading sequence

The liquid distillate fractions contain large concentrations of nitrogen and sulfur. These cracked-liquid distillates also contain aromatic species and, in the case of the naphtha fraction, substantial olefins. These distillates are therefore hydrotreated before being blended into marketable SCO. The four effects of hydrotreatment are as follows:

1. Olefin and diolefin saturation to provide a stable synthetic crude
2. Sulfur and nitrogen concentration reductions to levels suitable for downstream refining to finished products
3. Limited saturation of aromatic compounds to improve the cetane number and smoke point of diesel and jet fuels
4. A shift to more naphtha and distillates through hydrocracking of the gas-oil fractions

The sour offgas from hydroprocessing is combined with the offgas from primary upgrading, followed by H_2S removal and conversion to elemental sulfur. The resulting sweet gas is then available for use as fuel or as feed to hydrogen production plants.

The coke residue contains a large part of the sulfur and nitrogen and essentially all of the metals and ash. Unfortunately, the high sulfur content of this coke makes it generally unacceptable as a fuel without some form of sulfur removal—which would greatly increase operating costs.

Some of the byproduct coke is stockpiled and natural gas is typically used as the preferred refinery and boiler fuel. However, this practice is not sustainable in the long term. Decreased natural gas production, combined with steep price hikes, is forecasted to occur in the next decade, as conventional gas sources are depleted. Innovative ways to minimize coke production and maximize its utilization must be developed and implemented in oil sands refining operations.

There are six major bitumen-upgrading processes currently being used:

1. Delayed coking
2. Fluid coking
3. Flexicoking
4. LC-Fining
5. The H-Oil process
6. The CANMET hydrocracking process

The first three are thermal coking technologies, while the rest are hydrogen-based. In this chapter, only delayed and fluid coking and LC-Fining are considered, as these are the technologies currently used in large-scale commercial operations in the oil sands industry.

The delayed coking process (Figure 4) is used at Suncor (2007). In its process, the diluted bitumen coming from extraction is first distilled to recover the naphtha

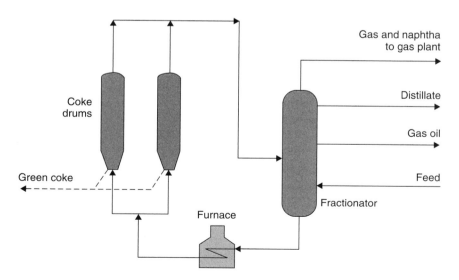

Figure 4 Delayed coking process flowsheet

and recycled back to the extraction plant. The bitumen is then preheated before entering high-temperature coking drums, where it resides for an extended period of time. Light hydrocarbons are vaporized in the coking drums and sent to a fractionating tower, where they are separated into four main streams:

1. Light gases, which are desulfurized and used as fuels
2. Naphtha, which can be upgraded to gasoline
3. Distillate for the production of jet fuels
4. Gas-oil used as a heating fuel or for diesel production

The severe conditions inside the delayed-coker produce a significant amount of coke. This, coupled with low conversion efficiencies, are the major issues of this process.

The largest bitumen fluid coking units in the world belong to Syncrude Canada Ltd. Each one of these units process more than 50,000 tons per day (Malherbe et al. 1983), The process schematic of fluid coking technology is shown in Figure 5. Diluent-free bitumen is fed to the fluid coker. Gas streams containing butanes, naphtha, and gas-oils are generated in the coking drum, along with pet-coke. The gases are sent to a fractionator and are blended into SCO downstream. The petcoke generated in the reactor is burned in a separate vessel, providing all the heat for the thermal cracking reactions. Unburned coke is withdrawn from the burner and stockpiled.

The liquid fraction conversion in fluid coking is higher than in delayed coking. However, coke coproduction is still an issue, although the net petcoke production of the former is lower than the latter. Coke removal and carryover are challenges in fluid-coking processes, and unscheduled shutdowns can occur.

The LC-Fining process is unique among hydrocracking technologies because it can handle the entrained solids in bitumen and operates at relatively low

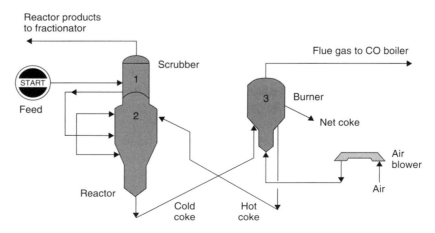

Figure 5   Fluid coking process flowsheet

pressures. Currently, Syncrude Canada and Shell Canada employ this technology in their upgrading processes. The former is a low-conversion process (60 percent), while the latter can reach up to 90 percent bitumen conversion rates (Lott and Lee 2003). For this reason, LC-Fining is Shell's only primary upgrading strategy, whereas the upgrading scheme of Syncrude involves fluid coking in addition to LC-Fining.

The process, shown in Figure 6, employs an expanding-bed reactor. Bitumen and hydrogen react in the presence of the catalyst, which is intermittently added and/or withdrawn to control product quality. The stream leaving the reactor is flashed in two steps to recover unreacted hydrogen before arriving at a gas fractionator. The final products include high-quality distillate gases, naphtha, and gas-oil. The heavy unconverted fraction is a feed suitable for coking or solvent deasphalter.

The unreacted hydrogen is recovered and purified at low pressure, before being mixed with make-up hydrogen and recycled to the reactor. This reduces capital and hydrogen production costs.

An advantage of the LC-Fining process over coking processes is that, aside from the high possible bitumen-to-liquids conversion levels, deep sulfur and metal removal rates can be simultaneously achieved in the reactor. In coking processes, the recovered oil fractions require more severe hydrotreatment downstream to achieve product specifications.

1.4   Energy Production

The extraction of bitumen and upgrading to SCO consume vast quantities of energy. Each process stage consumes energy in different forms. Of all the energy

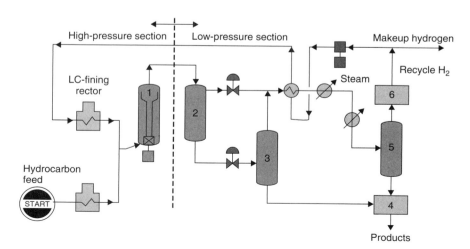

Figure 6   LC-Fining process flowsheet

commodities, steam, power, and hydrogen are produced in auxiliary units in oil sands operations. Diesel and process fuel (natural gas) are either produced internally or purchased. Steam, hot water, power, and hydrogen, are usually produced internally for all oil sands producers. In the following subsections, power and hydrogen production processes are briefly reviewed. Boiler technology for hot water and steam generation is a comparatively undemanding process that is covered elsewhere (Harrell 2002) and, thus, will not be reviewed here.

Hydrogen Production

Bitumen upgrading to synthetic crude consumes sizable quantities of hydrogen. The hydrogen requirements to produce refined petroleum products from bitumen are estimated to be 5 to 10 times larger than those to produce the equivalent refined products from conventional crude. It is anticipated that the projected expansion in oil sands upgrading operations over the next decade will quadruple the current western Canadian hydrogen production capacity from about 500 million SCFD to approximately 2,000 million SCFD (Thambimuthu 2003).

Currently, the most prevalent method of hydrogen production for bitumen upgrading and refining operations in the oil sands is the steam reforming of natural gas. Figure 7 shows a schematic of a typical steam methane reforming (SMR) plant. The gas is first treated to remove poisons such as sulfur and chloride to maximize the life of the downstream reformer and catalysts. Steam is produced in a boiler, and natural gas reacts with it over a catalyst in the reformer. The CO in the hydrogen-rich gas leaving the reformer is shifted with additional steam to produce CO_2 and more H_2 in the shift reactor. The shifted gas is treated in an amine absorption column, where bulk removal of CO_2 occurs. The clean H_2 gas is then refined in a pressure swing adsorption (PSA) system, to a purity of 99.99 percent. The tail gas from the PSA is compressed and recycled to the reformer, where it is burned as a fuel. CO_2 and H_2 are the final products of the plant.

The SMR plant in Figure 7 can operate in both CO_2 capture and no CO_2 capture modes. In the former, the CO_2 recovered in the amine unit is dehydrated and compressed, whereas in the latter, the CO_2 is vented to the atmosphere.

An emerging hydrogen production technology is the gasification of hydrocarbons. *Gasification,* in essence, refers to the reaction of hydrocarbons with oxygen and steam to yield a hydrogen-rich synthetic gas. This synthetic gas (or *syngas*)

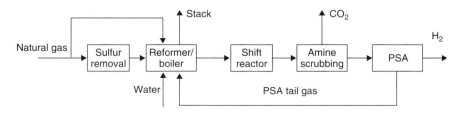

Figure 7   Natural gas steam reforming plant with CO_2 capture

can be used directly as fuel in a power plant or as a feed to synthesize other gaseous or liquid chemicals. Currently, about one-fifth of the hydrogen in the world is produced by this route (Kruse et al. 2002).

Figure 8 depicts a flow diagram of a typical gasification plant using coal as a feedstock. The coal is pulverized and slurried with water before being injected to the gasifier, where it reacts with steam and oxygen. The raw syngas is then cleaned of particulate matter by water quenching. The solids-free coal gas is then shifted with steam on a high/low (temperature) catalytic reaction, consisting of two reactors. The shifted gas, containing mainly hydrogen and CO_2, is cooled prior entering a physical absorption system. H_2S is removed first. Sulfur is recovered in a Claus/SCOT plant. The sulfur-free coal gas enters a CO_2 absorption system, where the bulk of the carbon dioxide is captured. When operating as a CO_2 capture hydrogen plant, the captured CO_2 is dried and compressed, and is ready for export. Otherwise, the CO_2 is removed from the syngas and vented to the atmosphere when the solvent is regenerated. The hydrogen-rich gas leaving the CO_2 absorption system is purified in a PSA unit, thus generating hydrogen with a purity of 99.99 percent. The PSA purge gas is burned in a combined cycle to generate electricity and steam for internal plant consumption.

Chiesa et al. (2005) provide an excellent coverage of the performance, emissions, and costs of hydrogen production via gasification with and without CO_2 capture.

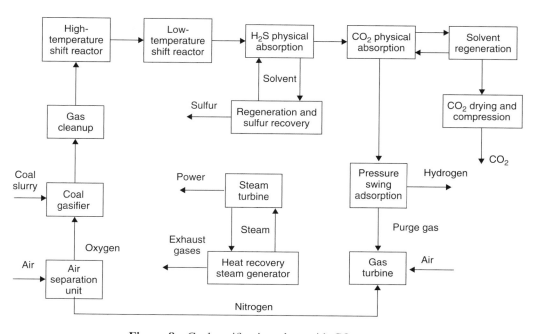

Figure 8   Coal gasification plant with CO_2 capture

Although other techniques, such as water electrolysis, exist for hydrogen production, their limited scope of application and high cost (Kruse et al. 2002) precluded their application in oil sands operations. Likewise, hydrogen production via thermonuclear processes (Schultz et al. 2003), although promising, is currently at the early conceptual stages of its development. The reader must note that the scope of this chapter is predominantly on technologies that have commercial status and, thus, have the potential to be readily implemented in the oil sands industry within a two-decade timeframe.

Power Production

A number of power generation technologies are employed in oil sands operations: natural gas combined cycle (NGCC), supercritical pulverized coal (PC), integrated gasification combined cycle (IGCC), and oxyfuel. These power plants can have CO_2 capture or not. In the former case, there can be three categories, according to the CO_2 capture mode: postcombustion, precombustion, and oxycombustion. The plants with postcombustion CO_2 capture include NGCC and PC units. The precombustion plants are IGCCs. The oxyfuel plants are fueled by coal or natural gas. Schematics of these plants and capture processes are depicted in Figure 9.

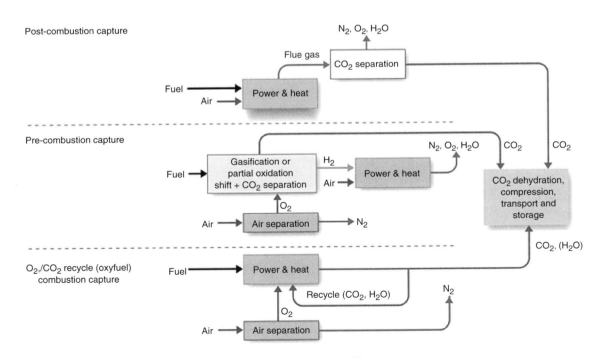

Figure 9   Power plants with CO_2 capture

NGCC and PC plants are the most common types of power plants (Rubin et al. 2004). PC plants burn coal in a boiler to raise steam, which drives a turbine, thus generating electricity. NGCC plants burn gas in a turbine, where a portion of the power is produced. The hot combustion gases exiting the turbine are then used for steam production in a heat recovery steam generator. The steam is subsequently used in a steam turbine, where additional power is generated. In conventional NGCC and PC plants, the flue gases are vented to the atmosphere through stacks.

When operating in CO_2-capture mode, NGCC and PC plants employ a chemical solvent to wash the flue gas downstream of the turbine in a scrubber. The most commonly used solvent is monoethanol amine (MEA). The CO_2 dissolves in the MEA, and is then thermally recovered in a stripping column, where the solvent is regenerated. The recovered CO_2 is then dried and compressed for export. The MEA is recycled to the scrubber and the cycle is repeated.

IGCC power plants are quickly gaining popularity, as they offer higher efficiencies than conventional coal-fired plants with lower overall emissions. More important, IGCC plants produce syngas streams with high CO_2 concentrations at high pressure. This reduces the overall volume of gas to be treated when CO_2 capture is contemplated, which positively impacts the costs of CO_2 removal.

A flow diagram of an IGCC power plant with and without CO_2 capture is shown in Figure 10. This IGCC plant operates in an almost identical way as

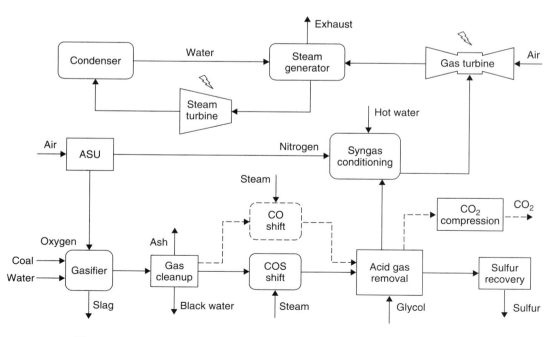

Figure 10   IGCC power plant with and without CO_2 capture flow diagram

the hydrogen production plant shown in Figure 8 and described earlier. When operating without CO_2 capture, the CO in the syngas is not steam-shifted after particulate removal. Hence, only H_2S is removed in the acid gas removal system and the syngas (containing H_2, CO, and CO_2) is burned in the gas turbines. In CO_2-capture mode (represented by dotted lines), all the CO_2 in the syngas is shifted to CO_2, yielding a syngas composed mostly of H_2 and CO_2. The CO_2 and H_2S are then removed separately in the acid gas removal unit and the hydrogen-rich syngas is burned with oxygen in a turbine for power generation.

The acid gas removal section of an IGCC plant typically removes CO_2 and H_2S from the syngas in a two-step process, as seen in Figure 10. An alternative configuration in which CO_2 and H_2S are co-captured simultaneously is possible, however. This results in the elimination of the sulfur recovery step and in a much simpler acid gas removal process, which greatly reduces the capital costs of the IGCC plant. Hence, in this work, both *separate* CO_2 and H_2S and $CO_2 + H_2S$ *co-capture* power and H_2 plants are included. A comprehensive technoeconomic study of IGCC plants without CO_2 capture, with CO_2 capture, and with CO_2 and H_2S co-capture can be found in Ordorica-Garcia et al. (2006).

Oxyfuel plants, as the name suggests, burn fuel (coal or natural gas) with pure oxygen instead of air in boilers (coal-fired) or turbines (gas operation). The resulting high-combustion temperatures necessitate that a portion of the flue gas be recycled back to the boiler/turbine, to moderate the temperature and prevent damage to materials. The flue gas of this process is composed mostly of CO_2 and H_2O. Once dehydrated, the resulting stream has an elevated CO_2 purity, suitable for underground storage or other uses. If the CO_2 is utilized, the atmospheric emissions of oxyfuel plants are negligible.

In principle, any fossil feedstock can be used in oxyfuel combustion, which makes this technology attractive in refinery applications, where fuel gas and other low-value feedstocks are available (Jordal et al. 2004). In practice, however, most of the research into oxyfuel technology focuses on coal- and natural-gas-fired power plants. More information on oxyfuel plants can be found in Davidson (2007).

2 MODELING OF OIL SANDS OPERATIONS

2.1 Introduction

This section discusses the modeling of the energy demands and GHG emissions associated with the oil sands industry (Ordorica-Garcia et al. 2007). The modeling is based on plant- and process-specific data for bitumen extraction and upgrading to SCO. The modeling framework quantifies the demands for power, H_2, steam, hot water, natural gas, and diesel fuel of the oil sands industry for given production levels of SCO and bitumen. Alongside the demands, the modeling process can estimate the resulting CO_2 emissions and CO_2 emissions intensity of SCO and bitumen production, using current energy-production technologies. We will illustrate the modeling of oil sands operations on a case study from Alberta oil sands.

As was discussed earlier, bitumen can be extracted by surface or in situ techniques. Surface techniques involve mining the oil sands and separating the bitumen by using the hot-water process (Meyers 1984). In situ techniques involve injecting an external agent in the underground reservoir, thus forcing the bitumen (and other substances) out of the basin. Steam-assisted gravity drainage (SAGD), which uses steam to extract bitumen from oil sands, is currently the prevalent in situ technology in the oil sands industry.

The bitumen produced by mining or in situ methods can be upgraded to SCO or diluted with naphtha solvent before being sold to refineries. To reflect such variety of commercial oil sands operations in Alberta, the OSOM includes three products:

1. Mined bitumen, upgraded to SCO
2. SAGD bitumen, upgraded to SCO
3. SAGD bitumen, diluted

Likewise, several technologies are employed in the upgrading process. In what follows, three technologies are considered for all mass and energy balances:

1. LC-Fining (LCF) + Fluid coking (FC) + Hydrotreatment (HT) – Syncrude
2. Delayed coking (DC) + Hydrotreatment – Suncor
3. LC-Fining + Hydrotreatment—Shell-Albian sands

The bitumen upgrading schemes correspond to the three leading oil sands operators, which currently extract and upgrade bitumen commercially to SCO in the Athabasca region (Alberta Energy and Utilities Board 2004). The modeling is based on published information for upgrading processes 1-3 as well as plant-specific data, where available.

The oil sands producers discussed are grouped by bitumen-extraction technology and upgrading scheme and are shown in Table 1.

SCO production from mined bitumen (producers A1–A3) is accomplished in four stages: (1) mining, (2) conditioning/hydrotransport, (3) extraction, and (4) upgrading. In stage 1, the oil sand is mined out of the ground by hydraulic shovels and transported by truck to the next process stage. In stage 2, the sand is mixed with hot water and chemicals and agitated to separate the bitumen from the sand. In stage 3, the resulting slurry is washed with hot water in separation cells; in which air and steam addition cause the bitumen to rise to the surface. The bitumen froth is then mechanically separated from the mostly water and sand slurry, deaerated and diluted with naphtha. The diluted bitumen is then centrifuged to remove traces of sand and water and is then ready for upgrading. In stage 4, the naphtha solvent is distilled from the bitumen and sent back to stage 3. The bitumen is then processed in a vacuum distillation unit where the lighter

Table 1   Aggregate Production Estimates for All OSOM Producers

Producer	Description	2003 Daily* Barrels (1000 bbl/d)
A1	Mined bitumen upgraded by LCF+FC+HT	231
A2	Mined bitumen upgraded by DC+HT	213
A3	Mined bitumen upgraded by LCF+HT	94
A	Total mined SCO production	538
B1	SAGD bitumen upgraded by LCF+FC+HT	0
B2	SAGD bitumen upgraded by DC+HT	0
B3	SAGD bitumen upgraded by LCF+HT	0
B	Total SAGD SCO production	0
A+B	Total SCO production	538
C	Total SAGD diluted bitumen production	350

*This column shows the production of a typical oil sands operation that will be considered in the case study of Section 3.

oil fractions are recovered. The bottoms are then cracked either thermally or by hydrogen addition processes or by a combination of both. The resulting products include naphtha, light and heavy gas oils, and petroleum coke, depending on the cracking method. Finally, all the oil fractions are sent to hydrotreaters in which sulfur and nitrogen compounds are removed by hydrogen addition. The treated fractions are blended together into a SCO, with a high API number and low sulfur content.

The production of SCO from SAGD bitumen (producers B1–B3) involves two stages: (1) in situ extraction and (2) upgrading. In stage 1, steam is injected into the underground bitumen reservoir through an injection well. The heated bitumen alongside condensates, and solution gas is collected in a second well (parallel to the injection well) and pumped out of the reservoir. Once condensate and solution gas have been removed from the bitumen, diluent naphtha is added, and the diluted bitumen is sent to upgrading. Stage 2 for producers B1–B3 is much the same as stage 4 for producers A1–A3, as the upgrading schemes considered in this study are the same.

The production of bitumen via SAGD (producers C) only involves stage 1 as described for producers B1–B3.

We will illustrate the calculation of mass and energy balances of each of the process stages for producers A1–A3, B1–B3, and C. The CO_2 emissions associated with SCO and bitumen production are calculated by determining each stage's net demand for a variety of fossil-fuel-intensive inputs, as shown in Figure 11. Each one of these "energy commodities" is produced in specific plants or units, which are modeled based on their real-world counterparts. For instance; all of the H_2 required for bitumen upgrading for producer A2 is produced in natural gas steam reforming plants. We will calculate the amount of natural gas required to produce the H_2 and also the associated CO_2 emissions, among other pertinent model outputs. Likewise, the demands for all of the other commodities

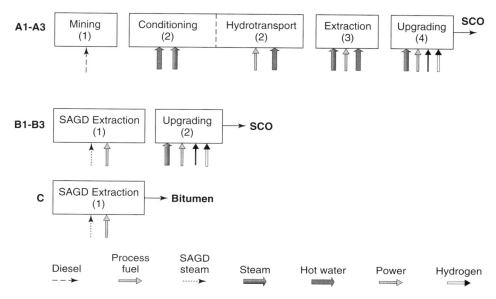

Figure 11   Energy commodities to process stages for producers A1–A3, B1–B3, and C

are calculated, along with their corresponding fossil feedstock/fuels consumption and resulting CO_2 emissions. The calculations hence include the CO_2 emissions per process stage, per SCO producer, as well as their CO_2 intensity. The CO_2 intensity is defined as the amount of CO_2/GHG emitted per unit of SCO or bitumen produced (given in tonnes of CO_2 eq./bbl).

2.2   Process Flow Diagram and Assumptions

All energy production operations are assumed to be based on conventional technologies. Consequently, natural gas (NG) and steam methane reforming (SMR) are the fuel and technology of choice for H_2 production, respectively. All the steam and hot water are produced in natural-gas-fired boilers (SB and SSB). The steam produced in SB boilers is used in mining-based SCO production and in bitumen-upgrading processes, whereas the steam produced in SSB boilers is destined for SAGD bitumen extraction alone. NG is also employed for power generation in combined cycle (NGCC) plants and as process fuel in upgrading. The resulting process diagram is shown in Figure 12.

The right side of the diagram represents the energy demand side. The large boxes correspond to bitumen and SCO producers, which require certain amounts of each one of the seven energy commodities, symbolized by circles in Figure 12. These commodities are produced in the units shown on the left side, which together represent the supply side. We will determine the energy demands of producers A1–A3 and C, based on oil sands mining and bitumen-extraction rates

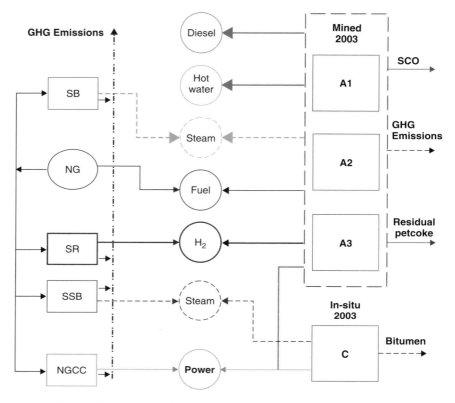

Figure 12 Process diagram of a typical oil production operation

and mass/energy balances particular to each producer. We will then compute feedstock consumption, as well as associated CO_2 emissions on the supply side, based on mass and energy balances for all energy-producing units.

On the demand side, each SCO and bitumen producer requires different energy inputs depending on that particular producer's *modus operandi*. For instance, producer A2 uses surface mining to extract bitumen from the sand and upgrades it by delayed coking, followed by hydrodesulfurization. By contrast, producer C uses SAGD technology to extract bitumen from underground reservoirs and dilutes it with naphtha—without upgrading it. Oil producers are characterized by the technologies they use to extract and upgrade bitumen and thus, their operations are considered to take in place in distinct process stages. These stages, for modeling purposes, are defined based on the operations of actual oil sands producers, as shown in Figure 13.

The modeling task is greatly facilitated by partitioning the operations of all producers into stages. An important approach for modeling these stages is that only the energy inputs specified in Figure 12 are considered. This implies that, for example, the presence of additives in the bitumen hydrotransport or extraction

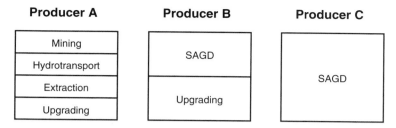

Figure 13 Assumed process/modeling stages per bitumen/SCO producer

stages (producers A1–A3) is excluded from the mass balances for these stages. Only bitumen, hot water, steam, and power are included in such balances. The goal of the model is to quantify energy requirements and CO_2 emissions associated with SCO/bitumen production. Hence, substances/processes that neither consume energy nor cause GHG emissions are not modeled. Nevertheless, substance handling and processing, if it involves energy consumption, is accounted for (e.g., power required to mix additives or to pump the bitumen slurry in mining operations).

2.3 Mining

The energy demands in this stage consist solely of diesel fuel for shovels and trucks, which are used to mine the oil sands. Historical data of oil sands mining rates, as well as the oil sand's bitumen saturation, taken from Alberta Energy and Utilities Board (2004), are used to calculate the fuel demands of these units.

A hypothetical fleet consisting of a variety of diesel-powered mechanical shovels and trucks is shown in Table 2. This fleet includes several commercial shovel and truck models, all of which have different mining capacities and varying mechanical specifications. These specifications were taken from each unit's product brochure available on their respective manufacturer's website. The type and number of units in service was specified to yield an oil sands mining fleet which is generic enough to adequately represent the performance of a real mining fleet, such as the ones used by Syncrude or Suncor (Kreutz et al. 2005).

In the mining stage, the shovels' and trucks' mining rates are determined based on their specified volumetric/mass capacities. The total reference mining fleet's fuel consumption (D) is then calculated, based on each unit's specified engine and individual fuel consumption (f) values as shown in the following equations:

$$D_t = \sum_t f_t N_t \tag{1}$$

$$D_s = \sum_s f_s N_s \tag{2}$$

Table 2   Reference Mining Shovel and Truck Fleet

Type	Model	Units	Engine	Capacity (tonnes)*	Net Power (kW)	Fuel Consumption (l/h)
Shovel	LeTourneau-L2350	4	Cummins QSK 60	72	1,715	375
Shovel	CAT-994D	2	CAT 3516B EUI	63	1,011	248
Shovel	LeTourneau-1850	1	Cummins QSK 60	45	1,492	330
Shovel	Terex-RH400	4	2 Cummins QSK 60C	85	3,280	740
Truck	CAT-793C XQ	20	CAT 3516B HD EUI	218	1,611	406
Truck	Komatsu-930E	11	Komatsu SSDA16V160	290	1,902	447
Truck	Terex-MT5500B	6	Cummins QSK 78	327	2,445	580
Truck	CAT-797B	4	2 CAT 3512B	345	2,513	579
Truck	Liebherr-T282B	2	Cummins QSK 78	363	2,445	580

*Assumed oil sand density of 2.095 tonnes/banked cubic meter (Syncrude 2006)

where N_t and N_S are the number of trucks and shovels in the reference fleet, respectively. Individual truck and shovel utilization for each producer in the OSOM-BC are calculated based on the producers' mining rates. The fuel demands of shovel and trucks corresponding to the calculated mining rates are then determined by calculating a shovel and truck utilization factor (uf) for each producer with respect to the reference mining fleet.

$$D_{P_i} = uf_{t,P_i} D_t + uf_{s,P_i} D_s \qquad (3)$$

While the utilization factors are defined as follows:

$$uf_{t,P_i} = \frac{OS_{P_i}}{OS_{t,ref}} \qquad (4)$$

$$uf_{s,P_i} = \frac{OS_{P_i}}{OS_{s,ref}} \qquad (5)$$

OS is the oil sand mining rate of each producer, and OS_{ref} is the mining rate of shovels/trucks of the reference fleet. Aside from diesel demands, the oil sand composition for each producer is determined on the basis of the specified oil sand mining rate and its bitumen saturation. The reason for this is that individual mass flows of sand and bitumen are required in the mass balance of downstream stages.

2.4   Hydrotransport

In this stage, the mined oil sand is slurried with hot water and pumped to the extraction stage. Hence, energy demands include hot water and power. The slurry has an assumed solids content (SCS) of 70 percent (mass) (Meyers 1984; Syncrude 2006). The water temperature is 35°C. The hot water demands (W) per producer based on the above parameters are given by:

$$W_{H,P_i} = \frac{OS_{P_i}(1 - SCS)}{SCS} \qquad (6)$$

The requirements for pumping bitumen slurry are determined by simulating the pumping of slurries of oil sands with different bitumen saturations in Aspen Plus. The resulting power demands were plotted as a function of bitumen saturation and head requirements. The following empirical pumping power factor (PF) is used to calculate the power demands for pumping bitumen slurries of varying qualities.

$$PF_{H,P_i} = 0.026x_{P_i} + 0.0007 \tag{7}$$

where x is the bitumen content of the oil sand. The power demands (P) for each producer are calculated based on the total slurry produced and user-supplied head values (h) according to:

$$P_{H,P_i} = PF_{H,P_i}(W_{H,P_i} + OS_{P_i})h_{P_i} \tag{8}$$

2.5 Bitumen Extraction

All producers use the two-stage hot water process (HWP) for bitumen extraction, as outlined in Meyers (1984) and shown in Figure 1. In primary extraction, bitumen is separated from the oil sand slurry as froth, using hot water and steam. In secondary extraction, the bitumen froth is diluted with naphtha and centrifuged to remove traces of sand and water.

Energy demands of this process include hot water, steam, and power. The hot (wash) water demands are determined based on the mass balances for each producer as:

$$W_{E,P_i} = OS_{P_i}WWE \tag{9}$$

where W_{E,P_i} is the hot water demand for producer i in the extraction stage and WWE is a model parameter that represents the washwater requirements for primary extraction. The composition of bitumen froth produced in primary extraction is determined using the following empirical correlations:

$$BF_{P_i} = 2.75x_{P_i} + 0.6 \tag{10}$$

$$WF_{P_i} = 0.6738x_{P_i} + 0.0204 \tag{11}$$

where BF and WF are the bitumen and the water mass fraction of the froth, respectively and x is the mass percentage of bitumen in the oil sand. Equations (10) and (11) are empirical models, derived from mass balances for oil sands with different bitumen saturations, presented in Meyers (1984). The bitumen froth (FR) produced in primary extraction is given by the expression:

$$FR_{P_i} = WF_{P_i}(W_{E,P_i} + W_{H,P_i}) + BF_{P_i}OS_{P_i}x_{P_i} + OS_{P_i}(1 - x_{P_i})SF \tag{12}$$

where SF is the sand content of primary froth, a model parameter.

Aside from determining the amount of froth produced and its composition, the percentage bitumen recovery in primary extraction, as well as the quantity of primary tailings produced, can also be calculated. The tailings are mostly composed of sand and water, with traces of bitumen. These tailings are treated for additional bitumen recovery prior being pumped (together with secondary tailings) to storage ponds.

In secondary extraction, the naphtha diluent requirements and steam demands (S) are calculated based on mass balances from Meyers (1984). The steam demands are a function of the bitumen froth.

$$S_{E,P_i} = FR_{P_i} SSE \qquad (13)$$

where SSE is a model parameter required to compute steam requirements in bitumen extraction. The overall power demands (P) in this stage comprise of the power required to pump tailings to disposal (PTA) and the power for diluted bitumen centrifugation (PC):

$$P_{E,P_i} = PTA_{P_i} + PC_{P_i} \qquad (14)$$

Similarly to the hydrotransport stage, the power requirements for tailings transport were determined by simulating the pumping of oil sands tailings with varying bitumen saturations in Aspen Plus. The following empirical equation was thus obtained:

$$TP_{P_i} = 0.001x_{P_i} + 0.0013 \qquad (15)$$

where TP is the pumping power factor and x is the bitumen content of the oil sand. The power demands for each producer are calculated based on their pumping power factor as calculated here and the overall tailing production and user-supplied head values.

The power demands for diluted bitumen centrifugation (two stages) are based on the specifications of an Alpha Laval CH-36B centrifuge (Alfa Laval 2007). The diluted bitumen product can either be sold to refineries, or upgraded to SCO, as is the case with producers A1–A3.

2.6 SAGD Extraction

The thermal extraction of bitumen by in situ methods is modeled according to the data corresponding to Opti-Nexen's Long Lake project (OPTI Canada 2002). The main energy demands are steam and electricity. The steam raised for SAGD extraction has a quality of 80 percent and a pressure of 8000 kPa. After separation, the resulting saturated steam is injected underground at a SOR (steam-to-oil ratio) of 2.4 (OPTI Canada 2002), which is a typical value of an economically viable SAGD operation.

Mass and energy data from the aforementioned source are used to compute the steam demands of producers B and C, as well as the amount of solution gas produced and the power demands for SAGD extraction. The solution gas is assumed to be burned in the steam boilers; thus, the amount of natural gas required for SAGD steam production is lowered accordingly in all calculations.

$$SS_{P_i} = B_{P_i} SOR \tag{16}$$

$$PS_{P_i} = B_{P_i} ERS \tag{17}$$

$$SG_{P_i} = B_{P_i} SOL \tag{18}$$

where SS and PS are the steam and power demands per producer and SG is the solution gas produced during bitumen extraction. B is the bitumen production rate from SAGD operations.

2.7 Bitumen Upgrading

The diluted bitumen from mining and/or thermal operations is upgraded to SCO in the upgrading stage. The modeling of this stage is complex, as it encompasses three possible upgrading routes for bitumen, which are shown in Figure 14.

Energy consumption in this stage is significant. The upgrading of bitumen to SCO requires vast amounts of hydrogen, steam, and power. Additionally, certain upgrading technologies also consume process fuel for heating. The energy demands for each one of the upgrading schemes shown in Figure 14 are based on their individual amounts of bitumen processed.

The first step in the upgrading process is the recovery of naphtha solvent in a distillation column. The recovered naphtha is sent back to the bitumen-extraction stage. The steam requirements for the diluent recovery unit (DRU) are based on mass and energy balances obtained from an ASPEN Plus model of the DRU (Ordorica-Garcia et al. 2007).

The products from the DRU are naphtha, light gas-oil (LGO) and atmospheric-topped bitumen (ATB). The LGO is sent to an LGO hydrotreater for sulfur and nitrogen removal. The ATB can either be sent to the vacuum distillation unit (VDU) or split its flow between the VDU and the LC-Finer. The bottoms of the VDU, called vacuum-topped bitumen (VTB), together with any ATB from the DRU are then sent to the LC-Finer (producers A–B1) or to the delayed coker (producers A–B2) while the LGO and HGO proceed to hydrotreatment for further processing.

The LC-Finers use hydrogen to convert the feed (ATB and/or VTB) into LGO, HGO, and naphtha products. The LC-Finer in upgrading scheme 1 has a lower liquid yield that that of scheme 3 (60 percent versus 90 percent). The specifications for the former LC-Finer were taken from Sunderland (2001), Schumacher (1982), and Van Driesen et al. 1979. The high-conversion LC-Finer was modeled based on data from Meyers (1984) and Sadorah (2005). Total electricity and fuel demands are modeled based on the specifications found in the *Hydrocarbon Processing Handbook* (2007).

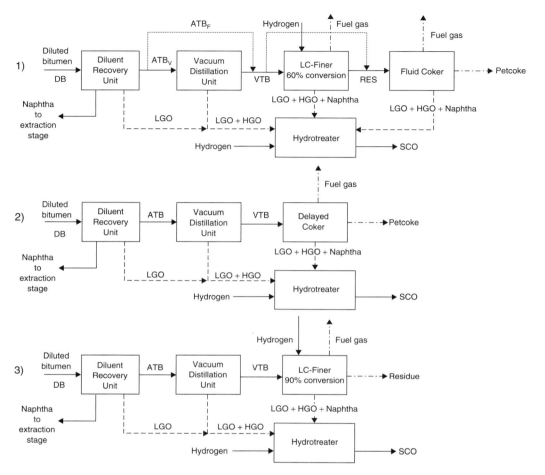

Figure 14 Bitumen upgrading schemes for producers A1–A3 and B1–B3

The product streams from the LC-Finer (LGO, HGO, and naphtha) are sent to hydrotreatment. Some fuel gas is also generated in this unit. This fuel gas is collected and after scrubbing with MEA (mono-ethanolamine), it can be used in the hydrogen plant as fuel (optional). Finally, the bottoms of the low-conversion LC-Finer are sent to the fluid coker (A–B1). In the case of the high conversion LC-Finer (A–B3), the residuum is considered to be a saleable byproduct and is not further processed.

The cokers process the bottoms of upstream units, yielding further LGO, HGO, and naphtha, together with petroleum coke and sour coker gas byproducts. Although the delayed coker consumes a fraction of the total petcoke for fuel, total petcoke production from the cokers is substantial. The petcoke byproduct is not used for energy production, due to its high sulfur and metals content. This petcoke is stockpiled until a better use can be found for it. The sour coker gas, similarly to the LC-Finer fuel gas, is treated with MEA and can be used as fuel in the hydrogen plants.

The fluid coker is modeled on yield data from Sunderland (2001) and Schumacher (1982). The energy demands for steam, power, and process fuel were taken from Meyers (1984). The modeling data for the delayed coker include yields from ConocoPhillips (2007) and energy demands (electricity/process fuel) found in the *Hydrocarbon Processing Handbook* (2007).

The last step in the upgrading process is the hydrodesulfurization of the oil fractions. This is accomplished in individual hydrotreaters for LGO, HGO, and naphtha. The flows of these fractions from all units upstream of the hydrotreaters are fed to their corresponding hydrotreater.

The hydrogen demands for hydrotreatment in the model are calculated based on yield data presented in Sunderland (2001). Additional density data for the oil fractions are taken from Yui and Chung (2001). After hydrotreatment, the treated fractions are blended together into SCO. The properties of the SCO, such as composition and density (which are a function of the upgrading scheme used) are calculated, together with the overall bitumen conversion to SCO.

The sulfurous gas removed from the fractions in the hydrotreaters is sent to the MEA plant for treatment. Aside from the sulfur balance, the model calculates the total sweet fuel gas (from LC-Finer and coker gases) that is available for use as fuel in the hydrogen plants can also be calculated.

The breakdown of the energy demands of the upgrading stage per unit is shown in Table 3 for the different upgrading processes.

Accordingly, the following equations determine power demands (P) for each upgrading scheme:

$$P_{U-1,P_i} = PLL(ATB_F + VTB)_{P_i} + PFC \cdot RES_{P_i} \qquad (19)$$

$$P_{U-2,P_i} = PDC \cdot VTB_{P_i} \qquad (20)$$

$$P_{U-3,P_i} = PLH \cdot VTB_{P_i} \qquad (21)$$

where PLL, PLH, PFC, and PDC are model parameters used to compute power demands of high conversion LC-Finer, low-conversion LC-Finer, fluid coking,

Table 3 Energy Demands Distribution for Upgrading Schemes 1 to 3

Commodity	Units	1	2	3
Power	kW	LC-Finer Fluid coker	Delayed coker	LC-Finer
Hydrogen	tonne/h	LC-Finer Hydrotreater	Hydrotreater	LC-Finer Hydrotreater
Steam	tonne/h	DRU VDU Fluid coker	DRU VDU	DRU VDU
Process fuel	GJ/h	LC-Finer	Delayed coker	LC-Finer

and delayed coking, respectively. ATB, VTB, and RES are feed streams to the above processes.

The steam demands (S) of each producer are given by:

$$S_{U\text{-}k,P_i} = \frac{DB_{P_i} \cdot HDR + (ATB - ATB_F)_{P_i} \cdot HVD + RES_{P_i} \cdot HFC}{\Delta HS} \tag{22}$$

DB is the diluted bitumen entering the DRU. HDR, HVD, and HFC are model parameters representing the heat requirements of the DRU, VDU, and fluid coker, respectively. ΔHS is the enthalpy of the steam consumed in upgrading.

Hydrogen demands for upgrading are divided into hydrogen for hydrodesulfurization (H_T, equation (23)) and hydrogen for hydrocracking (H_L, eq. 24 and 25). The former hydrogen is consumed in the hydrotreaters while the latter is a feedstock to LC-Fining units.

$$H_{T,P_i} = \left(\sum_j \frac{LGO_{P_i,j}}{\rho_{LGO_{P_i}}} \right) \cdot HLG + \left(\sum_j \frac{HGO_{P_i,j}}{\rho_{HGO_{P_i}}} \right) \cdot HHG$$

$$+ \left(\sum_j \frac{NAP_{P_i,j}}{\rho_{NAP_{P_i}}} \right) \cdot HNP \tag{23}$$

$$H_{L\text{-}1,P_i} = (VTB + ATB_F)_{P_i} \cdot HLL \tag{24}$$

$$H_{L\text{-}2,P_i} = VTB_{P_i} \cdot HLH \tag{25}$$

LGO, HGO, and NAP represent the flow rates of each oil fraction from individual process units (j) in upgrading. HLG, HHG, and HNP are model parameters that specify the hydrogen consumption of LGO, HGO, and naphtha hydrotreaters. Two other parameters, HLL and HLH, denote the hydrogen requirements of low-conversion and high-conversion LC-Finers, respectively. ρ is the density of each one of the oil fractions entering the hydrotreaters.

Process fuel is consumed in certain units in the upgrading stage. The specific fuel demands (F) of each producer, according to upgrading scheme are calculated with equations (26) to (28). FRL and FRD are parameters that set the fuel requirements of LC-Fining and delayed coking units, respectively.

$$F_{U\text{-}1,P_i} = (VTB + ATB_F)_{P_i} \cdot FRL \tag{26}$$

$$F_{U\text{-}2,P_i} = VTB_{P_i} \cdot FRD \tag{27}$$

$$F_{U\text{-}3,P_i} = VTB_{P_i} \cdot FRL \tag{28}$$

2.8   Total Energy Demands

The energy demands are computed by adding individual energy commodities for all process stages. The index k is used to differentiate specific energy demands of individual upgrading schemes. These equations summarize the total energy demands:

$$W = \sum_{P_i} (W_{H,P_i} + W_{E,P_i}) \tag{29}$$

$$P = \sum_{P_i} \left(P_{H,P_i} + P_{E,P_i} + PS_{P_i} + \sum_{k=1}^{3} P_{U\text{-}k,P_i} \right) \tag{30}$$

$$D = \sum_{P_i} D_{P_i} \tag{31}$$

$$S = \sum_{P_i} \left(S_{E,P_i} + \sum_{k=1}^{3} S_{U\text{-}k,P_i} \right) \tag{32}$$

$$SS = \sum_{P_i} SS_{P_i} \tag{33}$$

$$H = \sum_{P_i} \left(H_{T,P_i} + \sum_{k=1}^{3} H_{L\text{-}k,P_i} \right) \tag{34}$$

$$F = \sum_{P_i} \sum_{k=1}^{3} F_{P_i} \tag{35}$$

2.9 Energy Supply

The aggregate energy demands from all stages, represented by circles in Figure 12, for all the SCO and bitumen producers are satisfied by the energy-producing plants represented by boxes on the left side of the process flow diagram. SB and SSB represent boilers generating steam at 6300 kPa and 500°C and 80 percent steam at 8000 kPa, respectively. The former steam (S) is used in bitumen extraction, as well as in upgrading operations. The latter (SS) is employed exclusively for SAGD operations. The steam production in boilers SB and SSB is given by the expressions:

$$S_b = \frac{HHV_{NG} \cdot PSP \cdot \eta_{B_b}}{\Delta HS_b} X_b \qquad \forall b \in SB \tag{36}$$

$$SS_b = \frac{HHV_{NG} \cdot \eta_{B_b}}{\Delta HS_b} X_b \qquad \forall b \in SSB \tag{37}$$

where η_B represents the thermal efficiency of the boiler, ΔHS is the enthalpy of the steam, HHV_{NG} is the high heating value of the natural gas fuel and *PSP* equals the percentage of boiler capacity used for steam production. X_b is the natural gas consumption of the boiler.

Boilers SB also produce hot water (W), required in bitumen hydrotransport and extraction:

$$W_b = \frac{HHV_{NG} \cdot (1 - PSP) \cdot \eta_{B_b}}{\Delta HW} X_b \qquad \forall b \in SB \qquad (38)$$

and ΔHW is the enthalpy of the hot water. Since the OSOM-BC is based on the current manner of operation in the Athabasca region oil sands industry, the power demands (P) are met by natural gas combined cycle power (NGCC) plants, according to equation (39). HRP is the specified heat rate of the NGCC plant and X_p is its natural gas consumption.

$$P_p = \frac{HHV_{NG}}{HRP_p} X_p \qquad \forall p \in NGCC \qquad (39)$$

All the hydrogen (H) required for upgrading is produced in steam methane reforming (SMR) plants according to:

$$H_h = \frac{HHV_{NG}}{FCH} X_h \qquad \forall h \in SR \qquad (40)$$

where FCH equals the fuel consumption of SMR plants per unit of hydrogen produced. The steam reforming plant used in the model corresponds to the one described by Simbeck (2004). The power plants are modeled after the NGCC plant also described in Spath and Mann (2000). The steam boiler specifications are derived from Harrell (2002). The natural gas used in this study is western Canadian, with an HHV of 38 MJ/Nm^3.

Power demands in excess of those calculated for oil sands operations and the option to use internally produced refinery gas for hydrogen production can also be accommodated.

2.10 CO$_2$ Emissions

The CO$_2$ emissions (E) due to natural gas and diesel fuel use for energy production in oil sands operations are computed in the OSOM, based on the emissions factors (FEF) for each fossil fuel (Nyober 2003). Equations (41) to (44) show the CO$_2$ emissions of boilers (SB and SSB), hydrogen plants, and power plants, respectively.

$$E_b = \frac{FEF_{NG}}{\eta_{B_b} HHV_{NG}} (W_b \Delta HW + S_b \Delta HS_b) \qquad \forall b \in SB \qquad (41)$$

$$E_b = \frac{FEF_{NG}}{\eta_{B_b} HHV_{NG}} SS_b \Delta HS_b \qquad \forall b \in SSB \qquad (42)$$

$$E_h = HEF_h \cdot H_h \qquad \forall h \in SR \qquad (43)$$

$$E_p = FEF_{NG} X_p \qquad \forall p \in NGCC \qquad (44)$$

Table 4 CO_2 Emissions Breakdown for Hydrogen and Power Production Plants

CO_2 Emissions Source	NGCC (%)	SMR (%)
On-site hydrogen production	-	82.3
On-site electricity generation	84.4	2.5
Upstream NG production and distribution	15.0	14.8
Upstream construction and decommissioning	0.6	0.4
Total	100	100

HEF is a model parameter, denoting the CO_2 emissions of SMR plants. The total CO_2 emissions from oil sands operations for the base case are given by:

$$E = \sum_{b \in SB} E_b + \sum_{b \in SSB} E_b + \sum_{h \in SR} E_h + \sum_{p \in NGCC} E_p$$

$$+ FEF_D \sum_{P_i} D_{P_i} + FEF_{NG} \sum_{P_i} \sum_{k=1}^{3} F_{k,P_i} \tag{45}$$

where the last two terms in equation (45) represent the CO_2 emissions of diesel and process fuel, respectively.

CO_2 emissions from power and hydrogen production are reported according to their life-cycle components, as shown in Table 4. The aforementioned emission components were reported in life-cycle emissions and performance studies (Spath and Mann 2000 and 2001).

The calculation of other GHG emissions in addition to CO_2, such as CH_4 and N_2O can be performed in a similar fashion. These two gases are linked to NGCC and SMR plants operations. The user has the option to include methane and nitrous oxide in the calculation of GHG emissions corresponding to hydrogen and power production, or else, calculate only CO_2 emissions in these plants. The results for the 2003 case presented in Chapter 4 include CO_2, CH_4, and N_2O.

3   CASE STUDY

The case study considered here represents the SCO and bitumen production operations in Alberta in 2003 (column 3 of Table 1). This year was selected because it conforms to the Oil Sands Technology Roadmap production estimates (Thambimuthu 2003) and also because sufficient SCO and bitumen production data for that year are readily available (Van Driesen et al. 1979).

3.1   Energy Demands

The energy demands for oil sands operations for this case study are calculated for all producers and for all production stages, according to energy commodity. Table 5 shows the energy demands for the mining stage for producers A1–A3.

Table 5   Energy Demands for Mining for the Case Study

Variable	Units	A1	A2	A3
Oil sands mined	tonne/h	17,402	17,405	5,126
Bitumen saturation	percent	11.4	11.3	12.4
Diesel shovels	L/h	3,969	3,969	1,169
Diesel trucks	L/h	14,981	14,985	4,413
Total Diesel Demands	L/h	18,950	18,954	5,582

The diesel consumption of trucks is higher than that of shovels because there are more trucks than shovels in the reference mining fleet. This reflects the fact that a single shovel can load multiple trucks in the same amount of time that it takes for a truck to deliver its payload. Moreover, the shovels travel much shorter distances than trucks do.

In the bitumen conditioning/hydrotransport stage, the energy demands consist of hot water, steam, and electricity, as shown in Table 6. In this case study, only producer A1 uses bitumen conditioning, whereas the others hydrotransport the mined sand. This mirrors the way the industry operated in 2003, where the majority of the oil sands were subjected to the more energy-efficient hydrotransport process.

Table 7 summarizes the energy demands of the bitumen-extraction stage. Extraction is executed in two stages. In primary extraction, the oil sand slurry

Table 6   Energy Demands for Conditioning/Hydrotransport for the Case Study

Variable	Units	A1	A2	A3
Hot water—Conditioning	tonne/h	1,030	N/A	N/A
Hot water—Hydrotransport	tonne/h	4,291	5,222	1,538
Total Hot Water Demands	tonne/h	5,321	5,222	1,538
Steam—Conditioning	tonne/h	112	N/A	N/A
Power—Hydrotransport	kW	38,577	65,176	20,417

Table 7   Energy Demands for Extraction for the Case Study

Variable	Units	A1	A2	A3
Hot water—Primary extraction	tonne/h	7,139	7,140	2,103
Steam—Secondary extraction	tonne/h	130	132	43
Naphtha—Secondary extraction	tonne/h	1,299	1,324	428
Power—Secondary extraction	kW	12,944	10,079	3,256
Power—Tailings disposal	kW	100,771	126,975	38,681
Total Power Demands	kW	113,715	137,053	41,937

from conditioning/hydrotransport is diluted with hot water, causing the bitumen to rise to the surface as a froth. In secondary extraction, the froth is deaerated and diluted with naphtha and centrifuged to remove traces of sand and water. In all, the extraction processes consumes generous amounts of hot water, steam, and power, in addition to requiring naphtha solvent. The breakdown of these commodities is shown in Table 7.

From Table 7 it is evident that the power required to pump tailings to disposal ponds is very substantial. The production of tailings in the hot water process is elevated; in fact, the mass balances show that the ratio of tailings produced to bitumen recovered from the oil sand is 16, on a mass basis. Therefore, reductions in water requirements for bitumen extraction have a great potential to simultaneously cut energy demands in oil sands operations, and ultimately, their energy and emissions intensities.

The energy demands of the bitumen upgrading stage are different for all producers, as it is a function of their particular upgrading scheme. Hence, the breakdown of these energy demands varies widely from producer to producer, as seen in Table 8.

The power demands of producer A1 are more pronounced than the rest. This is due to the fact that its upgrading scheme involves coking and hydrocracking, both of which consume power. In contrast, producers A2 and A3 use only one primary upgrading step, followed by hydrocracking, which requires less power overall. The upside of the upgrading scheme of A1 is that it achieves a higher overall bitumen conversion and produces less petcoke byproduct than that of A2.

One noteworthy difference in the upgrading stages among producers is the process fuel demands of producer A2, which are roughly nine times greater than

Table 8 Energy Demands for Upgrading for the Case Study

Variable	Units	A1	A2	A3
Steam—Diluent recovery	tonne/h	900	925	304
Steam—Vacuum distillation	tonne/h	71	108	36
Steam—Fluid coking	tonne/h	326	N/A	N/A
Total steam demands	tonne/h	1,298	1,033	340
Power—LC-Fining	kW	21,061	N/A	34,858
Power—Fluid coking	kW	38,325	N/A	N/A
Power—Delayed coking	kW	N/A	24,347	N/A
Total Power Demands	kW	59,386	24,347	34,858
Process fuel—LC-Fining	Nm^3/h	3,135	N/A	5,190
Process fuel—Delayed coking	Nm^3/h	N/A	25,103	N/A
Total Process Fuel Demands	Nm^3/h	3,135	25,103	5,190
	GJ/h	119	995	197
Hydrogen—LC-Fining	tonne/h	3.0	N/A	7.0
Hydrogen—Hydrotreatment	tonne/h	26.5	24.5	10.8
Total Hydrogen Demands	tonne/h	29.5	24.5	17.8

Table 9 Energy Demands for Thermal (SAGD) Bitumen for the Case Study

Variable	Units	C
Steam—SAGD	tonne/h	5,642
Power—SAGD	kW	45,120
Process fuel—Lift gas	Nm^3/h	37,616

those of A1. The former uses delayed coking, a process requiring more heat than either fluid coking or LC-Fining. Natural gas is used as process fuel, which increases the cost of upgrading on a per barrel basis.

In addition to SCO produced from mined oil sands, the base case includes production of bitumen via SAGD. The energy demands for this process consist of power and steam. Table 9 summarizes the demands for producer C.

The steam demands for SAGD bitumen production are calculated in the base case on the basis of a steam-to-oil ratio (SOR) of 2.4. Although the value of SOR is reservoir specific and may change over its productive life, 2.4 is a figure representative of an economically feasible SAGD operation, not uncommon in commercial calculations (OPTI Canada 2002).

Table 9 includes the amount of natural gas required for injection into the reservoir, which aids in pumping the produced bitumen out of the reservoir. The reader must note that this gas is later used in boilers for steam production, and is thus lumped with the natural gas demands for SAGD steam production in the OSOM.

The overall energy demands of each producer, according to energy commodity, are computed and presented in Table 10. The breakdown of these energy demands by commodity, expressed in GJ/h, is shown in Table 10.

The results reveal that the unitary energy demands for SCO production are approximately 1.5 GJ/bbl. The energy intensity of producers A3 are the highest of all (1.55 GJ/bbl), followed by A1 and A2 (1.53 and 1.49 GJ/bbl). Published values of energy intensities in oil sands operations/heavy oil upgrading for the year 2001 are 1.38 GJ/bbl (Nyober 2003). The upgrading scheme used in this study is unknown, thus direct comparison is not possible. Nevertheless, the derived values in this chapter are comparable.

The upgrading scheme has a strong influence over the energy intensity. For instance, although the energy requirements for mining and extraction of producers A1 and A3 are lower than those of A2, the latter's energy for upgrading is lower than that of the former producers, which results in overall lower energy intensity. In other words, the base case results suggest that the energy intensity of SCO production from mined oil sands is proportional to the magnitude of the energy demands for the upgrading process.

In terms of energy consumption according to commodity, for the base case, the results reveal that steam generation is responsible for roughly half the energy

Table 10   Energy Demands: Producer Comparison

Variable	Units	A1	A2	A3	C
Diesel	L/h	18,950	18,954	5,582	N/A
	L/bbl	1.96	2.14	1.43	N/A
Hot water	tonne/h	12,460	12,362	3,640	N/A
	tonne/bbl	1.29	1.39	0.93	N/A
Steam—Process	tonne/h	1,539	1,166	383	N/A
	tonne/bbl	0.16	0.13	0.10	N/A
Steam—SAGD	tonne/h	N/A	N/A	N/A	5,642
	tonne/bbl	N/A	N/A	N/A	0.39
Power—All stages	kW	211,678	226,577	97,212	45,120
Power—Ancillary	kW	21,168	22,658	9,721	4,512
Total Power Demands	kW	232,845	249,235	106,933	49,632
	kWh/bbl	24.1	28.1	27.4	3.4
Hydrogen	tonne/h	29.5	24.5	17.8	N/A
	MMSCF/h	12.5	10.4	7.5	N/A
	MMSCF/bbl	1,293	1,169	1,929	N/A
Process fuel	Nm^3/h	3,135	25,103	5,190	N/A
	GJ/h	119	995	197	N/A
	GJ/bbl	0.012	0.108	0.051	N/A

Table 11   Energy Demands by Commodity

Variable	Units	A1	A2	A3	C	Fleet
Diesel	GJ/h	726	726	214	0	1,665
Hot water	GJ/h	1,822	1,437	423	0	3,682
Steam	GJ/h	5,268	3,989	1,311	11,175*	21,742
Power	GJ/h	1,720	1,841	790	367	4,717
Hydrogen	GJ/h	5,157	4,282	3,114	0	12,553
Process fuel	GJ/h	119	955	197	0	1,272
Total per producer	GJ/h	14,812	13,229	6,049	11,541	45,631
Oil production	bbl/h	9,650	8,868	3,907	14,583	37,008
Energy Intensity	GJ/bbl	1.53	1.49	1.55	0.79	1.23

*Used for SAGD extraction, steam quality is different than that of process steam for mining/extraction operations

demands of the fleet. Hydrogen, power, and hot water combined account for 46 percent of the total energy demands of the fleet while the share of process fuel and diesel fuel is only 6 percent.

For bitumen production, the bulk of the energy demands is due to SAGD steam generation (97 percent). The energy intensity of producer C is 0.79 GJ/bbl bitumen produced. Although the above intensity value is lower than those corresponding to producers A1–A3, the reader must keep in mind that the product of the latter is upgraded bitumen, whereas the product of C is diluted bitumen (not upgraded).

3.2 GHG Emissions

Figure 15 presents the calculated GHG emissions intensity for SCO and bitumen according to producer and process stage. In mining operations (producers A1–A3), the bulk of the emissions come from the bitumen upgrading step, which accounts for approximately 70 percent of the process emissions for producers A1 and A2. The total calculated GHG intensity for these producers is 0.083 and 0.080 tonne CO_2 eq/bbl SCO, respectively. The upgrading emissions of producer A3 represent 80 percent of its total GHG intensity, which was found to be 0.087 tonne CO_2 eq/bbl SCO. Its hydrogen-intensive upgrading scheme is the reason for such high value. The emissions distribution of the other process stages shows that roughly half of the non-upgrading emissions belong to the bitumen-extraction stage. The first two process stages and the balance of the plants cause 15 percent of the total GHG emissions, on average. The mining and hydrotransport stages are less energy-intensive than extraction and upgrading, consuming only diesel fuel, hot water, and moderate amounts of electricity and steam. In contrast, bitumen extraction and upgrading require large quantities of steam, hot water, and power, and—in the case of the latter stage—also a good deal of H_2, as shown in Table 12.

GHG emissions intensities reported in other studies (McCulloch et al. 2006; Alberta Chamber of Resources 2004) correlate well with the values calculated here for producers A1–A3. For instance, published values for mined bitumen upgraded to SCO range from 0.080 to 0.118 tonne CO_2 eq/bbl SCO (Alberta Chamber of Resources 2004), whereas the corresponding values are 0.080—0.087 tonne CO_2 eq/bbl SCO.

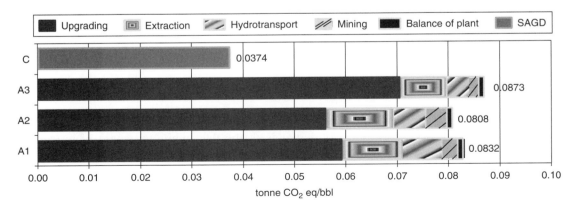

Figure 15 GHG emissions intensity by process stage

Table 12   GHG Emissions for All Producers in Tonne CO_2 eq/h for the Case Study

Producer	Hot water	Steam	Power	Hydrogen	Diesel	Process	CH4+N2O	Total
A1	86	248	81	281	51	6	51	803
A2	68	188	87	233	51	45	46	716
A3	20	62	37	169	15	9	29	341
C	0	526	17	0	0	0	3	546
Total	173	1023	222	683	116	60	129	2406

The GHG emissions intensity of thermal bitumen production is relatively high, when compared to that of mined bitumen. In mining operations, diluted bitumen is available at the end of the extraction stage. On the one hand, the average GHG intensity of the mining, hydrotransport, and extraction stages is 0.022 tonne CO_2/bbl bitumen, for producers A1–A3. This value excludes fugitive emissions of tailing ponds or those generated during overburden removal. The emissions intensity associated with producing a barrel of bitumen via SAGD, on the other hand, is 0.0374 tonne CO_2/bbl bitumen, or 68 percent higher than the intensity of mined bitumen, excluding fugitive and overburden emissions.

The GHG emissions intensity for thermal bitumen production predicted by the OSOM (producer C) is lower than those from other studies (McCulloch et al. 2006; Alberta Chamber of Resources 2004). The published figures range from 0.052 to 0.060 tonne CO_2/bbl bitumen. It is unknown if these values include fugitive emissions. Also, the steam quality and SOR from the aforementioned studies are unknown. Hence, direct comparison is not possible.

Of all the energy commodities consumed during SCO and thermal bitumen production, steam, hydrogen, and power are the most GHG-intensive. From Table 12, it can be seen that steam, H_2, and power are responsible for 80 percent of all GHG emissions in the Athabasca region oil sands industry in 2003. The individual contribution of each is 42 percent, 28 percent and 10 percent of the total GHG emissions, respectively. The GHG emissions resulting from hot water production are the fourth largest, accounting for 7 percent of the total.

The results also show that 95 percent of all emissions are CO_2, the rest being methane and nitrous oxide. These substances are emitted by power plants and H_2 plants and, for the most part, represent the GHG emissions upstream of the process, due to natural gas and coal production and transport. Their contribution to the total GHG emissions for the OSOM-BC is almost as significant as that of hot water. These non-CO_2 emissions are calculated based on the emissions factors presented in Nyober (2003) and Spath and Mann (2001). Furthermore, the combined calculated emissions from process fuel consumed during upgrading and diesel fuel were found to be relatively low, accounting for just 7 percent of the total, for all producers.

In mining operations alone, hydrogen production for upgrading is the leading source of GHG emissions, accounting for 37 percent of the total emissions. Steam and power represent 27 and 11 percent of total GHG emissions, respectively. The combined contribution of hot water production, diesel and process fuel use and non-CO_2 emissions is a quarter of the total GHG emissions of mining operations, which is estimated to be 1,860 tonne CO_2 eq/h.

The results suggest that the GHG abatement efforts in oil sands operations should be focused on the upgrading stage. In other words, considerable opportunities exist for less CO_2-intensive production of steam, hydrogen, and power in the oil sands industry in Alberta.

4 SUMMARY

Bitumen is becoming a predominant source of energy for North American markets. The bitumen-extraction and upgrading processes in the oil sands industry require vast quantities of energy, in the form of power, H_2, steam, hot water, diesel fuel, and natural gas. These energy commodities are almost entirely produced using fossil feedstocks/fuels, which results in significant CO_2 atmospheric emissions.

CO_2 capture and storage (CCS) technologies are recognized as viable means to mitigate CO_2 emissions. Coupling CCS technologies to H_2 and power plants can drastically reduce the CO_2 emissions intensity of the oil sands industry. The CO_2 streams from such plants can be used in enhanced oil recovery, enhanced coal-bed methane, and underground CO_2 storage.

This chapter introduced the relationship between energy demands, energy costs and CO_2 emissions associated with current and proposed oil sands operations using various energy production technologies. Accordingly, a modeling framework has been introduced that can serve as an energy planning and economic optimization tool for the public and private sectors. This framework-first model is an industrywide mathematical model that can serve to quantify the demands for power, H_2, steam, hot water, process fuel, and diesel fuel of the oil sands industry for given production levels of bitumen and synthetic crude oil (SCO), by mining and/or thermal extraction techniques.

The modeling framework was illustrated on a case study from the oil sands of Alberta. In 2003, steam, H_2, and power were found to be the leading sources of CO_2 emissions, accounting for approximately 80 percent of the total emissions of the oil sands industry. The CO_2 intensities calculated ranged from 0.080 to 0.087 tonne CO_2 eq/bbl for SCO and 0.037 tonne CO_2 eq/bbl for bitumen. The energy costs in 2003 were \$13.63/bbl and \$5.37/bb for SCO and bitumen, respectively.

REFERENCES

Alberta Chamber of Resources (2004), *Oil Sands Technology Roadmap Unlocking the Potential*, Alberta Chamber of Resources, Edmonton, Canada.

Alberta Energy and Utilities Board (2004), *Alberta Oil Sands Annual Statistics for 2003*. Report ST 2003-43, Alberta Energy and Utilities Board, Calgary, Canada.

Alfa Laval, http://www.alfalaval.com/digitalassets/2/file10080_0_PD41493e.pdf. Accessed on January 2007.

Chiesa, P., Consonni, S., Kreutz, T., and Williams, R. (2005), "Co-production of Hydrogen, Electricity, and CO_2 from Coal with Commercially Ready Technology. Part A: Performance and Emissions," *International Journal of Hydrogen Energy* 30: 747.

ConocoPhillips, http://www.coptechnologysolutions.com/thruplus_delayed_coking/thruplus/tech_specs/index.htm. Accessed in August 2007.

De Malherbe, R., Doswell, S. J., Mamalis, A. G., and de Malherbe, M. C. (1983), *Synthetic Crude from Oil Sands*, VDI-Verlag GmbH. Düsseldorf, Germany.

Davison, J. (2007),. "Performance and Costs of Power Plants with Capture and Storage of CO_2," *Energy,* Vol. 32, p. 7.

Harrell, G. (2002), "Steam System Survey Guide," Report ORNL/TM-2001/263, Oak Ridge National Laboratory, Oak Ridge, TN.

The Hydrocarbon Processing Handbook (2007). http://www.hydrocarbonprocessing.com/handbooks//contents2.html. Accessed in August 2007.

Jordal, K., Anhede, A., Yan, J., and Strömberg, L. (2004), "Oxyfuel Combustion for Coal-fired Power Generation with CO_2 Capture—Opportunities and Challenges," in *Proceedings of the GHGT-7 Conference* (Vancouver, Canada), September.

Kreutz, T., Williams, R., Consonni, S., and Chiesa, P. (2005), "Co-production of Hydrogen, Electricity, and CO_2 from Coal with Commercially Ready Technology. Part B: Economic Analysis," *International Journal of Hydrogen Energy*, Vol. 30, p. 769.

Kruse, B., Grinna, S., Buch, C. (2002), *Hydrogen—Status and Possibilities*. The Bellona Foundation, Oslo, Norway.

Lee, S. (1996), *Alternative Fuels*, Taylor & Francis, Washington.

Long Lake. http://www.longlake.ca/default.asp. Accessed in August 2007.

Lott, R., and Lee, L. K. (2003), "Upgrading of Heavy Crude Oils and Residues with (HC)3 ™ Hydrocracking Technology," Paper presented at the 2003 American Institute of Chemical Engineers annual conference (New Orleans, LA), April 2, 2003.

McCulloch, M., Raynolds, M., and Wong, R. (2006), *Carbon Neutral 2020: A Leadership Opportunity in Canada's Oil Sands*. The Pembina Institute, Drayton Valley, Canada.

Meyers, R. A. (1984), *Handbook of Synfuels Technology*, McGraw-Hill Inc., California.

Nyober, J. (2003), "A Review of Energy Consumption in Canadian Oil Sands Operations, Heavy Oil Upgrading 1990, 1994 to 2001," Canadian Industry Energy End-use Data and Analysis Centre. Vancouver.

OPTI Canada (2002), *OPTI Canada Long Lake Project Application for Commercial Approval. Volume 1 Sections A, B, and C.* Application submitted to the Alberta Energy Utilities and Board, Calgary, Canada.

Ordorica-Garcia, G., Douglas, P., Croiset, E., and Zheng, L. (2006), "Technoeconomic Evaluation of IGCC Power Plants for CO_2 Avoidance," *Energy Conversion and Management*, Vol. 47, p. 2250.

Ordorica-Garcia, G., Douglas, P., Croiset, E., and Zheng, L. (2007), "Modeling the Energy Demands and Greenhouse Gas Emissions of the Canadian Oil Sands Industry," *Energy and Fuels*, Vol. 21, pp. 2098–2111.

Rubin, E. S., Rao, A. B., and Chen, C. (2004), "Comparative Assessments of Fossil Fuel Power Plants with CO_2 Capture and Storage," in *Proceedings of the GHGT-7 Conference* (Vancouver, Canada), September, 2004.

Sadorah, R. (2005), Shell Canada, personal communication.

Schumacher, M. (1982), *Heavy Oil and Tar Sands Recovery and Upgrading International Technology*, Noyes Data Corporation, Park Ridge, NJ.

Schultz, K. R., Brown, L. C., and Besenbruch, G. E. (2003), "Large-scale Production of Hydrogen by Nuclear Energy for the Hydrogen Economy," *Proceedings of the National Hydrogen Association Annual Conference* (Washington, DC), April 6–9.

Simbeck, D. R. (2004), "Hydrogen Costs with CO_2 Capture," *Proceedings of the GHGT-7 Conference* (Vancouver), September 2004.

Spath, P. L., and Mann, M. K. (2000), "Life Cycle Assessment of a Natural Gas Combined-Cycle Power Generation System," Report NREL/TP-570-27715. National Renewable Energy Laboratory (Colorado).

Spath, P. L., and Mann, M. K. (2001), "Life Cycle Assessment of Hydrogen Production via Natural Gas Steam Reforming," Report NREL/TP-570-27637, National Renewable Energy Laboratory, Colorado.

Suncor. http://www.suncor.ca. Accessed in August 2007.

Sunderland, B. (2001), "Hydrogen Production, Recovery and Use at Syncrude," Presentation at the EFI Conference on Technologies to Enable the Hydrogen Economy. April 22–24, Tucson, AZ.

Syncrude, http://www.syncrude.ca. Accessed in August 2007.

Syncrude, "Operations Overview," http://www.syncrude.ca/users/folder.asp?FolderID=5726. Accessed in July 2006.

Thambimuthu, K. (2003), "Clean Coal Roadmap 'Strawman,'" Presentation at the 1st Canadian Clean Coal Technology Roadmap, Calgary, Alberta, March 21–21.

Van Driesen, R., Caspers, J., Campbell, A.R., and Lunin, G. (1979), "LC-Fining Upgrades Heavy Crudes," *Hydrocarbon Processing,* Vol. 58, p. 5.

Yui, S., and Chung, K.H. (2001). "Processing oil sands bitumen is Syncrude's R&D focus," *Oil & Gas Journal,* Vol. 99, p. 17.

CHAPTER 3

ENVIRONMENTALLY CONSCIOUS COAL MINING

R. Larry Grayson
Department of Energy and Mineral Engineering
The Pennsylvania State University
University Park, Pennsylvania

1 INTRODUCTION

The world's total consumption of coal is over 5.9 billion metric tons today, with China (at 2.4 billion metric tons) and the United States (at 0.9 billion metric tons) accounting for nearly 57 percent of it (IEA 2008). The world's estimated recoverable coal reserves are 930 billion metric tons, with the United States (at 260 billion metric tons), Russia (at 177 billion metric tons), and China (at 130 billion metric tons) accounting for 61 percent of them. Coal continues to be a cost-effective choice for electric power generation, and as the global demand for energy continues to increase dramatically, it is projected for even greater roles in conversion to liquids and gas. However, the contribution of coal to an apparently growing increase in atmospheric carbon dioxide, which may be warming the Earth's climate, is now a concern embraced by many nations. Coal's future role is thus uncertain, at least without a transition to the use of new technologies to clean up coal-fired emissions. Other effects on the environment have occurred historically and continue to occur, including impacts on air, land, and water at mining operations. In the foreseeable future, though, coal is projected to be a mainstay in the global energy picture, in which coal accounts for 24 percent of the current energy mix and is projected to account for 27 percent by 2030 (IEA 2008).

In her 1994 paper, Nancy Bingham outlined what the public thought about mining. In a nutshell, based on extensive interviews, the following results were found:

- The public believed mining harms the environment.
- The public believed mining harms people in nearby communities.
- The public believed mining exploits workers.
- The public believed mining has little personal benefit to the individual.

These perspectives were reinforced by previous articles in the popular media that focused on "The Death of Mining" (Anon. 1984) and "The Curse of Coal" (Gup 1991).

Globally, historical environmental impacts are well known. In the author's travels to China, Russia, and Ukraine, for example, waste piles and impoundments dotted the landscape in coal regions, and pollution of air, land, and water was evident. In the United States, the historical experience is similar, although mitigation of impacts began much sooner than elsewhere, largely in the 1960s, 1970s, and 1980s. Thus, today in the United States, dramatic improvements have been realized in the emissions of sulfur dioxide, nitrogen oxides, and particulate matter. Similarly, damage from mine subsidence, and acid mine drainage and pollution of streams and rivers have been significantly abated by enforcement of strict environmental mining laws. A program also exists to rehabilitate abandoned mine lands where environmental damage poses a significant threat to public health. The progress in the United States, like in many other countries, began with public outrage over incidents such as the Buffalo Creek impoundment failures in 1972 where 118 citizens died and over 4,000 of them were left homeless (West Virginia Division of Culture and History 2008). Today's issues involve mountaintop removal surface coal mining, carbon dioxide emissions from coal-fired power plants, mercury emissions, emissions of other trace elements, and longwall subsidence. The industry is more proactive today in dealing with the issues, particularly from a technological standpoint, and government is heavily involved much more quickly today that ever before in seeking resolution of the issues. This chapter discusses historical environmental performance, the evolution of environmental regulations impacting coal mining, noteworthy environmental achievements, and progress toward environmentally conscious coal mining and adoption of sustainable development principles into coal mining business.

2 HISTORICAL ENVIRONMENTAL PERFORMANCE

As mining was promoted in the nineteenth century to open and develop the territories west of the United States, there were no environmental performance standards in a vast, open land. Riches sought in the territories eventually were realized, and the United States grew westward as economic development progressed. The General Mining Act of 1872, passed by Congress, enhanced the growth by authorizing and governing prospecting and mining of minerals such as gold, platinum, and silver on federal public lands. However, no major environmental laws were in effect in the United States until much later, especially regarding mining.

Beginning in the 1960s and 1970s, environmental damage and indifference to its impacts began taking the forefront as a key issue marring the coal industry's image (Grayson and Warneke 2006).

Large gob piles dotted the landscape in Appalachia particularly, and acid mine drainage fouled rivers and streams, often robbing the public of fish and recreation. With the cascading impoundment failures at Buffalo Creek, West Virginia, which evoked public outrage, and the growth of surface coal mining in Appalachia, the U.S. Congress passed legislation to control mining's impact on the environment. In 1977, the Surface Mining Control and Reclamation Act (Title 30, United States Code, Chapter 25), called SMCRA for short, was passed as the first serious federal action taken to coordinate improvement of environmental performance in the coal industry. Again, because of public sentiment, this time regarding the effects of acid rain, the Clean Air Act Amendment of 1990 was passed, which regulated the emissions of sulfur dioxide. At this time, the United States was inexorably marching toward tighter regulation of the coal mining industry itself and its downstream impacts on air, land, and water. The major legislation evolving from this point onward is summarized next.

3 CURRENT ENVIRONMENTAL REGULATIONS

The Surface Mining Control and Reclamation Act of 1977 was intended "to ensure that coal mining activity is conducted with sufficient protections of the public and the environment, and provides for the restoration of abandoned mining areas to beneficial use" (Title 30, United States Code, Chapter 25). The Act established the Office of Surface Mining and Reclamation Enforcement (OSM for short), which generally gives primacy for enforcement to state authorities, but also oversees the effectiveness of the state enforcement programs. A state may determine its own requirements under the Act, provided the requirements are at least as stringent as those contained in the Act.

For example, in 2005, the Virginia Division of Mined Land Reclamation established a guide to bond reduction/release, noting that a "permittee may submit an application for a Phase I, II, or III release of bond for the entire permit or increment(s). The permittee does not have to request a Phase I or II bond reduction before requesting a Phase III release." The guide established criteria to be used, as follows:

Phase I. After completion of the required backfilling, regrading (which may include the replacement of topsoil depending on the weather), and drainage control of the subject area, up to 60 percent of the bond or collateral for the area may be released. However, the minimum bond of not less than $10,000 must be retained through Phase II. Requesting a Phase I release is at the option of the permittee.

Phase II. After two growing seasons and the required reestablishment of vegetation on the regraded mine lands, an additional amount of bond may be released. The division must ensure that any reduction will leave a sufficient amount to cover any remaining reclamation work, such as structure removal, grouting of wells, and so forth. For future revegetation costs the division requires an amount of $125 for each disturbed acre or $10,000, whichever is greater. A vegetative survey should be conducted by the permittee at this time to serve as a record to determine if the standards of success are being met during the applicable liability period. The information obtained from the survey would alert the permittee if replanting of the woody plants (20 percent) and follow-up seeding are necessary to avoid an extension of the liability period. The permittee could request a phase II reduction without first requesting a phase I release.

Phase III. Once the permittee has successfully implemented the reclamation plan and met the applicable liability obligation, an application for the release of the remaining bond may be submitted to the division. The bond may not be fully released until the reclamation requirements of the Act and the permit have been fully met. The permittee's bond release application must include a final vegetative survey to document that the standards of success have been and are being met.

Most other states with primacy similarly specify such bond reduction/release requirements, which must be consistent with SMCRA requirements.

SMCRA also manages the Abandoned Mine Land (AML) tax fund, which was renewed by the Surface Mining Control and Reclamation Act Amendments (U.S. House of Representatives 2006). In this act, the surface coal mine tax was decreased from 35 to 31.5 cents/ton, the underground coal mine tax from 15 to 13.5 cents/ton, and the lignite mine tax was fixed at 9 cents/ton. A lower limit for taxes is maintained by a percentage assessed on coal value. The taxes will be reduced again in 2013 to 28, 12, and 8 cents/ton, respectively. The fund is used to reclaim abandoned mine land problems which had historically accumulated across the United States.

The U.S. Congress passed the Comprehensive Environmental Response, Compensation, and Liability Act (CERCLA) in 1980, which is commonly known as Superfund (Title 42, United States Code, Section 9601). CERCLA created a tax on the chemical and petroleum industries and provided broad federal authority to respond directly to releases or threatened releases of hazardous substances that may endanger public health or the environment. CERCLA may apply to mining operations as well, for example, when cyanide is used in ore processing.

The Clean Air Act (CAA) of 1963, as extended in 1970 and amended in 1977 and 1990 (Title 42, United States Code, Section 7622), defined "the U.S. Environmental Protection Agency's responsibilities for protecting and improving the nation's air quality and the stratospheric ozone layer." Amendments to the CAA have forced reductions in sulfur dioxide emissions from coal-fired power plants and, more recently, in mercury emissions.

The Clean Water Act of 1972 (Title 33, United States Code, Section 1251) had the objective "to restore and maintain the chemical, physical, and biological integrity of the Nation's waters." A National Pollution Discharge Elimination System (NPDES) permit is required to mine. The NPDES permit is embedded in the OSM permitting system administered by a state authority with primacy for enforcement.

There are myriad other environmental protection-related acts that impact mining. Some of them are listed here:

- Endangered Species Act (1973), which was passed to "protect critically imperiled species from extinction as a consequence of economic growth and development untendered by adequate concern and conservation" (Title 7, United States Code, Section 136, and Title 16, United States Code, Section 1531)
- Historic Landmark and Historic District Protection Act of 1978 (Title 17, United States Code, Section 470)
- Insecticide, Fungicide and Rodenticide Act (1973), which was passed "for controlling the sale, distribution and application of pesticides" (Title 7, United States Code, Sections 136–136y)
- Marine Protection, Research and Sanctuaries Act (1972), which regulates "the dumping of all types of materials into ocean waters and to prevent or strictly limit the dumping into ocean waters of any material which would adversely affect human health, welfare, or amenities, or the marine environment, ecological systems, or economic potentialities" (Title 33, United States Code, Sections 1401–1445)
- National Environmental Protection Act (1970), passed "for the protection, maintenance, and enhancement of the environment" (Title 42, United States Code, Section 4321)
- Resource Recovery and Conservation Act (1976), which was passed to control the "disposal of solid and hazardous waste" (Title 42 United States Code, Sections 6901–6992k)

Importantly, OSM's Applicant/Violator System (AVS) now provides state regulatory authorities with a central database of application, permit, ownership and control, and violation information (Office of Surface Mining 2006). OSM and state officials can review this data when evaluating an applicant's eligibility for new permits, thereby prohibiting the issuance of new permits to applicants who own or control operations with unabated or uncorrected violations.

4 ENVIRONMENTAL ACHIEVEMENTS

OSM reports its accomplishments, largely compiled from state data, in a quasi-annual report to the U.S. president and Congress, the last of which was published in 2006. In the last report, over 1.8 million hectares were permitted for mining in the United States, while over 77,300 hectares were newly permitted that year, and nearly 20,200 hectares met Phase III requirements for complete bond release. Another 37,200 hectares have Phase I or Phase II bond reductions/releases. A total of over 37,000 operations/activities were *inspectable,* with 294 of them newly permitted entities in 2006. In the 2006 annual report, OSM notes that 91.5 percent of active sites are free of offsite impacts, which is a remarkable achievement, clearly demonstrating the effectiveness of the federal-state program. Each year, OSM recognizes industry excellence in environmental reclamation, and Figure 1 shows the national award-winning operation using before and after results.

OSM is also responsible for the reclamation of abandoned mine lands. In 2006, approximately 2,800 hectares of land were reclaimed or mitigated from the effects of degradation from past mining. Additionally, approximately 111 stream-kilometers of streams degraded by surface mining were improved, and approximately 13 hectares of degraded water were improved. One such site is shown in Figure 2, where the contractor was recognized for excellence in reclamation of abandoned mine land.

AML high-priority projects protect public health, safety, and general welfare, as well as address emergency projects. Overall since 1978, high-priority projects

Figure 1 2006 National Award, Peabody Energy Seneca Coal Company, Seneca II West Mine, Routt County, Colorado (OSM 2006)

Figure 2   Pennsylvania Luciana Bottoms West, Abandoned Mine Reclamation Project, Huntington County, Pennsylvania (OSM 2006)

of the AML program have reclaimed 882 kilometers of clogged streams, 1.0 million meters of dangerous highwalls, over 1,700 dangerous impoundments, and nearly 22,000 hectares of land involving surface burning, underground mine fires, subsidence, industrial/residential waste, dangerous slides, dangerous piles and embankments, and clogged stream areas. Environmental restoration projects, priority 3, have protected over 19.3 million liters per minute of water, reclaimed another 51 thousand meters of highwalls, closed more than 700 abandoned mine openings, and improved more than 17,400 hectares of degraded land. It has proven to be a program of excellence for 30 years.

Electric power generation and coal use continue to grow across the world and in the United States. A schematic of how coal is used in a traditional coal-fired power plant is given in Figure 3. Emissions from such plants have decreased

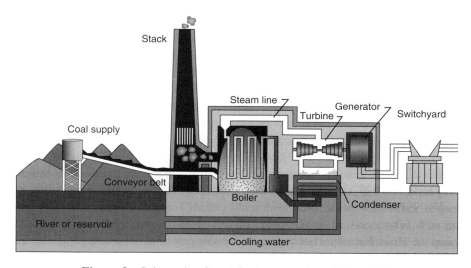

Figure 3   Schematic of coal-fired power plant (TVA 2006)

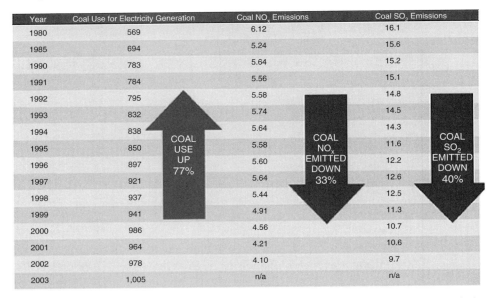

Year	Coal Use for Electricity Generation	Coal NO$_x$ Emissions	Coal SO$_2$ Emissions
1980	569	6.12	16.1
1985	694	5.24	15.6
1990	783	5.64	15.2
1991	784	5.56	15.1
1992	795	5.58	14.8
1993	832	5.74	14.5
1994	838	5.64	14.3
1995	850	5.58	11.6
1996	897	5.60	12.2
1997	921	5.64	12.6
1998	937	5.44	12.5
1999	941	4.91	11.3
2000	986	4.56	10.7
2001	964	4.21	10.6
2002	978	4.10	9.7
2003	1,005	n/a	n/a

COAL USE UP 77%

COAL NO$_x$ EMITTED DOWN 33%

COAL SO$_2$ EMITTED DOWN 40%

Figure 4 NO$_X$ and SO$_2$ emissions reductions vs. coal use (National Mining Association 2006)

dramatically in the United States since the 1990 Clean Air Act (see Figure 4), in spite of a dramatic increase in the use of coal for electricity, and emissions will continue to decrease dramatically with adoption of new technology to clean the coal even more, or to transform the coal into gas or liquid form and then use it for power generation or other markets. A schematic of a generic integrated gasification combined cycle (IGCC) power generation plant is shown in Figure 5. Although no full-scale IGCC plants have been built yet, several plants are in the permitting stage in the United States in 2008 (e.g., Florida, Indiana, Ohio, and West Virginia). They will likely come on board in three to five years. The pilot IGCC Polk Plant in Florida achieved a reduction of SO$_2$ to less than 1.35 lb/MWh (0.17 g/KJ), kept NO$_x$ below 0.52 lb/MWh (0.06 g/KJ), reduced particulate below 0.04 lb/MWh (0.005 g/KJ), and removed over 98 percent of total sulfur. The ultimate reduction in adverse emissions from new clean-coal technology appears optimistic, and makes a convincing case that air quality does not have to be compromised when coal is used for electric power generation. Progress in using abundant coal for clean power generation can also better protect public health.

Ultimately, public awareness of various treatments and their ultimate effects provides the comprehensive information needed to build confidence in coal as the primary source for power generation (Grayson and Warneke 2006). For example, the 1,600-MW (1600 MJ/s), mine-mouth Prairie State Energy Campus in Illinois with state-of-the-art clean-coal technology (http://www.prairiestateenergycampus.com/index-nn.html) will employ low-NO$_x$

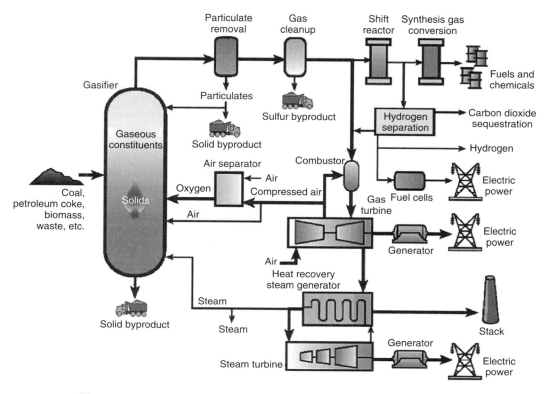

Figure 5 Schematic of a generic IGCC plant (Ratafia-Brown et al. 2002)

burners, limestone injection, selective catalytic reduction, dry precipitators, wet scrubbers, and wet precipitators, projected to attain the following results:

- Removal of SO_2 at approximately 98 percent to 0.167 lbs/MMBtu (0.072 g/KJ)
- Removal of NO_x to 0.09 lbs/MMBtu (0.039 g/KJ)(greater than 80 percent removal)
- Particulate removal of greater than 99 percent

The ultimate goal of Peabody Energy and other major coal companies is to not only use the latest proven clean-coal technology for power generation but also to capture and sequester carbon dioxide emissions in deep geologic formations. This should occur over the next decade.

A stronger commit to demonstrating clear improvements in reducing coal waste disposal, mine effluents, and subsidence are important as well. In the quest of capturing public support for mining, the adoption of sustainable development principles and transparency of data and information used to show improvements are critical. Success stories in restoring land, helping to plan post-mining land use, and investment in communities will help. For all good deeds, verifiable figures should be made available.

5   MOVEMENT TOWARD ADOPTION OF SUSTAINABLE DEVELOPMENT PRINCIPLES

Grayson and Warneke (2006) gave a brief overview of how the concept of sustainable development and its principles grew from early writings on environmental transgressions and land ethics, evolving from the early environmental writings of Leopold (1949), Darling (1955), and Carson (1965). As a growing number of countries responded to the issues, international conferences began to focus on them, as described by Moffatt (1996). The most famous definition of sustainable development came from the Brundtland Commission (World Commission on Environment and Development 1987), stating it as "...a notion that there may be limits to growth and that society must be reorganized to protect the interest of future generations." This is a simple definition, but it is easily understood by the global public. Nonetheless, many government organizations as well as nongovernment ones and industries, have different, sometimes widely divergent definitions.

Not long after the Brundtland Report, the environment was no longer viewed as an externality to economic matters. Today, more and more businesses have adopted sustainable development and its principles as guiding aspects in their planning and execution of operations. Sustainable development has also galvanized on the political agenda of numerous countries, and it has been at the forefront of global gatherings ever since (e.g., the 1992 Rio Earth Summit and its Agenda 21, and the 2002 World Summit on Sustainable Development in Johannesburg).

By 1998, nine of the largest mining companies embarked on a new initiative aimed at achieving a serious change in the way their industry deals with ongoing problems, dubbing it the Global Mining Initiative (IIED and WBCSD 2002). The joint research project executed by the International Institute for Environment and Development (IIED) focused on finding "a more informed understanding of the industry's actual and potential contribution to sustainable development," since the industry partners agreed that "it is no longer enough to argue that because society needs its products, it must tolerate whatever occurs in their production." Further, it is not "enough to play to a public audience conditioned to expect the worst from mining or other minerals industries." Finally, the initiative's partners noted, "We need to define what higher performance amounts to, and create sanctions and incentives to achieve it."

A set of guiding principles for the four pillars of sustainable development were specified early in the study and used as a framework for issues to be addressed. The four pillars, each with three or more subprinciples, included the following major spheres, which are to "be applied in an integrated manner in decision-making":

- Economic sphere
- Social sphere
- Environmental sphere
- Governance sphere

The report specified important key challenges that must be met in order to achieve the major goal. The nine key challenges include: (1) Viability of the Mineral Industry; (2) Control, Use, and Management of Land; (3) Minerals and Economic Development; (4) Local Communities and Mines; (5) Mining, Minerals, and the Environment; (6) An Integrated Approach to Using Minerals; (7) Access to Information; (8) Artisanal and Small-Scale Mining; and (9) Sector Governance: Roles, Responsibilities, and Instruments for Change. With the industry's approach to sustainable development outlined, the point must now be made that mining cannot have sustainable development of itself; rather, mining plays an important role in sustaining society by provision of raw materials. That role must be pursued in consonance with sustainable development principles.

6 THE FUTURE

6.1 An Integrated Approach to Using Coal

Data and information exchanged on the various aspects discussed thus far are helpful, but if provided alone there are gaps in understanding by the average citizen. Without an integrated linking of the economic, environmental, and social impacts, gaps in knowledge and understanding ultimately occur and lead to gaps in acceptance and mistrust. Thus, presenting information in a way that cultivates full understanding of the issues could pay huge dividends.

Material flows analysis is a way of presenting interrelations among physical aspects of coal flows in the U.S. economy and their potential impacts. The impacts may be quantified as well in an associated analysis of flows to the environment. Economic impacts may be quantified by linking the physical flows with associated monetary flows. In essence, a material flows accounting system is an analog of a monetary flows system, which has been in existence for decades to help guide governmental fiscal policy making. Linking material flows accounts with associated financial flows accounts expands the understanding of nonfinancial impacts.

A general economywide material-flows-accounting system is shown in Figure 6. The key focuses are on physical inputs to and outputs from the economy, but diagrams also track the build-up of materials in stock (in infrastructure primarily). Imports, exports, both used and unused extractions, and indirect (or hidden) flows associated with material extractions are shown. Such diagrams may be modified to show exactly where flows go to the environment (air, land, and water). Of course, environmental disturbances generally occur on the materials extraction and storage side, as well.

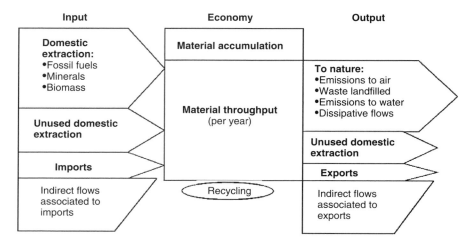

Figure 6   General scheme for economywide material flows accounting illustration (Eurostat 2001)

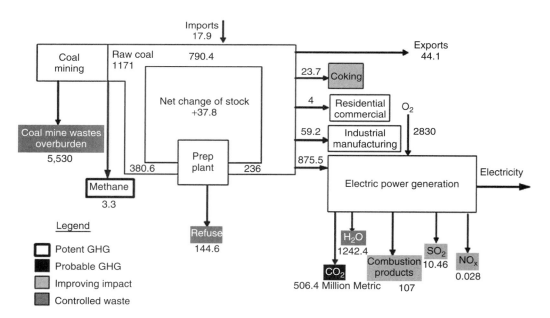

Figure 7   Coal material flows analysis for 2001, in MMT, as modified by Lusk (2007)

Figure 7 shows an early (1993) material flows analysis for the U.S. coal industry done by Ayres and Ayres (1998), as modified by Warneke (2004). It shows the paths of consumption, as well as various emissions that occur. Added to the diagram are colors representing potential impacts on the environment. Much of the data are generally available through the U.S. Energy Information Administration, but much of it is not. For example, considerable work is required to make a reasonably accurate calculation of the coal mine overburden disturbed, other emissions from extraction, the amount of refuse generated from coal cleaning, and the emissions from coking and power generation. In fact Warneke (2004) noted several discrepancies with the previous 1993 analysis.

Beyond the material flows analysis picture, which the public generally will understand, is a description of the ultimate impacts the flows have on people. These impacts can be illustrated through the use of driving force-pressure-state-impact-response models (Petruzela et al. 2000). Figure 8 (Warneke 2004) shows one possible model for emissions from coal utilization. The driving force in this instance is the production of coal itself, while pressures are exerted on the environment by acidifying, global warming, and other polluting (CO, NO_x, SO_2, particulate, etc.) emissions. The state resulting from the pressures relates to the total amount of substances in the air as well as the mean temperature and other effects of the various emissions. The ultimate impacts people and society care about relate to their survivability, their health, availability of food and potable water, and aesthetically pleasing recreational areas. As negative impacts occur at an unacceptable level, the public response eventually influences government policy, and laws are passed or regulations revised to reflect the desire for reduced negative impacts.

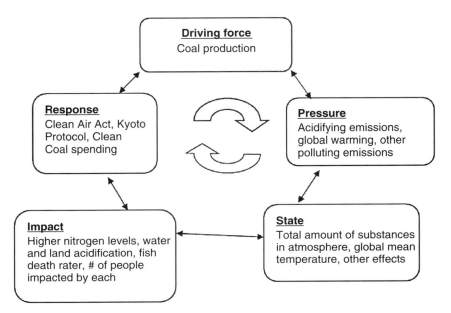

Figure 8   Driving Force-Pressure-State-Impact-Response model on emissions from coal production

6.2   Transparency of Information

For the entire integrated approach to work in building trust among the public constituencies, information must be available for anyone to examine. One of the best examples of this approach is the commodity flow information developed and shared by the U.S. Geological Survey (USGS 2005a; USGS 2005b). These data are open to public access, but at the same time it protects proprietary information of data providers. Companies wishing to build public trust and indicate social responsibility as stewards of the communities in which they work, the environment, and of sustainable development principles will need to share their data and information, too. Transparency of data equates to improved trust that the data and information will tell the true state of affairs.

7   CONCLUSION

Looking at Nancy Bingham's article once again, all is not lost, according to the public. She noted, "The good news is that the most common beliefs about mining are not strongly held and are quite susceptible to change." It was clear from her study that "the information people do have comes mostly from movies and entertainment television, rather than textbooks or documentaries." This situation opens the door for good, integrated communication about our industry, which could lead to a change in public attitudes about mining.

Although mines cannot be sustainable unto themselves, mining will long play a major role in sustaining society in many ways. Pursuing the principles of sustainable development has already begun to develop a new integrated message for industry that the public and government will understand, and they will often buy-in. The historical stigma attached to the coal industry is now slowly being reversed, as the industry expands in the twenty-first century. Through a concerted effort using a well-integrated message, the coal industry could make the change a permanent one. Those who work at mines or supporting operations are helping to sustain the massive U.S. economy, nurturing communities and workers, and are strong stewards of the environment in a direct way. Hopefully, one day the coal industry will be self-enforcing as well, and the need for more and more regulation will diminish because of prevailing good-faith efforts. In the meantime, the key stakeholders of the industry must be diligent in pursuing the principles of sustainable development and place strong peer-pressure on all operators to do so.

REFERENCES

Anon. (1984). "The Death of Mining," *Business Week* (December 17), pp. 64–70.

Ayres, R. U. and Ayres, L.W. (1998), *Accounting for Resources, 1: Economy-Wide Applications of Mass-Balance Principles to Materials and Waste*, Edward Elgar, Cheltenham, 245 pp.

Bingham, N. (1994), "Mining's Image—What Does the Public Really Think?" *Mining Engineering,* Vol. 46, No. 3, pp. 200–203.

Carson, R. (1965), *Silent Spring*, Penguin, Harmondsworth.

Darling, F. F. (1955), *The West Highland Survey: An Essay in Human Ecology*, Oxford University Press, Oxford.

EPA (2006), "Integrated Gasification Combined Cycle (IGCC): Background and Technical Issues,". State Clean Energy-Environment Technical Forum, June 19, 25 pp.

Eurostat (2001), *Economy-Wide Material Flow Accounts and Derived Indicators: A Methodological Guide*, Office for Official Publications of the European Communities, Luxembourg, 92 pp.

Grayson, R. L., and Warneke, J. (2006), "Coal's Role in Sustaining Society—An Integrated-Message Approach," *Mining Engineering.* Vol. 58, No. 10, pp. 23–29.

Gup, T. (1991), "The Curse of Coal," *Time* (November 4).

International Energy Agency (2008), *World Energy Outlook 2008*. Organization of Economic Cooperation and Development/International Energy Agency, Paris, 578 pp.

International Institute for Environment and Development and World Business Council for Sustainable Development (2002), *Breaking New Ground: Mining, Minerals, and Sustainable Development*, Earthscan Publications Ltd., London and Sterling, VA.

Leopold, A. (1949), *A Sand County Almanac and Sketches Here and There*, Oxford University Press, New York.

Lusk, S. (2007), "Analysis of the Holistic Impact of the Hydrogen Economy on the Coal Industry," Dissertation, University of Missouri–Rolla, 122 pp.

Moffatt, I. (1996), *Sustainable Development: Principles, Analysis and Policies*, The Parthenon Publishing Group, London.

National Mining Association (2006), "Clean Coal Technology: Current Progress, Future Promise," at http://www.nma.org/technology/environmental.asp.

Office of Surface Mining (2006), "2006 Report to the President and Congress, Office of Surface Mining Reclamation and Enforcement," Department of Interior, Washington, DC.

Petruzela, L., Lizner, M., Lippert, E., and Rencova, L., eds. (2000), "Report on the environment in the Czech Republic in 1999," Ministry of the Environment, at http://www.env.cebin.cz/publikace/3_report99_e/aobsah.htm.

Prairie State Energy Campus (2008) At http://www.prairiestateenergycampus.com/index-nn.html..

Ratafia-Brown, J., Manfredo, L., Hoffmann, J., and Ramezan, M. (2002), "Major Environmental Aspects of Gasification-Based Power Generation Technologies," National Energy Technology Laboratory, Department of Energy. http://www.netl.doe.gov/technologies/coalpower/gasification/pubs/prev-reports.html.

Title 7, United States Code, Section 136, and Title 16, United States Code, Section 1531.

Title 7, United States Code, Sections 136–136y.

Title 17, United States Code, Section 470.

Title 30, United States Code, Chapter 25.

Title 33, United States Code, Section 1251.

Title 33, United States Code, Sections 1401–1445.

Title 42, United States Code, Section 4321.

Title 42 United States Code, Sections 6901–6992k.

Title 42, United States Code, Section 7622.

Title 42 United States Code, Section 9601.

TVA (2006), at http://www.tva.gov/power/images/coalart.gif.

USGS (2005a), *Minerals Yearbook*. http://minerals.usgs.gov/minerals/pubs/myb.html.

USGS (2005b), *Mineral Commodity Summaries*. http://minerals.usgs.gov/minerals/pubs/mcs.

U.S. House of Representatives (2006), Public Law 109-432, Surface Mining Control and Reclamation Act Amendments of 2006, Washington, DC.

Virginia Division of Mined Land Reclamation (2005), "Virginia Coal Surface Mining and Reclamation, Permanent Regulatory Program: A Guide to Bond Reduction/Release," Commonwealth of Virginia, Department of Mines, Mineral, and Energy, Big Stone Gap, VA.

Warneke, J. R. (2004), *The Development of a Coal Material Flow Analysis, Indicators, and Models to Facilitate the Creation of Public Policy*, Dissertation, University of Missouri–Rolla, 157 pp.

West Virginia Division of Culture and History (2008), http://www.wvculture.org/history/buffcreek/bctitle.html.

World Commission on Environment and Development (1987), *Our Common Future* (Brundtland Report), Oxford University Press, Oxford.

CHAPTER 4

MARITIME OIL TRANSPORT AND POLLUTION PREVENTION

Sabah A. Abdul-Wahab
Department of Mechanical and Industrial Engineering
Sultan Qaboos University
Muscat, Sultanate of Oman

1 INTRODUCTION

Oil spills are really a major problem in most countries because of their bad impact on the environment and economy. These spills are usually spread rapidly over wide areas of the sea, which results in the deterioration of the water quality and the destruction of the marine life. In this chapter, three case studies will be presented. The objective of the first case study is to illustrate the development of a real time in situ device for earlier detection of oil pollution in seawater. The operation of the device is based on the concept of the electrical conductivity. This device is very useful, as it can be used to monitor the seawater pollution continuously and to send an alarm if the oil level exceeds a certain limit. The results of the study are very important, as they may contribute to the development of advanced practices in oil spill detection.

Work is also needed to reduce ship accidents. Therefore, it is very important to detect ruptures in ship hulls at an early stage to prevent propagation and so as to avoid disastrous impacts. Hence, in the second case study, an innovative technique is presented to design a smart ship that has the ability to predict ruptures in ship hulls as early as possible to prevent oil spills. The technique is a simple solution to the problem of fissures and can be applied to all ships, especially those that are used to transport oil. The results of this study are very important in preventing oil spills and in protecting of the marine environment and human life.

Moreover, emissions from ships at berth seriously threaten the environment and contribute to the poor air quality in many coastal and port regions. They are considered one of the major sources of air pollution that negatively impacts public health and communities in coastal and far inland. Since nitrogen oxides (NO_x) are considered the main pollutants emitted from ships during the berth mode, the third case study of this chapter is presented to assess the impact of NO_x emissions originated from ships at berth on a nearby community. The output results of this case study can help to derive recommendations for ships operators and environmentalists to better protect workers and the surrounded environment from pollution.

2   FIRST CASE STUDY: DETECTION OF OIL SPILL IN SEAWATER

2.1   Introduction

Most of the produced oil is transported by sea from one part to the other, so water bodies are used as the primary means of transport of oil. Hence, significant amounts of oil spill during transportation. This exposes the marine environment to the dangers of oil spillage. Oil spills are, therefore, a major problem in the world. Oil, with its aromatic compounds, is categorized as a toxic material that can destroy the marine life, as well as damage habitat for terrestrial animals and humans. The main impacts of the oil spills include damage to marine organisms; problems to human health due to the introduction of oil in the food chain; and physical prevention of birds from flying when smeared with oil.

Oil spills enter the seawater from a wide range of sources that include accidents or grounding of tankers; oil exploration and production; cargo tank washings at sea; and discharge of ballast water by ships into the sea. Generally, a significant amount of oil pollution of the world's oceans is caused by shipping accidents. The world has seen many oil spill accidents in different countries and for different reasons (Marine Emergency Mutual Aid Centre 2003). Due to the relative volumes of discharges, illegal emissions from ships represent a greater long-term source of harm to the environment than infrequent large-scale accidents. Monitoring illegal discharges is thus an important component in ensuring compliance with marine protection legislation and general protection of coastal environments (Inggs and Lord 2006).

An oil spill that occurs near a coastline will always impact on more living organisms than one which occurs in the open ocean. This is simply because coastal areas are home to a much more concentrated and diversified population of marine life than the open ocean (IMO/UNEP 2002). Nevertheless, all oil spills have an impact on marine organisms, and oil from open ocean spills can end up contaminating beaches hundreds of miles away. This has highlighted the need for authorities to work together, especially in the coastal areas that suffer from oil spills. During the last 30 years, therefore, pollution of the oceans of the world has become a matter of increasing international concern.

In spite of rigorous controls, oil spills do still occur, and so deterioration of water quality continues at a high rate. Earlier and more reliable detection of oil spills means faster containment and reduced clean-up costs. Hence, the rapid detection of oil pollution in seawater is an essential part of oil pollution prevention. It is important for minimization of environmental and financial impacts. An alternative for detecting oil spills in seawater is by designing an innovative in situ technique that can be used for earlier oil spill detection. This earlier spill detection will prompt slick containment, which accordingly results in a significant reduction in clean-up costs. For successful operation of this technique, it must be easy to carry and install at different locations and depths of marine water. In addition, it must be safe and its use should not create any additional environmental hazards.

In this first case study, an operative real time in situ device for oil spill detection in seawater is developed. The device can monitor the seawater pollution continuously and send an alarm if the level exceeds a certain limit. It can also be used for the surface measurement of oil spill thickness. This case is based on the work of Abdul-Wahab (2006). It is focused on minimizing the environmental and financial impacts of oil pollution by designing a real time, environment friendly and cost effective in situ device to detect the oil pollution in seawater. The technique is based on using the concept of the electrical conductivity to characterize and to measure the thickness of an oil layer in seawater.

2.2 Electrical Conductivity

Oil possesses a number of thermal properties that make it distinguishable from seawater (e.g., heat capacity and thermal conductivity). Electrical conductivity is the ability of an aqueous solution to conduct an electric current, and it is a measure of how well a material accommodates the transport of electric charge. The electrical conductivity of oil is much lower than that of seawater. The electrical current flows through the seawater (Figure 1), whereas it does not flow through

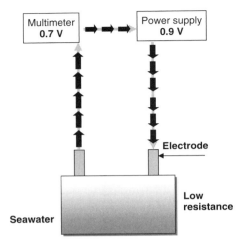

Figure 1   The flow of the electrical current through the seawater

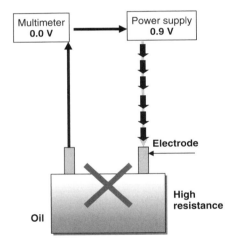

Figure 2   The resistance of the oil layer to the flow of the current

the oil layer (Figure 2). Hence, the oil layer acts as a layer of resistance to the flow of the current.

2.3   Preliminary Design System

A preliminary design was first constructed to measure the conductivity of seawater and compare it to that of the oil polluted seawater. Figure 3 depicts in detail the whole system of this preliminary design. The design consists of five main components: glass container (1 liter), which holds the seawater; DC power supply (0 to 15 volts) to provide electrical current or voltage difference; two

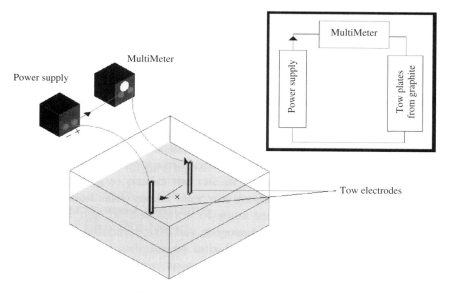

Figure 3   The electrical conductivity measuring device

electrodes (conductors); wires; and digital multimeter to measure current that was given by the DC power supply.

The oil layer detection in seawater was examined by slowly adding a small amount of crude oil to the glass container that contained the seawater so as to form an oil layer. The reading of the multimeter (i.e., electrical conductivity) was compared to that of the seawater. The results indicated an abrupt decrease in the multimeter readings, when the seawater was covered by an oil layer. This decrease in conductivity between the two electrodes was due to the high resistance of the oil layer, which acted as an insulator.

2.4   Final Operative Real-Time Design System

An innovative technique was then designed to develop an operative real-time system for oil spill detection. The system provides an in situ technique for detection of oil pollution in seawater. It can work in such a way that monitoring of seawater pollution can simultaneously detect the oil layer and send an alarm if the oil level exceeds a certain limit. Figure 4 shows the electrical circuit of the real-time system for oil spill detection in seawater. The design consists of seven main components: glass container (1 liter), which holds the seawater; two electrodes (conductors); potentiometer; computer program; wires; motor (i.e., moving the plates up and down); and DC power supply (0 to15 volts).

The system was designed to send an alarm if the level of the oil increased to a certain threshold. Also, the system was designed to data log the status information about seawater conditions every 15 minutes, or at any time interval selected on

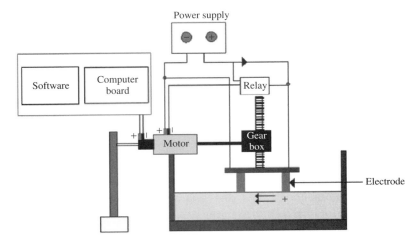

Figure 4 The electrical circuit of the real-time system for oil spill detection in seawater

the computer. This helps not only in finding the thickness of the oil layer, but also for determining the speed of the oil layer (since the downward distance and the time between the two steps are known). Finally, the data are displayed in the program screen with the thickness and the speed of the oil layer.

The designed system is easy to use. It may be installed at different locations in the marine water and at different depths, as well. In addition, it has the ability to be connected locally with a data logger. The system in reality will be constructed in a small box that will float in seawater (Figure 5). It will consume only a small

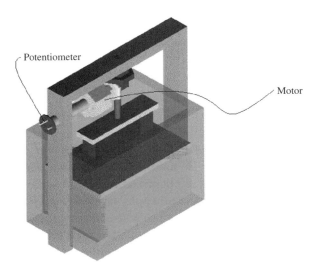

Figure 5 Construction of real-time system for oil spill detection in seawater

amount of electricity. Hence, the total cost will be very low compared with other techniques that have been used to detect oil pollution. Moreover, safety concerns have been taken into account because the voltage supply is very low.

Figure 6a shows the current flow when there is no crude oil in the seawater whereas Figure 6b depicts the current flow when there is oil pollution in the seawater. It can be seen that the electrical circuit in both figures consisted of two paths. The first path was the combination of the power supply, DC-motor and a variable resistance (R), whereas the second path consisted of the power supply,

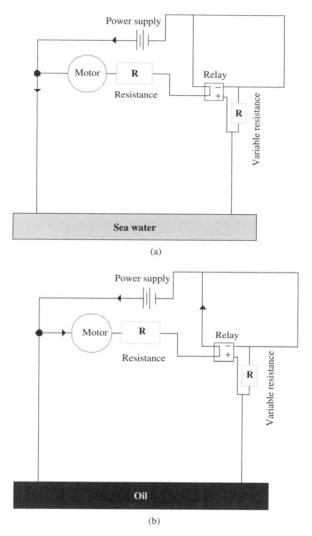

Figure 6 The current flow (a) when there is no crude oil in the seawater and (b) when there is oil pollution in the seawater

seawater, and a variable resistance (R). It should be noted that the first and the second variable resistances were used for the purpose of controlling the current flow through the motor and the relay, respectively.

If there is no oil in the seawater (Figure 6a), then the electrical current would flow in the seawater side. No flow would pass through the motor side because of the higher resistance on this side (i.e., motor and variable resistance) as compared with that of the seawater side. When the relay is opened, then no current would flow through the motor side. When oil is present in the seawater (Figure 6b), the direction of the electrical current in the circuit will be changed. Since the relay circuit is closed, the current would flow in the motor. Therefore, the motor rotates, and as a result, the two electrodes move downward into the polluted seawater. This downward movement of the two electrodes continues as long as the water is polluted with oil. This downward movement continues until the electrodes meet a nonpolluted seawater layer (beneath the oil layer). The direction of the flow of the current would change again so it passes through the seawater. The motor then stops its rotational movement. As a result, the electrodes stop their downward movement. The distance that the motor moved could be calculated.

Calibration of the designed system was carried out in order to find out the relationship between the downward distance (thickness of the oil layer) and the voltage coming from the potentiometer (i.e., volt signal received by the computer through the data logger). Each full revolution of the potentiometer corresponds to a downward distance of 2 mm, which is equivalent to 10 volts. The results of the calibration are illustrated in Figure 7.

2.5 Preliminary Experimental Design

In carrying out the experiment, an amount of seawater (300 ml) was placed in a glass container. The electrical circuit was prepared in such a way that it consisted of two electrodes that were connected by wires with the power supply and the

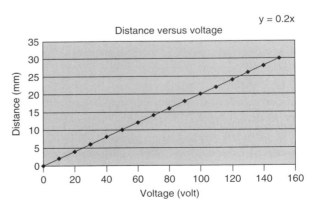

Figure 7 Calibration curve between the downward distance and the measured voltage

multimeter (Figure 3). All the components were connected in series. The electrical circuit consisted of two electrodes that protruded into the seawater. A constant voltage (V) was applied across the electrodes by using the DC power supply. An electrical current (I) flowed through the seawater due to this voltage and was proportional to the concentration of dissolved ions in the seawater (i.e., the more ions, the more conductive the seawater), resulting in a higher electrical current, which was measured electronically by the multimeter. The oil layer detection in seawater was then examined by slowly adding a small amount of crude oil in the glass container so as to form an oil layer. The new reading of the electrical current of the multimeter was recorded. This reading was compared with that of the seawater.

2.6 Final Experimental Design (Real-Time System)

In carrying out the experiment, an amount of seawater (300 ml) was placed in a glass container. The electrical circuit was prepared in such a way that it consisted of two electrodes that were connected by wires with the power supply and the motor. The power supply was connected in parallel with the electrodes, as well as with the DC motor. The whole system is depicted in detail on Figures 4 and 5. The electrical circuit consisted of two electrodes that protruded into the seawater (i.e., the two electrodes were adjusted to touch the surface of the seawater in the container), power supply, and the motor. A constant voltage (V) was applied across the electrodes by using the DC power supply. Due to this voltage in the electrical circuit, an electrical current (I) flowed through the electrical circuit. All the generated electrical current (I) in the circuit was expected to flow through the seawater side (less resistance) and therefore there no electrical current would flow through the motor side. This is because of the existence of the dissolved ions in the seawater that makes the seawater more conductive in comparison to the motor side. Hence, the electrical resistance was much higher in the motor side than in the seawater side. This caused all the electrical current (I) to flow through the seawater side across the two electrodes. As a result, the motor does not rotate and so the two electrodes do not move downward in the seawater. The two electrodes stayed in their position at the surface of the seawater. The readings of the electrical current that passed through the seawater side across the two electrodes were taken and saved by the computer.

The oil layer detection in the seawater was then investigated by slowly pouring an amount of oil (150 ml) into the glass container so as to form an oil layer. A constant voltage (V) was again applied across the electrical circuit by using the DC power supply. Due to this voltage in the electrical circuit, an electrical current (I) flowed through the electrical circuit. In this case, all the generated electrical current (I) in the circuit was expected to flow through the motor side (less resistance), and therefore, there was no electrical current flowing through the seawater side (high resistance due to the presence of the oil layer). Hence, there was no electrical current passing across the two electrodes. This was due to the high resistance of the oil layer, which worked as insulator. The current flowing through the motor was used to rotate the motor (rotational movement), and the gearbox was used to convert the rotational movement of the motor into downward movement. This resulted in the movement of the two electrodes downward into the polluted seawater. Hence, the two electrodes continued to move downward within the oil layer as long as the water was polluted. This downward movement of the two electrodes continued until the electrodes met a clean layer of seawater beneath the oil layer. The signal then was given to the computer to stop rotation of the motor and so the downward movement of the two electrodes. The computer program measured this downward movement of the two electrodes and considered it as a measure of the thickness of the oil layer. It should be noted that a potentiometer was connected with the motor. The function of this potentiometer was to convert the rotational movement of the motor into a volt signal that the computer could understand.

In summary, the case study demonstrated that the designed system for early detection of oil pollution in seawater made it possible to not only detect oil in seawater, but also to determine the thickness of the oil layer. The design was experimentally tested by using different types of oil and was used to determine the average speed of the oil spill. The design was simple, cost-effective, automated, compact, with low energy consumption, and could be constructed at any time and at any location. The result of the study is very important since it can be used to tackle the oil spill problem and hence minimize damage caused.

3   SECOND CASE STUDY: PROTECTION OF THE MARINE ENVIRONMENT

3.1   Introduction

Research into previous ship accidents indicates that the major cause of any ship damage was a fissure or crack in the body of the ship. Cracks may begin as imperfections caused by welding and thermal cutting of edges. In the beginning, the problem usually starts with some small cracks that appear in certain locations of the ship body. Then, these cracks propagate because of the ship load and the huge amount of force that is generated by sea waves. The cracks grow in size, causing damage to the whole ship. Subsequent crack propagation may cause failure of the primary structural members, leading to catastrophic consequences such as massive oil pollution or loss of the whole ship (Garwood 1997). Even when fissures or cracks do not lead to complete failure, the cost of inspections and repairs and the cost of the consequences of oil pollution due to oil leakage can be high (Kirkhope et al. 1997).

Damage to ship hulls, especially in ships that are used to transport crude oil, may harm the marine population in the area of the oil spill. Oil pollution can harm marine life in three different ways: by poisoning after ingestion, by direct contact, and by habitat destruction (Baird and Hayhoe 1993; Brown et al. 2000; Al-Obeidani 2004, Al-Obeidani et al. 2008). Oil pollution creates many problems that can lead to loss of lives, damage to property and the environment (Wang et al. 2003). As the oil spreads along the coastline, it covers sea animals, birds, and plants. It transforms previously clean areas with significant biological resources into areas of ecological disaster (Al-Obeidani 2004; Al-Obeidani et al. 2008). Many plants and animals suffer or are killed within a short time after the oil pollution occurs. Many people spend their time and money cleaning up the oil. Scientists also spend their time and government resources trying to find different types of technology for treating oil pollution.

In addition to the usual toxic effects on marine life and shore damage, oil pollution can threaten desalination plants that provide the fresh water supply in some areas. Conventional seawater desalination plants cannot operate if the feed seawater contains oil. In the case of an oil spill disaster, the desalination plant must be stopped until the oil is removed from the feed seawater. The cleanup process can take a long time. Therefore, most countries of the world suffer from oil spills due to their detrimental impact on the environment and economy.

With the increasing demand for safety at sea and protection of the environment, it is of great interest to be able to predict ruptures in ship hulls as early as possible to prevent oil spills. Hence, there is a need to detect any fissures in ship hulls before they cause serious problems. In this second case study, an innovative technique for designing a smart ship is developed to detect ruptures in ship hulls at an early stage. This smart ship has the ability to both detect the existence, and determine the location, of a crack that might occur in its body. The study is based on the work of Abdul-Wahab et al. (2008). It deals specifically with finding an innovative solution to the problem of fissures that can be applied to all ships, especially those that are used to transport oil. It can also be applied for detecting cracks in operational pipe lines, such as those used to transport gas or oil. The designed system is simple and based on the resistance theory technique. It can be inserted into the structure of a ship in order to detect and specify the location of a fissure.

3.2 Basic Principles of the Designed System

Figure 8 shows a copper wire with thickness of 0.01 mm and a resistance of 0.20 ohm. When there is no crack on the wire, the multimeter reads approximately 0.2 ohm. However, if there is a crack in some section of the wire, it will suddenly break down. A zero indication will appear in the multimeter (Figure 9).

3.3 Switch Device

Figure 10 shows the top view of the switch device. It is a box that consists of a motor, a shaft, and a pointer. Also, it contains eight metal knobs. To explain the function of the switch device, the circuit will be connected without the switch

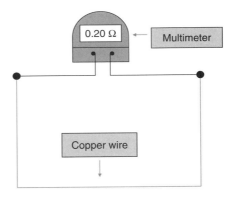

Figure 8   Testing without fissure

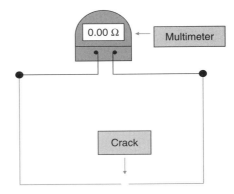

Figure 9   Testing with fissure

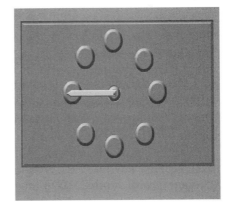

Figure 10   Top view of the switch device

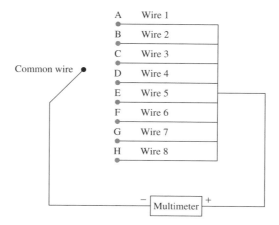

Figure 11   The circuit without the switch device

device as shown in Figure 11. The objective of this circuit is to measure the resistance of the wires, each one individually. For example, if the resistance of wire 1 is required to be measured, then the common wire (negative side of the multimeter) should be connected to point A (left end of the wire 1). This closes the circuit. Therefore, the multimeter will read only the resistance of wire 1. Again, the same procedure will be used to measure the resistance for each one of the remaining wires. As you have seen the procedure of measuring the resistances of wires is simple. However, the electrical connection needs to be made manually at each time the wire is measured. Hence, if there are 100 wires, this means that we should connect the electrical circuit manually in each wire individually in the circuit (i.e., to measure its resistance). This is impractical.

Therefore, a switch device is fabricated in such a way that it works as a multiplexer to switch between the wires. This switch device will automatically measure the resistance of each wire separately. Figure 12 shows the same circuit described in Figure 11 but with a switch device. In this case, the common wire

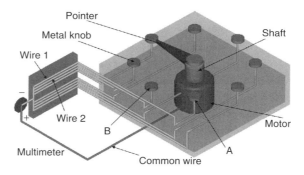

Figure 12   The circuit with the switch device

of the multimeter (negative side) will be connected to the shaft of the motor (and so it is connected with the pointer). The wires (ends A to H) will be connected to the metal knobs. Each wire will be connected to one metal knob. Then all the wires leaving the metal knobs will be combined together in one wire that will be connected to the other end of the multimeter (positive side). If the resistance in wire 1 is required to be measured (i.e., end A), then the pointer of the motor will be automatically connected to wire 1 through point A. This makes the circuit close automatically, and accordingly, the resistance of wire 1 will be measured by the multimeter. If the resistance of wire 2 is required to be measured (i.e., end B), then the pointer of the motor will be automatically connected to wire 2 through point B. This makes the circuit close automatically, and accordingly, the resistance of wire 2 will be measured by the multimeter, and so on for the remaining wires.

3.4   Experimental Procedure

The bench scale consists of the following main components: hard paper sheet, plastic tape, copper wires, a small ship, personal computer with Data Acquisition Card (DAC), Lab View program, alarm system, switch device, converter (i.e., to convert the resistance to voltage so the computer can understand it), and power supply.

The hard paper sheet was covered on both sides by plastic tape to resist water. Then, the hard paper sheet was divided into 15 sections (three horizontal rows and five vertical columns) as shown in Figure 13. The copper wires were arranged in the hard paper sheet in a horizontal direction. In each horizontal section only one wire was used, as shown in Figure 14. Figure 15 shows the wires in the three horizontal rows. The spaces between the wires were small (approximately 4 mm), and there was no connection between the wires in different rows. The side of the hard paper sheet that included the wires was then covered by the plastic tape, as shown in Figure 16. The copper wires were then arranged in

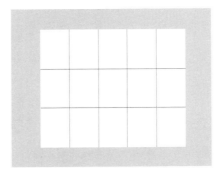

Figure 13   Hard sheet paper with its sections

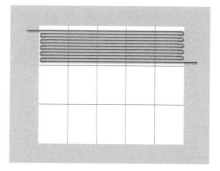

Figure 14   Wire arranged in one horizontal row

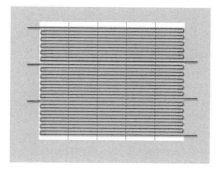

Figure 15   Wires arranged in horizontal rows

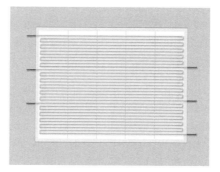

Figure 16   Wires covered with plastic tape

the hard paper sheet, but in a vertical direction. In each vertical section, only one wire was used, as shown in Figure 17.

Figure 18 shows the wires in all the five vertical columns. The spaces between the wires were again small (approximately 4 mm), and there was no connection between the wires in different columns. The side of the hard paper sheet that

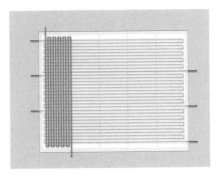

Figure 17   Wire arranged in one vertical column

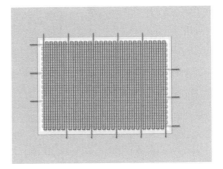

Figure 18   Wires arranged in all vertical columns

included all the wires in the horizontal and vertical sections was then covered by the plastic tape, as shown in Figure 19. The hard paper sheet was then stuck into the bottom side of the ship as shown in Figure 20. The inlets of the horizontal and the vertical wires of the paper sheet (I1 to I8) were combined together in one wire that was connected to the converter. The converter was used in order to read the resistance signal of the combined wire and convert the reading into a voltage signal. The outlets of the horizontal and the vertical wires of the paper sheet (O1 to O8) were connected to the switch device in such a way that each wire was connected into one metal knob. A common wire was connected between the shaft of the motor of the switch device and the converter. The output signal from the converter (i.e., volt signal) was taken into the computer through the interface board. The data analysis of the whole circuit was achieved by using the Lab View program that controlled the whole circuit. The designed circuit is depicted in detail in Figure 21, whereas the converter is shown in Figure 22. An alarm system was also connected to the computer. The ship was tested and the readings were recorded every few seconds.

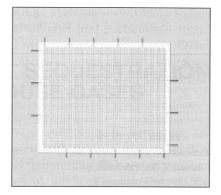

Figure 19   Copper wires covered with plastic tape (final sheet that it will be inserted into the ship)

Figure 20   Sheet paper that was inserted into the ship

3.5   Results of the Study

The hard paper sheet described in Figure 19 (with all its horizontal and vertical wires) was stuck into the bottom side of the ship or any other location in the structure of the ship. The whole circuit was connected as depicted in Figure 21. The switch device was controlled by the computer program so that the switch between wires happened every few seconds. Also, the DAC supplied the switch device with voltage (the maximum output voltage is 10 volts). The power of the converter was provided by the power supply. The designed system was tested

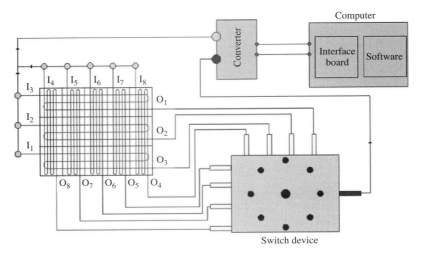

Figure 21 The designed circuit in detail as connected to the switch device, converter, and computer

under two conditions. In the first condition, there was no crack in the ship hull. The readings of the computer associated with this case were equal to 2 volts, indicating that the circuit was closed. In the second condition, the system was tested when there was a crack present in the ship hull. The readings of the computer associated with this case were equal to 10 volts, indicating that the

Figure 22 Resistance to voltage converter

circuit was opened. Simultaneously, the computer program sent a signal to the alarm system to be turned on. The computer program was also used to specify the location of the crack. All this information was displayed on the computer screen. The readings associated with each individual section of the hard paper sheet were also displayed on the computer screen (Figures 23 and 24). The whole circuit (which includes the ship, hard paper sheet with its wires, switch device, converter, alarm, and the computer) is depicted in Figure 25.

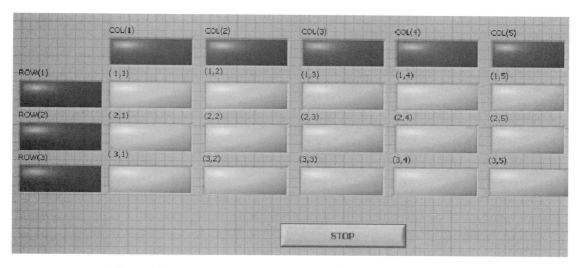

Figure 23   The computer screen when there is no fissure in the ship

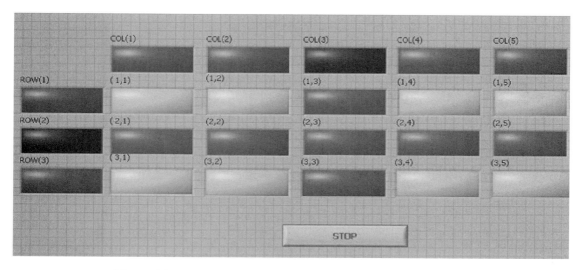

Figure 24   The computer screen when there is a fissure in the ship

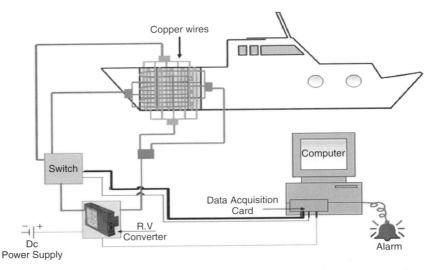

Figure 25 The smart ship, hard paper sheet with wires, switch device, converter, alarm, and the computer as connected together in our designed system

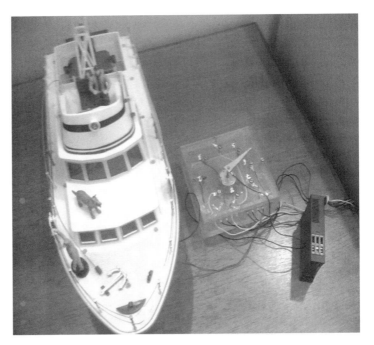

Figure 26 Top view of the designed system

The designed system will be inserted into the structure of the ship (Figures 26 and 27) to detect any fissures in the structure of the ship at an early stage and to prevent fissure propagation. With this system, the ship will then have the ability to detect the crack directly when it occurs—and its location as well. Thus, costs

Figure 27   Side view of the designed system

resulting from ship accidents and damage will be reduced. In addition, predicting fissures in ships that transport crude oil as early as possible will prevent oil from entering the marine environment. This results in protection of marine and human life. In the long run this will have a positive impact on the environment and the economy, as this will save time and money spent in cleaning up oil. As we have seen, construction of the designed system requires only hard paper sheet with wires, a switch device (i.e., multiplexer), lab view program, DC power supply, converter, alarm, and computer. All these components are cheap. However, the multiplexer was relatively expensive in the market. Therefore, a multiplexer (i.e., switch device) that cost only $15 was designed and fabricated. The designed system is cost-effective. In addition, it has a low operating cost and it is very low on maintenance costs. It also does not require any extra space on the floor of the ship, since it is inserted into the body of the ship. In addition, the designed system has almost no moving parts and it should last for an indefinite period (subject to normal maintenance). Further, it requires a very low level of skill for operation.

Regarding the environmental impacts, the designed system is clean, presents no risk and is environmentally friendly. All of the construction materials do not generate any air, water or land pollution. In terms of social impacts, the designed system is helpful and efficient because it has the ability to detect the location of a rupture in a ship's hull as early as possible. The system does not conflict with any logical or ethical rules. As for safety, the system does not present any danger to humans because it relies on the resistance theory. Therefore, there is no current or voltage applied in the designed system. By using the resistance method, any short circuit that may occur in the ship is eliminated.

4 THIRD CASE STUDY: OXIDES OF NITROGEN EMISSIONS FROM SHIPS AT BERTH

4.1 Introduction

Ships are powered at sea by diesel engines (i.e., main and auxiliary engines) to provide propulsion and energy power. While in port, most of the time auxiliary engines are operated to supply power for lighting and other domestic services. Operation of these engines can cause an array of environmental impacts that can seriously affect local communities and marine and land-based ecosystems (Bailey and Solomon 2004). Emissions from shipping seriously threaten the environment and are considered one of the major sources of air pollution that affect the health of people working onboard ships and living in nearby communities. According to Corbett and Fischbeck (1997), ships are among the world's highest polluting combustion sources per quantity of fuel consumed. Therefore, the air pollution threat caused by diesel engines cannot be neglected (Lin and Huang 2003).

Diesel engines are considered a major source of air pollution in port and urban areas because of their release of black smoke, hydrocarbons (HC), nitrogen oxides (NO_x), particulate matter (PM), carbon monoxide (CO), carbon dioxide (CO_2), sulphur dioxide (SO_2), and other toxic chemical compound contaminants into the air (Lin and Huang 2003; Lin and Pan 2001). Marine emissions are confirmed to cause the ecological environmental problems such as the ozone layer destruction, enhancement of the greenhouse effect, and acid rain production (Lin and Huang 2003). Emissions from ships also contribute significantly to global warming and to regional air pollution problem. In addition, studies on air pollution and human health indicate that human exposure to diesel exhaust causes cancer, numerous respiratory diseases, and premature death (International Maritime Organization 2005). Although those most threatened are children, communities living near port-loading terminals, shipyard workers, crews, and passengers in the immediate vicinity of commercial marine vessel engines are also threatened (International Maritime Organization 2005).

Nitrogen oxides are considered the main trace gases that emit from ships during the berth mode (i.e., auxiliary engines). Emissions of nitrogen oxides from ships increase tropospheric ozone and hydroxyl radical concentration over their background levels, thus increasing the oxidizing capacity of the atmosphere. According to Murphy et al. (2003), marine shipping is considered as the largest unregulated source of oxides of nitrogen emissions. It represents, therefore, a significant long-term obstacle to achieve ozone standards in coastal areas. Further, air pollutants such as NO_x, and O_3 accelerate the rate of deterioration of a large number of various materials.

In this third case study, the dispersion of NO_x originating from ships at berth is illustrated at Said Bin Sultan Naval base in the Sultanate of Oman. The study is based on the work of Abdul-Wahab et al. (2008). It is focused on assessing the impact of NO_x emissions origination from ships at berth on a nearby community. For this purpose, the Industrial Source Complex Short Term (ISCST) model has been used to determine the spatial distribution of the ground-level concentration of NO_x in and around the port. The results have been analyzed to determine the affected area and the level of NO_x concentrations. The highest concentration points in the studied area are also identified.

4.2   Ship Characteristics

The studied area in Said Bin Sultan Naval Base includes a total of 20 ships at berth that are responsible for NO_x generation (Table 1). A detailed database of the characteristics of the ships and performance of auxiliary diesel engines were gathered from Royal Navy of Oman. This includes information on ship names, arrival and departure dates and direction, and ship type. Table 2 shows the activities of the ships used in the study during 2005 in terms of the number of days and hours the ships spend at sea and port, according to the month.

Table 1   Characteristics of the Ships (Auxiliary Engines or Diesel Generators) Used in the Study

Seq.	Name of Ship	Type of Ship	Cross Tonnage (tons)	Number of DG[*]	DG Output (kW)
1	Al-Mabrukah	Training class	2147	3	235
2	Qahir Al- Amwaj	Corvette class	1384	4	350
3	Al-Muzzar	Corvette class	1384	4	350
4	Dhofar	Province class	543	3	145
5	Al-Sharqiyah	Province class	543	3	145
6	Al-Batnah	Province class	543	3	145
7	Mussandam	Province class	543	3	145
8	Al-Bushra	Gun boat class	490	3	220
9	Al-Mansoor	Gun boat class	490	3	220
10	Al-Najah	Gun boat class	490	3	.220
11	Seeb	Inshore patrol boat class	110	2	45
12	Shinas	Inshore patrol boat class	110	2	45
13	Sadh	Inshore patrol boat class	110	2	45
14	Khassab	Inshore patrol boat class	110	2	45
15	Nasr Al-Bahr	Landing class	3560	3	180
16	Neemran	Landing mechanism class	170	2	74
17	Sab Bahr	Landing mechanism class	170	2	74
18	Doghs	Landing mechanism class	170	2	74
19	Temsah	Landing mechanism class	170	2	74
20	Sultanah	Support class	2032	2	83

DG = Diesel generator

Table 2 Ships Activities during 2005

Seq.	Ships	Activity	Jan.	Feb.	Mar.	Apr.	May	Jun.	Jul.	Aug.	Sep.	Oct.	Nov.	Dec.	Total(days)	Total(hr)
1	Al-Mabrukah	Port	10	10	7	10	6	5	9	0	0	7	11	8	83	1992
		Sea	5	15	6	7	15	2	5	0	0	0	10	10	75	1800
2	Qahir Al-Amwaj	Port	20	21	13	12	0	5	16	7	0	0	0	0	94	2256
		Sea	11	7	8	7	7	13	7	3	0	0	0	0	66	1584
3	Al-Muzzar	Port	0	0	0	0	3	0	0	26	10	5	16	13	73	1752
		Sea	0	0	0	0	13	26	25	4	6	5	5	8	92	2208
4	Dhofar	Port	4	18	0	10	0	5	7	10	6	24	24	2	110	2640
		Sea	10	5	0	5	0	5	7	9	6	6	3	8	64	1536
5	Al-Sharqiyah	Port	25	15	6	0	18	11	0	2	11	11	3	26	128	3072
		Sea	3	6	10	0	5	8	0	3	12	5	9	5	66	1584
6	Al-Batnah	Port	0	0	10	8	15	15	9	26	3	4	0	5	94	2256
		Sea	0	0	6	6	12	4	8	4	7	7	0	0	47	1128
7	Mussandam	Port	20	0	15	10	9	14	26	20	5	6	18	13	156	3744
		Sea	5	0	0	8	13	7	5	8	10	2	7	7	73	1752
8	Al-Bushra	Port	0	20	18	16	0	0	0	0	11	6	12	23	106	2544
		Sea	0	8	8	2	0	0	0	0	12	4	7	5	46	1104
9	Al-Mansoor	Port	15	7	3	16	22	12	17	26	14	0	0	0	132	3168
		Sea	2	4	7	2	7	5	2	4	0	0	0	0	33	792
10	Al-Najah	Port	24	0	9	12	27	10	14	8	22	10	9	27	172	4128
		Sea	6	4	2	8	3	7	7	10	5	3	7	4	66	1584
11	Seeb	Port	0	0	0	0	0	13	9	27	3	31	20	17	120	2880
		Sea	0	0	0	0	0	5	4	0	2	0	5	0	16	384
12	Shinas	Port	10	10	10	21	12	3	17	17	4	12	15	3	134	3216
		sea	15	6	5	5	4	5	5	4	2	4	4	0	59	1416
13	Sadh	Port	5	14	14	15	9	15	15	29	2	5	28	8	159	3816
		Sea	2	0	6	6	11	7	5	1	3	0	1	9	49	1176
14	Khassab	Port	11	12	6	7	10	17	14	10	9	14	11	0	121	2904
		Sea	6	3	8	10	7	5	6	2	7	7	0	0	61	1464
15	Nasr Al-Bahr	Port	5	8	5	0	7	0	0	5	27	11	14	6	88	2112
		Sea	7	11	8	0	0	0	0	7	3	8	6	2	52	1248
16	Neemran	Port	10	26	16	7	21	23	11	8	6	13	18	18	177	4248
		Sea	2	1	5	5	3	2	5	9	0	7	4	6	49	1176
17	Sab Bahr	Port	24	13	15	7	21	14	16	5	15	22	18	13	183	4392
		Sea	2	6	5	0	2	6	6	4	5	4	5	5	50	1200
18	Doghs	Port	15	12	18	24	10	0	11	16	24	20	14	31	195	4680
		Sea	2	4	3	2	5	0	0	5	2	3	7	0	33	792
19	Temsah	Port	0	4	0	17	7	14	25	17	17	29	4	24	148	3552
		Sea	0	0	0	4	0	5	2	3	2	1	4	2	23	552
20	Sultanah	Port	7	0	9	12	10	13	11	16	15	8	12	13	123	2952
		Sea	4	0	7	3	6	4	1	8	2	5	7	6	53	1272

4.3   Emission Rate Calculation

The principal source of nitrogen oxides from ships at berth is the auxiliary diesel generators. The emissions of nitrogen oxides were calculated based on the prescribed methodology reported in National Pollution Inventory manual (NPI 1999).

4.4   Industrial Source Complex Short-Term (ISCST) Model

The Industrial Source Complex Short-Term (ISCST) model was used to predict the concentrations of nitrogen oxides. The emission source information that needs to be input into the model is restricted to the physical stack dimensions (height, location, internal diameter) as well as the velocity and temperature of the released gas and the NO_x emission rates. Table 3 shows the information of the stacks of the ships (auxiliary engines). This includes stack height, stack diameter, stack temperature, and stack exit velocity. Each ship was treated as a point source. The coordinates of each ship were specified in terms of X and Y coordinates. These coordinates were calculated to the reference ship *Al-Mabrukah*. In addition, the model requires the site-specific meteorological information as input data. The local meteorological information input into the model consisted of the Julian

Table 3   Information That Must Be Input into the ISCST Model

Seq.	Name of Ship	Stack Height (m)	Stack Diameter (m)	Stack Temperature (k)	Stack Exit Velocity (m/s)	X(m)	Y(m)
1	*Al-Mabrukah*	7.0	0.20	833	16.2	0	0
2	*Qahir Al- Amwaj*	11.8	0.22	823	18.4	100	0
3	*Al-Muzzar*	11.8	0.22	823	18.4	−186	−62
4	*Dhofar*	0.7	0.20	803	22.0	72	−50
5	*Al-Sharqiyah*	0.7	0.20	803	22.0	95	−50
6	*Al-Batnah*	0.7	0.20	803	22.0	72	−60
7	*Mussandam*	0.7	0.20	803	22.0	95	−60
8	*Al-Bushra*	6.0	0.30	793	19.5	70	−83
9	*Al-Mansoor*	6.0	0.30	793	19.5	90	−83
10	*Al-Najah*	6.0	0.30	793	19.5	70	−93
11	*Seeb*	0.5	0.08	736	9.0	65	−116
12	*Shinas*	0.5	0.08	736	9.0	85	−116
13	*Sadh*	0.5	0.08	736	9.0	65	−126
14	*Khassab*	0.5	0.08	736	9.0	85	−126
15	*Nasr Al-Bahr*	8.4	0.25	873	17.7	−186	−150
16	*Neemran*	5.5	0.08	783	8.4	65	−159
17	*Sab Bahr*	5.5	0.08	783	8.4	85	−159
18	*Doghs*	5.5	0.08	783	8.4	65	−169
19	*Temsah*	5.5	0.08	783	8.4	85	−169
20	*Sultanah*	10.0	0.20	753	10.2	−156	0

day of the year, the average wind flow vector, wind speed, height of the mixing layer, ambient air temperature, and the Pasquill stability category. The hourly meteorological data for the period of January 1, 2005, to December 31, 2005, was collected from Rustaq meteorological station. The ISCST model was set up to simulate the NO_x ground-level concentrations resulting from all 20 ships at all points covered by the receptor area.

4.5   Results of the Study

The spatial distribution of NO_x pollution in the modeled area during the period of study was predicted on an hourly and monthly base to assess the impact on air quality in and around the port. The NO_x concentration values were predicted using the ISCST model, while the NO_x emissions rate ships characteristics and meteorological data were used as input parameters to the model. The computations were done separately on each month. It was assumed that emission rates of NO_x were constant over the year and, therefore, it is expected that the pattern of pollutant concentration changed on a monthly basis depending only on variations in meteorological conditions. Generally, the results indicated that the highest NO_x concentrations were found to occur relatively close to the original position of the sources. This suggests that the air quality inside the ships and surrounding areas, in the port would be affected the most in terms of NO_x concentrations. The hourly and monthly distributions for total year of 2005 are shown in Figures 28 and 29, respectively. The highest hourly concentration, found close to the original point,

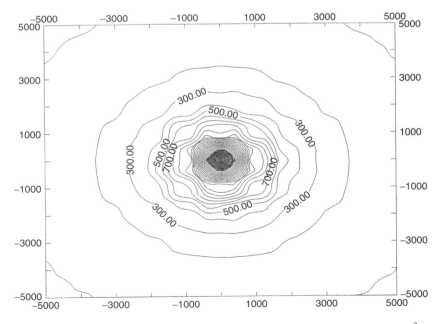

Figure 28   Predicted hourly average concentration contours of NO_x ($\mu g/m^3$)

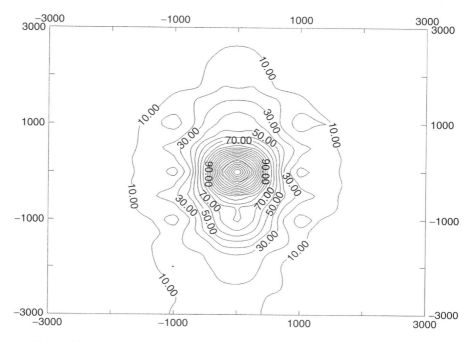

Figure 29   Predicted monthly average concentration contours of NO_x ($\mu g/m^3$)

was 4856.4 $\mu g/m^3$, while the monthly concentration was 256.9 $\mu g/m^3$and was also found at original point of the emission sources.

5   CONCLUSIONS

For the purpose of this chapter, three case studies on the maritime oil transport and pollution prevention were included and discussed. These cases are very important in minimizing the environmental and financial impacts of oil pollution. In the first case study, an innovative, real-time, and environment friendly in situ device for early detection of oil pollution in seawater was presented. The operation of the device was based on the concept of the electrical conductivity to characterize and to measure the thickness of an oil layer in seawater. The system was designed to monitor the seawater pollution continuously and to send an alarm if the oil level exceeds a certain limit. Hence, it is very important for early detection of oil pollution in seawater. The device was experimentally tested by using different types of oil and was used to determine the average speed of the oil spill. Based on the results of this case study, it was found that the device made it possible to not only monitor oil in seawater, but also determine the thickness of the oil layer. Furthermore, the device was very simple, cost-effective with low energy consumption, and could be constructed at any time and at any location.

The result of the study is very important because it can be used to tackle the oil spill problem and hence minimize damage caused.

In the second case study, an innovative technique to detect fissures in the structure of a ship was presented. The technique was simple and was based on the theory of resistance. The designed system can be simply inserted into the structure of a ship or an operational pipeline in order to detect and determine the location of a crack. The response of the system was accurate and rapid, even if the rupture was small. The design does not only solve the problem of crack detection, but it also has the ability to specify the exact location of a crack. In addition, the design is not complex and can be constructed anytime and anywhere. It does not present any negative legal, ethical, and social impacts. In addition, the invention is environmentally friendly and economical. The study is very important, as it will improve our common understanding of the fissures and ruptures in ship hulls and the environmental impacts of oil spills, as well. Furthermore, the concept of the designed system can be used in any operational pipelines, such as those used to transport gas or oil. The innovative design ensures a rapid system response and an ability to deal with different conditions, offering the potential to protect marine and human life.

In the third case study, the ground-level concentration of nitrogen oxides (NO_x) released from ships in berth was predicted. The study was undertaken at Said Bin Sultan Naval base in the Sultanate of Oman. The isopleths of NO_x concentrations into the port and beyond into its surrounding area were plotted. These isopleths indicated that the maximum average predicted ground level of NO_x concentrations were seen at a location close to the source of emission. The results of this study would have important implications to the ships operators and decision makers that are concerned about the health of ships crew and good atmospheric air quality. These will be a great reference for port authority and decision makers to take correct actions to improve the air quality in the port and the surrounding areas. As a result of this study, it is recommended that port authorities, in cooperation with local environmental agencies and ship crews, monitor and record ships emissions. This will help to identify ships that cause emission problems in the port and also will help ships' crews to take correct actions to reduce and control emissions from the main and auxiliary engines of the ships.

REFERENCES

Abdul-Wahab, S.A. (2006), "In situ Device for Detection of Oil Spill in Seawater," *Electroanalysis*, Vol. 18, No. 21, pp. 2148–2152.

Abdul-Wahab, S. A., Al-mammari, K. H., Al-Kindi, N. K., and Al-Sawafi, A. R. (2008), "Smart Ship System: Protection of the Marine Environment," *Environmental Engineering and Science*, accepted 2008.

Abdul-Wahab, S.A., Elkamel, A., Al Balushi, A. S., and Siddiqui, R. (2008), "Modeling of nitrogen oxides (NO_x) concentrations resulting from ships at berth," *Journal of Environmental Science and Health, Part A: Toxic/Hazardous Substances and Environmental Engineering*, under review.

Al-Obeidani, S.K.S., Al-Hinai, H., Goosen, M.F.A., Sablani, S., Taniguchi, Y., Okamura, H. (2008), "Chemical Cleaning of Oil Contaminated Polyethylene Hollow Fiber Microfiltration Membranes," *Journal of Membrane Science,* Vol. 307, No. 2, pp. 299–308.

Al-Obeidani, S. K. (2004), "Chemical Cleaning of Micro-Filtration Membranes Used for Pretreatment of Oil-Contaminated Seawater," A project report submitted in partial fulfillment of the requirements for the degree of Master of Sciences in Mechanical Engineering. Collage of Engineering, Sultan Qaboos University (SQU).

Bailey, D., and G. Solomon, G. (2004), "Pollution Prevention at Ports: Clearing the Air," *Environmental Impact Assessment Review*, Vol. 24, pp. 749–774.

Baird, S., and Hayhoe, D. (1993), *Oil Spills. Energy Fact Sheet.* Originally Published by the Energy Educators of Ontario.

Brown, A., Tikka, K., Daidola, J., Lutzen, M., and Choe, I. (2000), *Structural Design and Response in Collision and Grounding*, The Society of Naval Architects and Marine Engineers. Annual Meeting preprints, p. 28.

Corbett, J. J., and Fischbeck, P. (1997), "Emissions from Ships," *Science,* Vol. 278, pp. 823–824.

Garwood, S.J. (1997), "Investigation of the MV KURDISTAN Casualty," *Engineering Failure Analysis* Vol. 4, No. 1, pp. 3–24.

IMO/UNEP (2002), *Forum on Regional Arrangements for Co-operation In Combating Marine Pollution Incidents*, London.

Inggs, M. R., and Lord, R. T. (2006), *Applications of Satellite Imaging Radar*. Department of Electrical Engineering. University of Cape Town. South Africa.

International Maritime Organization (IMO) (2005), "Prevention of air pollution from ships," Marine Environment Protection Committee (MEPC 53/4/8).

Kirkhope, K. J., Bell, R., Caron, L., and Basu, R. I.. (1997), "SSC-400 Weld Detail Fatigue Life Improvement Techniques," Technical report 1997. Ship Structure Committee.

Lin, C. Y., and Huang, J. C. (2003), "An Oxygenating Additive for Improving the Performance and Emission Characteristics of Marine Diesel Engines," *Ocean Engineering*, Vol. 30, pp. 1699–1715.

Marine Emergency Mutual Aid Centre (MEMAC) (2003), Oil Spill Incidents in ROPME Sea Area, 1965– 2002. National Pollutant Inventory (NPI) (1999), Emissions Estimation Technique Manual for Aggregated Emissions From Commercial Ships/Boats and Recreational Boats, Commonwealth of Australia. ISBN: 0642 648102.

Wang, G., Jiang, D., and Shin, Y. (2003), *Consideration of Collision and Contact Damage Risks in FPSO Structural Designs*. Offshore Technology Conference. Houston. Texas.

CHAPTER 5

ACCIDENTAL OIL SPILLS BEHAVIOR AND CONTROL

M. R. Riazi
Department of Chemical Engineering
Kuwait University
Safat, Kuwait

1 INTRODUCTION

In general, an oil spill or oil slick refers to release of oil into a sea or land; however, the term *oil spill* usually refers to marine oil spills where oil is released into water at sea, ocean or coastal areas. The word *oil* mainly refers to liquid petroleum oils such as a crude oil or petroleum products (i.e., gasoline, kerosene, jet fuel, diesel fuel or any other petroleum derived fraction and refined products). This oil release could be accidental (intentional or unintentional) as a result of human activities during exploration, production and transportation in or near the sea. In addition, oil spill can be caused by war or natural disasters such as hurricanes. Terrorists may also cause oil spills by dumping oil into ocean to create attraction. Gasoline and other light fuels are more toxic, but due to volatility can vaporize more quickly. The heavier crude oil and fuel oil have less toxicity effects at the beginning but may remain longer on the water. Due to the major impact that oil spills have on coastal areas for tourism and fishing the oil companies and shipping industry have taken greater responsibility in preventing them.

When a major oil spill occurs, it usually involves contamination of coastal or inland shorelines and marshlands, which can result in serious environmental and economic damage. For example, killer whales can be poisoned if they eat oil or a fish dead by the oil (EPA 1998). Such damage can be significantly reduced if proper protection and cleanup actions are taken promptly. A good review of damages to coastal and marine resources has been published by the U.S. Department of Interior (1987). Problems of oil pollution in the Persian Gulf area became increasingly important after the Gulf War.

There are a number of different methods that might be used to deal with the oil floating on the seawater surface. Some of these methods are burning, skimming it off the surface, absorbing it with some absorbent and then removing the absorbent together with the oil, converting it into a gel and then skimming it from the surface, sinking it to the bottom, and emulsifying or dispersing it. These methods have been discussed by Breuel (1981a,b,c) and Zobell (1969). However, it is quite impossible to specify one single method of dealing with an oil slick approaching the coast or even after it has landed onto beach. In selecting an appropriate technique for oil removal, properties of oil such as density, viscosity, pour point, flash point, surface tension, boiling point, and oil solubility in seawater are quite important. For example, knowledge of density determines whether the oil floats or sinks. Surface tension determines whether the oil spreads. The velocity of an oil spill propagation, which depends on the velocities of wind and water, affects the rate of dissolution. The surface area of an oil spill depends on its volume and thickness.

At the coast when the oil reaches coastal waters, it wreaks far more damage on fragile ecosystems, some of them vital to local human economies. Sometimes with fuel oil, when it reaches shallow waters, quite a bit will pick up sand and sink. But every time there is a storm, fuel oil is released out of the sand again. It is possible for shellfish to be tainted by the toxicity of the oil over periods of years. This has severe implications for not only the shellfish populations themselves, but the creatures, including birds and humans, that feed on them (BBC 2002).

A good example of this is a relatively small spill of fuel oil in Buzzard's Bay, United States, in 1969. Shellfish were still tainted many years after the incident. Even salt marsh plants, if severely oiled, can take up to a decade to recover fully. According to some scientists, efforts to clean up sensitive areas such as this can be as damaging as the oil itself, and in some cases, it could be better simply not to intervene, and let nature do the repair work. It is human nature to want to do something, but sometimes it is better to leave it (BBC 2006).

The fate of a spill of crude oil or refined product in the marine environment is determined by spreading, evaporation, dissolution, dispersion, emulsification, sedimentation, and various degradation processes, as discussed by Kuiper and Van den Brink (1987). Therefore, analytical solutions to the problems associated with hydrocarbon oil spills are important. An ideal model for the oil fate simulates the following processes mathematically: evaporation of oil components, dissolution of oil components in water, dispersion of oil droplets in water, advection of slick and water masses, oxidation of oil components (particularly photooxidation), emulsification, biodegradation, sedimentation.

All of these processes are time-dependent and must be described by dynamic models. State-of-the-art models include some, but not all, of these processes at varying degrees of sophistication, but field or laboratory experiments designed to calibrate or test models usually focus on only one process. For example, Huang and Monastero (1982), Spaulding (1988), Stiver et al. (1989), Villoria et al. (1991), Riazi and Edalat (1996), Riazi and Alenzi (1999, 2002), Riazi and Roomi (2005, 2008) and ASCE (1996) review and propose some of these models.

Perhaps the cheapest and easiest way to remove an oil spill from seawater is to allow it to vaporize and disperse. Light petroleum fractions such as gasoline or kerosene can be completely vaporized with time. Crude oils, which contain some heavy compounds, may not be completely vaporized. Heavy compounds and residues tend to disperse into the water or sink to the bottom of sea.

The importance of evaporation in the bulk loss of oil has been noted by many researchers (Muller 1987). The rate at which a hydrocarbon dissolves in water is generally lower than the rate of evaporation under the same conditions (Kuiper and Van den Brink 1987). It is widely considered that, after volatility, the most significant property of oil components, from the point of view of their behavior in aquatic environments, is their solubility in water (Green and Trett 1989). One of the most complete list of references involving oils, occurrence, identification and monitoring, behavior, response techniques, cleanup methods, and modeling was provided by Broje while working at the School of Environmental Science and Management of University of California at Santa Barbara (Broje 2005).

2 NATURE OF OIL SPILLS

Most oil spills are from spilling of crude oil or its products such as gasoline, diesel, or fuel oil. Crude oil is produced at the production wells after a reservoir fluid is processed in a separator to reduce its pressure and temperature to the ambient conditions. During this process, light gases (mainly methane and ethane) are separated, and the liquid called a *crude oil* can be sent to a refinery or to a tanker for transportation and export. Crude oil is a mixture of hundreds of different hydrocarbon compounds. Hydrocarbons in a crude are mainly from paraffinic, naphthenic, and aromatic groups from methane to a very heavy hydrocarbon with carbon numbers greater than 50. As the number of isomers increases significantly for hydrocarbons with carbon number greater than heptane (C_7), all the compounds can be grouped as C_{7+} fraction. A typical composition of a Middle East crude oil is given in Table 1. Components or pseudocomponents in a crude may be specified by their properties such as boiling point, specific gravity, or molar mass. Once these specifications are known, various properties can be estimated from methods given in the ASTM Manual 50 (Riazi 2005).

Table 1   Composition and General Characteristics of a
Kuwaiti Crude Oil for Export (Riazi and Alenzi 1999)

Component	Mole (%)
Ethane (C_2)	0.2
Propane (C_3)	2.0
Iso-Butane (iC_4)	2.4
n-Butane (nC_4)	4.2
iso-Pentane (iC_5)	2.4
n-Pentane (nC_5)	4.1
Hexanes (C_6)	5.3
Heptane-Plus (C_{7+})	79.4
Total	100.0

For the crude:
Mol. Wt. = 210, API = 31, Sp. gravity = 0.8708
Sulfur weight percent = 2.4 percent.
For C_{7+} fraction:
Mol. Wt. = 266.6, Sp. gravity = 0.8910,
Average boiling point = 603.2 K

Once a crude oil is processed in a refinery, various products are produced. A list of typical products is given in Table 2. Each product is a mixture of various hydrocarbons from different families; however, their boiling ranges are much less than the boiling range of a crude, which includes many more compounds. Products with lower boiling point (e.g., gasoline) are lighter than those with higher boiling point (e.g., fuel oil). Light hydrocarbons with lower boiling points have greater vapor pressure and tend to vaporize more quickly. Similarly, heavier oils with lower vapor pressure have less tendency to vaporize. However, very heavy oils contain hydrocarbons, mainly from aromatic group, and contain more sulfur (a corrosive and undesirable element in any oil). Light hydrocarbons have

Table 2   Products and Composition of Alaska Crude Oil (Riazi 2007)

Petroleum Fraction	Approximate Hydrocarbon Range	Approximate Boiling Range		Vol (%)	Wt. (%)
		°C	°F		
Light gases	C_2-C_4	90 to −1	−130 to 30	1.2	0.7
Light gasoline	C_4-C_7	−1 to 83	30 to 180	4.3	3.5
Naphthas	C_7-C_{11}	83 to 205	180 to 400	16.0	14.1
Kerosene	$C_{11}-C_{16}$	205 to 275	400 to 525	12.1	11.4
Light gas oil (LGO)	$C_{16}-C_{21}$	275 to 345	525 to 650	12.5	12.2
Heavy gas oil (HGO)	$C_{21}-C_{31}$	345 to 455	650 to 850	20.4	21.0
Vacuum gas oil (VGO)	$C_{31}-C_{48}$	455 to 655	850 to 1050	15.5	16.8
Residuum	$>C_{48}$	655+	1050+	18.0	20.3
Total Crude	C_2-C_{48+}	−90 to 655+	−130 to 650+	100.0	100.0

Table 3   General Characteristics of Petroleum Products Derived from Crude Oil
of Table 1 (Riazi and Alenzi 1999)

SampleProperty	Naphtha	Kerosene	Diesel Fuel	Gas Oil
ASTM-D86 Distillation ($^\circ$C)	65.6	147.2	223.9	274.4
0	81.1	162.8	257.8	351.1
5%	83.3	167.8	268.9	371.1
10%	91.7	180.0	293.3	396.7
30%	100.6	195.6	310.0	434.4
50%	112.8	214.4	332.2	456.1
70%	128.9	235.6	364.4	492.2
90%	135.6	243.3	379.4	513.3
95%	152.8	257.2	388.9	530.0
100%				
Sp. Gr. (15.5 C)	0.7188	0.7898	0.8589	0.9259
Density (20 C)	0.7136	0.7867	0.8568	0.9225
Refractive Index (20 C)	1.4049	1.4410	1.4804	1.5262
Molecular Weight	101.3	155.7	243.7	394.3

higher heat content and produce more heat during combustion, while heavier
oils have lower heating value. Some specifications of petroleum products derived
from crude oil of Table 1 are given in Table 3. Further specifications of various
oils are discussed in the ASTM Manual (Riazi 2005).

3   MAJOR OIL SPILLS IN THE WORLD

According to Oil Tanker Statistics of 2005 published by the International Tanker
Owners Pollution Federation (ITOPF 2006), for the period of 1974–2005, a total
of 9,300 oil spills occurred, in which 7,800 were less than 7 tons and 343 were
more than 700 tons. The major cause of spill for those less than 7 tons was
loading and discharging (37 percent), followed by collisions (25 percent), while
by fires and explosions was only 1 percent. For larger spills (more than 700 tons),
both loading/discharging and explosions contributed about 8 percent, while the
cause by collisions was up to 8 percent.

Hundreds of oil spills occur every year in Mediterranean sea, which can be
monitored by SAR satellite system, as shown by Vogt and Tarchi (2004). EPA
gives the list of the biggest oil spills occurred so far (EPA 1999). The Gulf
War oil spill was one of the worst oil spills in history, resulting from actions
taken during the Gulf War in January 23, 1991. It did considerable damage
to wildlife in the Persian Gulf. The exact size of the spill remains unknown;
however, the estimate of 6 to 8 million barrels (about 1 million m^3) is most
often referenced.

On March 24, 1989, some 200 million liters (1,250,000 barrels) of crude oil
poured off the coast of Alaska, covering an area of 1,100 miles. This was the

worst oil spill in U.S. history (EPA 1999). This number was later revised, and the latest data indicate that the amount was about 11 million gallons (261,000 bbl), which polluted about 2,000 km (1,243 miles) of coastline (Petroleum Association of Japan 2005), covering 10,000 square miles. The disaster is estimated to have killed 250,000 seabirds, 2,800 sea otters, 300 harbor seals, 250 bald eagles, up to 22 killer whales, and an unknown number of salmon and herring. A court case started in 1994 in San Francisco by more than 32,000 fishermen, native Alaskans, and property owners is one of the longest noncriminal ones in U.S. history. After 20 years, this case has not yet been settled (BBC News 2006).

In 1978 off the coast of Brittany, France, some 260,000,000 liters (1,625,000 barrels) of oil spilled on the sea that covered an area of 400 km. The cleanup operations took 8 months. In 1997, an accident near Sharjah in United Arab Emirates split some 40,000 barrels of diesel oil, causing the shutdown of a 20-million-gallons-a-day desalination plant in Sharjah that supplies drinking water to some 500,000 people in the region (*Arab Times* 1997). Two more accidents in the same area occurred when tankers carrying illegal Iraqi fuel and gas oils were interrupted due to UN sanctions imposed on Iraq (ASCE 1996). Infoplease has a further list of disasters due to oil spills for reference (Infoplease website 2005).

In the last decade, there has been an increase in accidental oil spills in aquatic environments especially in the Middle East and the Persian Gulf area. Possible sources of spills include oil transportation, offshore oil exploration, and storage facilities. Only from 1988 to 1991, 28 oil spills of 500,000 liters (3,145 barrels) or more occurred (Arab Times 1997). Total volume of oil spilled during this period, excluding the Kuwaiti spill of 1991, was 1.84 million barrels. The Kuwaiti oil spill (as a result of the Persian Gulf War) in 1991 was estimated at 900 million barrels (*Arab Times* 1997). This was the largest oil spill in the human history.

In Europe, the worst pollution disaster was in 1991, in a disaster in which the *Haven* oil tanker loaded with 144,000 tons of crude oil sank, killing five people. For the next 12 years, the Mediterranean Coast of Italy was polluted, especially around Genoa (Draffan 2005). After the *Exxon Valdez* spill, scientists came to believe that oil is more toxic to fish than previously thought. The impact of that spill was felt even after 10 years. (*Exxon Valdez* Oil Spill 1994). However, the *Exxon Valdez* was actually only the thirty-fourth largest oil spill in history. The 10 largest oil spill according to Green (2008) are as follows:

1. **Kuwait, 1991: 520 million gallons** Iraqi forces opened the valves of several oil tankers in order to slow the invasion of American troops. The oil slick was four inches thick and covered 4,000 square miles of ocean.

2. **Mexico, 1980: 100 million gallons** An accident in an oil well caused an explosion, which then caused the well to collapse. The well remained open, spilling 30,000 gallons a day into the ocean for a full year.

3. **Trinidad and Tobago, 1979: 90 million gallons** During a tropical storm off the coast of Trinidad and Tobago, a Greek oil tanker collided with another ship and lost nearly its entire cargo.

4. **Russia, 1994: 84 million gallons** A broken pipeline in Russia leaked for eight months before it was noticed and repaired.

5. **Persian Gulf, 1983: 80 million gallons** A tanker collided with a drilling platform, which, eventually, collapsed into the sea. The well continued to spill oil into the ocean for seven months before it was repaired.

6. **South Africa, 1983: 79 million gallons** A tanker caught fire and was abandoned before sinking 25 miles off the coast of Saldanha Bay.

7. **France, 1978: 69 million gallons** A tanker's rudder was broken in a severe storm, and despite several ships responding to its distress call, the ship ran aground and broke in two. Its entire payload was dumped into the English Channel.

8. **Angola, 1991: more than 51 million gallons** The tanker exploded; the exact quantity of spill is unknown.

9. **Italy, 1991: 45 million gallons** The tanker exploded and sank off the coast of Italy. It continued leaking oil into the ocean for 12 years.

10. **Odyssey oil spill, 1988: 40 million gallons** 700 nautical miles off the cost of Nova Scotia.

Some other smaller oil spills are occurred at Cornwall, England (1967, 38 million gallons off the coast of Scilly Islands), North Sea (1977, 81 million gallons due to blowout of well in Ekofisk oil filed of Norway), Portsall, France (1978, 68 million gallons), Gulf of Mexico (1979, 140 million gallons), Persian Gulf, Iran (1983, 80 million gallons), Canary Islands (1988, 19 million gallons of crude oil spilled in Atlantic Ocean forming 100 square-mile oil slick), Uzbakistan (1992, 88 million gallons), and South Korea (2007, 3 million gallons). In recent years, due to advances in logistics and tanker hulls, spills have slowed down. Although single-hulled tankers are no longer being built, there are still many that have not yet been decommissioned, and as long as these tankers exist, there will be the potential for catastrophic accidents. In 2007 alone, there were three: one that spilled more than 3 million gallons of oil in South Korea (Green 2008).

4 OIL SPILL REPORTING AND RESPONSE PROGRAMS

Once an oil spill occurs, it is necessary to prepare a report and contact an environmental or a government agency. For example, in the United States, the Environmental Protection Agency (EPA) or its regional offices or U.S. Coast Guard Marine Safety Office in the area near the incident can be contacted. These agencies alternatively contact the National Response Center (NRC) to handle the case (EHSO 2008).

In the United States, almost 14,000 oil spills are reported each year, mobilizing thousands of specially trained emergency response personnel and challenging the best-laid contingency plans. Although many spills are contained and cleaned up by the party responsible for the spill, some spills require assistance from local and state agencies, and occasionally, the federal government. Whether or not it manages the response, EPA tracks all reports of oil spills. EPA usually learns about a spill from the responsible party, who is required by law to report the spill to the federal government, or from state and local responders (EPA 1996).

In reporting an oil spill incident, the following information should be reported:

- Name, location, organization, name and address of party responsible for the incident
- Date, time, and location of incident
- Source and cause of release of spill
- Types and materials released, quantity of material released
- Danger or threat posed by the release or spill
- Number and types of injuries, if any
- Weather conditions at the location of the incident
- Any other information that may help emergency personnel respond to the incident

In the United States, EPA maintains the oil spill program information line to answer questions and provide information to the public and owners or regulated facilities (ESHO 2008).

The Oil Pollution Act (OPA) was signed into law in August 1990, largely in response to the public concern following *Exxon Valdez* incident. The OPA improved the country's ability to prevent and respond to oil spills by expanding federal government's ability and providing money and people if necessary. The OPA also created the national Oil Spill Fund, which is available to provide up to $1 billion per incident. In addition, OPA increased penalties for regulatory noncompliance and broadened the response and enforcement authorities of the federal government. In 1994, EPA finalized a set of revisions to the OPA regulations that require facility owners or operators to prepare, and, in some cases submit, to the federal government, plans for responding to a major oil spill incident. For example, this law says the owners of oil tankers must have a detailed plan on what they will do if there was a spill. This law also says that all U.S. ships are required to have a double hull by 2015. The law says owners of a ship that spills oil will have to pay $1,200 for every ton they spill. The government collects money from companies that transport the oil so that when a spill occurs, it can pay for the cleanup (EHSO 2008).

The severity of an oil spill's impact depends on a variety of factors, including the physical properties of the oi, and the natural actions and receiving water on the oil. A number of advanced response mechanisms are available for controlling oil spills and minimizing their impacts on human health and the environment. The key to effectively combating spills is careful selection and proper use of the equipment and materials most suited to the type of oil and the conditions at the spill site.

Most spill response equipment and materials are greatly affected by such factors as conditions at sea, water currents, and wind. Damage to spill-contaminated shorelines and dangers to other threatened areas can be reduced by timely and proper use of containment and recovery equipments. In general, clean-up methods can be categorized into three groups of mechanical, chemical/biological and physical methods (EPA 1996). Containment and recovery equipment include a variety of booms, barriers, and skimmers, as well as natural and synthetic sorbent materials that are used to capture and store the spilled oil until it can be disposed of properly.

Chemical and biological methods can be used in conjunction with mechanical means for containing and cleaning up oil spills. Dispersants and gelling agents are most useful in helping to keep oil from reaching shorelines and other sensitive habitants. Biological agents have the potential to assist recovery in sensitive areas such as shorelines, marshes and wetlands. Physical methods used to clean up shorelines include natural processes such as evaporation, oxidation, and biodegradation, which can start the cleanup process, but are generally too slow to provide adequate environmental recovery. Physical methods, such as wiping with sorbent materials, pressure washing, and raking and bulldozing, can be used to assist these natural processes.

Scare tactics are used to protect birds and animals by keeping them away from oil spill areas. Such devices as propane scare-cans, floating dummies, and helium-filled balloons are often used, particularly to keep away birds. When an oil spill occurs, birds and marine mammals are often injured or killed by oil that pollutes their habitat. Without human intervention, many distressed birds and animals have no chance of survival.

Much has been learned about the care and treatment of oiled birds and animals through experience with recent oil spill incidents. First, the need for immediate response is essential for rescuing birds and marine mammals. Second, personnel must be trained. The rehabilitation of oiled wildlife is a complex medical and technical procedure, and volunteers must be properly trained. Third, a commitment must be made to reclaim oiled wildlife using proven, documented procedures, and avoiding shortcuts. Finally, open communication with other response agencies is crucial for any wildlife rescue operation to be successful. For areas that have been polluted by oil, rescuers must capture affected birds as quickly as possible in order to save them. Once birds have been captured, they are taken immediately to treatment centers, where they are given medical treatment and cleaned. After a bird is alert, responsive, stable, and its body's fluid balance restored to normal, detergent is gently stroked into its feathers to remove the oil. An oiled bird may require three or more washings to remove the oil entirely (EPA 1996).

5   OIL SPILL MONITORING

Oil spills can be monitored by a geographic information system (GIS), which is a computer system capable of integrating, storing, editing, analyzing, sharing, and displaying geographically referenced information. Oil-spill detection by Synthetic Aperture Radar (SAR) is based on the dampening effect oil has on capillary and short ocean surface waves. The use of SAR permits to detect oil pollution on the sea surface day and night and in most weather conditions (Automatic Oil Spill Detection 1999). SAR image classification can show three different classes for oil spill: spill area in the center, surrounded by high pollution area and the outer layers of low pollution area (Mansor et al. 2006). The weather condition has a good impact in use and effectiveness of SAR technique. At high wind speeds, the oil may be washed down into the sea, and no surface effect is observed in the SAR image. At very low winds, no SAR signal is received from the sea, so no slicks can be seen (Automatic Oil Spill Detection 1999). The principle of SAR technology is by sensitivity of radar backscatter to surface roughness as oil has a smoothing effect on the surface roughness of water and acts as a mirror. The unpolluted water surface shows higher surface roughness than the oil slick, which translates into an increased backscattered signal (Vogt and Tarchi 2004).

New techniques such as radar image processing have been used in order to make the proper distinctions between different thickness of oil spill. The results of this research show that with radar information, the signature of oil can be used to detect minute concentrations of hydrocarbon (oil spill) on the sea, and it can distinguish between different types of its thickness (Mansor et al. 2006). The scheme proposed by these researchers shows the following steps for oil spill detection and classification: (1) radar image, (2) preprocessing such as radiometric and geometric corrections, (3) postprocessing, such as dark slick detection and gamma filtering, and (4) dark slick classifications. Figure 1 shows a SAR image classification showing three different classes for an oil spill.

Montero et al. (2006) reports monitoring and forecasting an oil spill that occurred off the coast of Galicia (northwest Spain) in November 2002 spilling some 11,000 tons of oil in water. Several organizations, including Spanish Institute of Oceanography, were involved in monitoring this oil spill. A new Office of Nearshore Surveillance (UOP) (Unidade de Observación Próxima) was created. The spill's movement was monitored by overflights, and its path was forecast

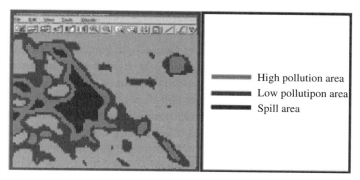

Figure 1 SAR image classification showing three different classes for oil spill (with permission from Mansor et al. 2006)

with different models. The main objectives of the UOP were to monitor the oil spill and operatively forecast its evolution near the Galician coast to help manage the crisis. Some agencies from other countries, such as Portugal, France, and the United Kingdom, were also involved in helping Spanish agencies.

Several helicopters and ships were used to monitor the spill with help of some volunteers and fishermen. Information obtained from overflights was entered in a GIS to obtain a picture of situation that was used to control the crisis and to forecast movement of the spill. Information such as maps, text, location, extend, patch type, and damaged area was made public twice a day through webpage. In order to forecast the spill movement different numerical models was used (Daniel et al. 1998). The track of the oil spill is obtained by considering that its speed is about 3 percent of the surface wind module with the same direction.

The aircrafts used for monitoring of an oil spill must have (1) good downward visibility (e.g., fixed-wing aircraft with an overfuselage wing, or helicopters), (2) radios that allow direct communications with vessels or ground personnel, if used in support of marine response or ground surveys. and (3) Global (Geographic) Positioning System (GPS), as discussed in the Oil Spill Monitoring Handbook (AMSA 2003). By knowing the speed of aircraft and the time taken to fly over the length and width of the spill, the length and size can be determined (*Note:* 1 knot = 0.5 m per second, or 1.8 km per hour).

From observing the color of the slick, approximate thickness and volume per square-km can be determined according to data given in Figure 2. In this figure the approximate percentage covered by the slick is also given based on the observation of slick. Monitoring of marine and shoreline environment is discussed in details by ASMA (2003). The personnel involved in monitoring of an oil spill must have sufficient training and skills—especially, they must have local knowledge.

Oil Color and Thickness		
Description/Color	Thickness (mm)	Volume (m³/sq km)
Silvery sheen	0.0001	0.1
Bright bands of rainbow color	0.0003	0.3
Dull colours seen	0.001	1.0
Yellowish brown slick	0.01	10
Light brown or black slick	0.1	100
Thick dark brown or black slick	1.0	1,000

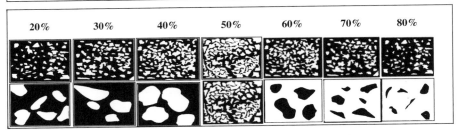

Figure 2   Visual method for approximate estimate of thickness and percent coverage area of an oil spill (Courtesy of AMSA 2003)

6   FATE AND BEHAVIOR OF OIL SPILL

The behavior of oil spills depend on the type of oil as well as the environment that it is spilled into, such as water and air temperature, water and wind speeds, and wave conditions. The processes that occur just hours following any oil spill incident are spreading, evaporation, and emulsification. However, in more general terms, the processes that marine type oil spills may go through include the following:

- Spreading
- Evaporation
- Dissolution
- Dispersion
- Emulsification
- Sedimentation
- Degradation processes

General behavior of a marine oil spill is also shown in Figure 3 (Riazi and Roomi 2008).

Spreading is a physical process in which oil spreads rapidly over a large area and breaks up in *windrows,* which are long, narrow slicks with the same orientation as the wind. Properties of oil such as density, viscosity, and interfacial tension (oil to water) play important roles in rate of spreading. Only 10 minutes after a spill of 1 ton of oil, the oil can disperse over a radius of 50 m, forming a

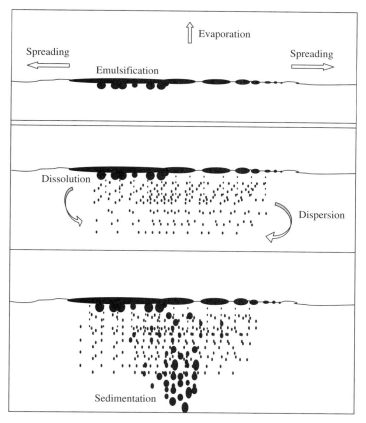

Figure 3 Behavior of an oil spill on seawater surface (Riazi and Roomi 2008)

slick of 10 mm thick. The slick gets thinner (less than 1 mm) as oil continues to spread, covering an area of up to 12 km². During the first several days after the spill, a considerable part of oil transforms into the gaseous phase. Besides volatile components, the slick rapidly loses water-soluble hydrocarbons. The rest—the more viscous fractions—slow down the slick spreading (Patin 1999). Figure 4 shows rate of spreading of the *Exxon Valdez* oil spill (Riazi and Roomi 2008), which is based on data published by Leber (1989) and Galt et al. (1991).

Further changes can take place under the impact of environmental conditions, which mainly depend on the power and direction of wind and water currents. The oil spill usually drifts in the same direction as the wind. After the slick thins to critical thickness of 0.1 mm, it disintegrates into separate fragments that spread over larger areas. Storm and turbulence speed up the dispersion of the slick and its fragments. A considerable part of spill disperses in the water as fine droplets that can be transported over very large distances away from the place of the spill. So a day after the oil spill, one may see a sticky oil mixed with all

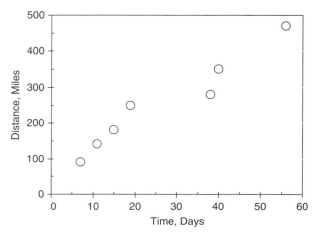

Figure 4   Rate of spreading of *Exxon Valdez* oil spill (Riazi and Roomi 2008)

types of floating debris, such as wood, cans, rope, plastic, and so on. Therefore, for a successful oil spill response we should have four elements:

1. Minimal response time
2. Efficient and fast concentration of the widely spread oil
3. Skimmers and pumps that can handle high viscosity emulsion and debris
4. Appropriate temporary storage capability (Patin 1999)

For light oils, which vaporize over a short period of time, only when oil has been trapped in a harbor may a quick response allow recovery of significant amounts of the oil.

After spreading, in which oil spill surface area increases (with decrease on slick thickness), *evaporation* becomes the most important process, especially for light oils. Evaporation is vaporization of components due to their vapor pressure or low boiling points, and sometimes it takes care of up to two third of oil. Light compounds with lower boiling points (higher vapor pressure) have higher tendency to vaporize more quickly and over a short period of time. For this reason, oil products such as naphthas and gasoline vaporize quickly, while heavier oil such fuel oil have less tendency to vaporize, leaving larger amount of oil on the water surface. Light products such as gasoline may completely vaporize within few hours. Evaporation enhances with increase in air temperature (higher vapor pressure) and increase in wind speed (higher rate of mass transfer in the air). In addition as oil spreads and the surface area increases (thinner oil), the rate of evaporation increases. As volatile components leave an oil spill, the remaining oil becomes more viscous, sticky, and heavier.

After evaporation, oil *dissolution* in water is perhaps the most important process. Most oil components are water-soluble to a certain degree, especially low-molecular-weight hydrocarbons (both aromatics and aliphatics). Components of oil that go through oxidation in the marine environment produce polar components that also dissolve in seawater. Dissolution is much slower than evaporation and approximately rate of oil dissolution is less than 0.1 percent of rate of evaporation. But it is important to know the amount of dissolution because of oil components toxicity. More toxic compounds are more soluble compounds. In general, lighter hydrocarbons have higher solubility and presence of salt in water reduces oil solubility. In addition, higher temperature increases oil solubility in water. While air temperature affects rate of evaporation, water temperature affects rate of oil dissolution. The components that dissolve in water are in contact with water, and the oil molecules that first evaporate are in contact with air in the outer layer of slick.

Table 4 shows some solubility data as measured for some products of a Middle East crude oil in which their characteristics were presented earlier in Table 3. A graphical presentation of solubility of some oil samples is presented in Figure 5. As shown in this figure, with increased salt concentration, solubility decreases.

Oil *emulsification* in water depends strongly on oil characteristics, its composition and water turbulence. In general oil emulsions are masses of oil (nonvolatile part) that contain water—as much as 30 to 80 percent. If the amount of water is

Table 4   Measured Solubilities of Different Oil Products in Water (Riazi and AlEnzi 1999)

Oil	Temperature(°C)	Salt Concentration(wt %)	$C_s(10^5 mol/l)$
Crude	20	0	2.5
	20	1.5	1.8
	20	2.9	1.0
	20	3.9	0.7
	50	0	11.9
	50	1.5	6.0
	50	2.9	3.8
	50	3.9	2.9
Naphtha	25	0	3.9
	25	2.9	3.7
	25	3.9	3.5
Kerosene	35	0	6.3
	35	3.9	3.5
	50	0	10
	50	3.9	6.5
Diesel	20	0	2.2
	20	2.9	0.8
	20	3.9	0.5
Gas Oil	25	0	1.8
	25	3.9	0.1
	50	0	4.9
	50	3.0	0.2

Other units of solubility: ppm $\simeq C_s \times$ Mol. Wt. $\times$ 1000 Mole fraction $\simeq C_S/55.5$

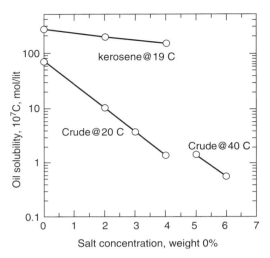

Figure 5 Measured and predicted oil solubilities in water (Riazi and Edalat 1996)

less or higher than these values, the oil-water emulsions are not very stable. Oil containing heavier hydrocarbons, especially asphaltenic components (molecular weights greater than 1000), have a higher tendency to emulsify. In addition, turbulence helps formation of larger water masses that promote the emulsification process. Stable emulsions can stay in the marine environment for more than 100 days, and their stability usually increases with decreases in temperature. Reverse emulsions as components of oil in water (oil droplets in water) are less stable. Emulsifiers accelerate the decomposition of oil products in water and promote oil dispersion (Patin 1999).

Although evaporation, dissolution, and emulsification are physical processes, *oxidation* is a chemical process in which oil components—especially those in the outer layer of slick and in contact with air—decompose due to the oxidative nature of these components. Products of oxidation process are phenols, carboxylic acids, ketones, and others. These products are more toxic than the original oil components. Oxidation processes involve photochemical reactions under the influence of ultraviolet waves of the solar spectrum, which may be enhanced by presence of elements such as vanadium in oil. Sulfur in the oil will decrease the oxidation process. Photooxidation and photolysis reactions under the influence of ultraviolet waves of sun cause the formation of solid oil aggregates and increase oil viscosity. Elements such as vanadium enhance degradation/oxidation process, while sulfur compounds slow the rate of degradation.

Sedimentation is another physical process in which heavier oil components may deposit in the bottom of the sea. Sedimentation can occur in two ways: First, those very heavy components of oil that have density greater than density of water (specific gravity > 1) fall to the bottom of the sea; and second, those components that are not denser than water but are adsorbed on suspended materials increase in density, becoming denser than water and then falling to the bottom. Depending on the crude oil composition, usually between 10 to 30 percent of oil may sediment, while for some products such as gasoline, there is no sedimentation. Sedimentation of oil, except for those very heavy oil fractions, is a very very slow process.

The other important chemical process is *microbial degradation,* in which determines the ultimate fate of an oil spill in the marine environment. Microbial degradation is decomposition of oil due to microbial activities of some 100 known species of bacteria and fungi in the marine environment. In areas of water that are highly polluted with oil, some 1 to 10 percent of oil may go through this process. In biochemical processes, oil decomposes with microorganism of several types of enzyme reactions. The rate of biodegradation depends strongly on the oil composition. In general, alkanes (paraffinic oils) biodegrade faster than aromatic or naphthenic oils. With an increase in molecular weight of the oil spill and the carbon number and chain branching (iso type molecules), the rate of microbial reaction decreases. Oil with higher rates of dispersion have higher rates of microbial degradation. In addition, temperature, concentration of nutrients and oxygen, as well as species composition, have strong impact on the rate of biochemical reactions (Patin 1999).

Finally, heavy oils may go through another physical process of aggregation. Oil aggregates in the form of petroleum lumps or tar balls, which may be found in the open water, beaches, and coastal waters. After lighter components of oil vaporize or dissolve in water, the heavier particles have greater tendency toward aggregation. The basis for oil aggregation is the presence of heavy molecules such as asphaltenes. Asphaltene molecules are high molecular weight (>1000) aromatic components that are present especially in heavy crude oils. Usually, oil aggregates have size of 1 mm to 10 cm and look like gray, brown, or black sticky lumps. Oil aggregates may exist for several years in open sea and several months in enclosed seas. Finally, as a result of these processes, oil in the marine environment rapidly loses its original properties and disintegrates into various fractions with different composition than the original oil spill. Through such self-processing, carbon dioxide and water are formed (Patin 1999).

In the case of *oil spilled on land,* the behavior of oil might be significantly different. For example, the light components in marine environment cause the least problems, but on the land, light components tend to promote more and faster polluting of groundwater, while heavier oil due to its higher viscosity penetrates slowly and its effect on soil contamination is less. In the case of spilled land, there is no spreading, and as a result, evaporation rate is limited. In addition there is no emulsification, but it may mix with gravel and soil making clean up operation more difficult. Oil spills on land are usually from product and finished petroleum products that are produced in refineries, and they contain additives (to improve fuel performances) that are sometimes very toxic and harmful to environment.

7   CLEANUP OPERATIONS

As already discussed, the seriousness of oil spill damage depends on factors such as type of oil (i.e., light or heavy), amount, and location, biological characteristics of the sea and the season (winter or summer). The most important factor is the composition of oil. The method of cleanup may increase or decrease the impact of the oil spill, and for this reason, it should be carefully selected. Decisions regarding the method of cleaning must be based on the environmental impact of the method selected.

Some cleanup methods include:

- Bioremedation and biostimulation
- In situ burning
- Skimming and mechanical recovery
- Use of chemical dispersants to emulsify oil
- Use of absorbent
- Use of solidifiers and to sink it to the bottom of sea
- Self-cleaning

Some of the equipments needed for such cleanup methods are:

- Booms
- Skimmers
- Pumps
- Storage
- Dispersants and spray systems
- Response vessels
- Absorbents
- Other spill response equipments such as aircrafts and boats

As it was discussed earlier, biodegradation is a process that includes evaporation, dissolution, dispersion, photochemical oxidation, adsorption onto suspended materials, sinking, and sedimentation. *Bioremedation* is to use microorganisms to remove pollutants from an environment. At its natural rate, it is called *biodegradation*; when affected by human manipulated conditions, it is called bioremediation. It is most often used for treatment of nontoxic liquid and solid wastes and grease decomposition. It is a cost-effective method to remove hydrocarbons from an area. Generally, about 70 to 100 microbial genera are known to contain organisms that can degrade hydrocarbon compounds. The microorganisms transform contaminants to less-harmful compounds through aerobic and anaerobic, fermentation, and reductive dehalogenation. Bioremediation agents are microbiological cultures, enzyme additives, or nutrient additives that are deliberately introduced into an oil spill. They significantly increase the rate of biodegradation and thereby mitigate the effects of the spill (U.S. Department of

Interior 2008). In general, bioremedation is more successful in land than in sea and the products of bioremediation process should be handled with special care.

The rate of natural biodegradation is limited by factors such as temperature and level of microbes (concentration of bacteria, fungi, unicellular algae, etc. in water), nutrients, and oxygen present in the immediate environment, as well as oil composition. To work effectively, microbes need a sufficient level of carbon, nitrogen, and phosphorous of C/N/P ratio. When a spill is occurred sufficient amount of carbon is available from oil itself leaving shortage of nitrogen and phosphorous levels. *Biostimulation* is a technique in which, by application of fertilizers, one can adjust C/N/P ratio and enhance the rate of biodegradation by the naturally occurring microbial community. Heavy oils with large quantities of resins and asphaltenes biodegrade very slowly if the process occurs at all (ITOPF 2008).

There are many ways to stop spread of oil in the sea. Workers can place a boom around the tanker that is spilling. Booms collect oil off the water. The workers can also use skimmers to remove the oil. Skimmers are boats that collect oil off the water. Sorbents are sponges that can collect the oil. An airplane can fly over the water, dropping chemicals into the ocean. The chemicals can break down the oil into the ocean. To clean oil off the beaches, high-pressure hoses may be used to spray the oil that is on the beach. Vacuum trucks may be driven on the beaches to vacuum up the oil. The method used for cleaning beaches depends on factors such as type and amount of oil, if people are in the area and in some cases it may be more harmful to use a cleaning method than doing nothing. This is a natural recovery that oil is left in place to degrade naturally without any response method also called *self-cleaning*.

Absorbent materials are those chemicals such as oleophilic material that have the capability of attracting oil and then removing the oil and absorbent together. Absorbents must be collected and removed finally, and overuse of sorbent should be avoided. Absorbent materials can be made as a pad with large surface area to increase the rate of absorption.

Boom is a floating physical barrier placed in water to contain, divert oil toward recovery area, deflect or exclude oil reaching sensitive areas until it can be removed. Maximum speed of current that flow to the boom is about 0.7 knot (0.35 m/s). Waves, wind, debris, and ice contribute to boom failure. Containment booming of gasoline is not usually attempted because of fire, explosion and inhalation hazards. However, when public health is at risk, gasoline can be boomed if foam is used and extreme safety procedures are applied (DAWG 2005). Spill berms are barriers other than booms that can be used to prevent oil from passing.

Skimming is to recover floating oil from the water surface using mechanized equipment such as skimmers. Skimmers are usually placed at the oil/water interface to recover oil and may be operated independently from shore or be mounted on the vessel. Skimmers can take oil at various rates mainly depending on the slick thickness. For example for slick of 1 mm thickness the recovery rate is about 35 tons/hr while for 6 mm thickness film the rate of oil removal is about 70 tons or 420 bbl/hr (OPEC 2006). Waves, currents, debris, seaweed, kelp, ice, and viscous oils will reduce skimmer efficiency (DAWG 2005). An oil *skimmer is a device that can be used to recover floating oil from (or near) the surface of water. A skimmer may be free floating, side mounted on a vessel*, built into a vessel, built into a containment boom, held by crane or by hand. It is most appropriate to have free-floating skimmer in sight of the operator. The closer it is, the better control and recovery efficiency. A free-floating skimmer is, in most cases, supposed to be within sight of the operator. The closer, the better control and recovery efficiency (Fleming 2008).

Dispersants are chemicals that are added to oil floating on the water surface to reduce oil/water interfacial tension and decreasing the work needed to break the oil into small particles and mix them with water column. This method is particularly useful when the oil spill is in a sensitive shoreline with habitats and animals. Dispersants are surface-active agents that are sprayed at the rate of 1 to 5 percent by volume of the oil from aircrafts or boats onto the slicks (DAWG 2005). To avoid losses due to winds large rain size droplets may be the most useful means (rather than fogs or mists) to spray the dispersants. The spraying of dispersant from aircraft has the significant advantages of rapid response, good visibility, high treatment rates, and optimum dispersant use. In addition, the aircraft allows treatment of spills at greater distances from shore than with vessels.

Some agitation is also needed to achieve dispersion and to send the oil from surface to the column of water. Dispersants are only useful for oils with low viscosity and in particular with viscosities of less that 2000 cSt which is equivalent to medium fuel oil at 20°C. In the upper part of water (about 10 meters from surface), the dispersants may adversely impact organisms. Use of dispersants should be carefully controlled and be followed by government regulations.

Generally, dispersants are based on hydrocarbon solvents with 15 to 25 percent surfactant. Typical dose rate is about 1 volume of dispersant for each 3 volumes of oil spill. Another group of dispersants are used at much lower doses the solvent based is alcohol or glycol (oxygenated solvent) with higher concentration of surfactant. These dispersants are used at doses of 1 volume surfactant to 10 to 30 volume oil spill (ITOPF 2008).

Another group of chemical agents used in oil spill cleanup operations is *solidifiers*, which cause a physical change of spilled oil from a liquid to a solid. These chemicals are polymer type (with hydrocarbon base) and are applied at the rate of 10 to 45 percent solidifying oil within few hours. This method is not generally used for very heavy oil that is already viscous and has high asphaltene content with a semisolid form. Oil treated with solidifiers usually has low toxicity and is typically disposed in landfills.

There are some chemical agents specifically used for shoreline clean up operations and oil removal from contaminated substrates. These chemicals help to mobilize the oil and lower the temperature and pressure of water used for flushing of oil. This method is especially useful when weathering has caused the oil to become viscous with low mobility.

One of pieces of equipment needed for cleanup operation is *vacuum pumps* used to remove oil pooled on a shoreline. The equipment has different sizes and may be mounted on a vessel for water-based operations, or on trucks or be carried by hand for remote areas affected by oil spills. Collected oil or oil/water mixture may need to be stored temporarily before permanent recycling or disposal. In such cases, large amount of water are often recovered which requires separation and treatment (DAWG 2005).

In situ burning is the term given to the process of burning oil slicks at sea, at or close to the site of a spill. Burning may be seen as a simple method with the potential to remove large amounts of oil from the sea surface. In reality, there are a number of problems that limit the viability of this response technique: the ignition of the oil; maintaining combustion of the slick; the generation of large quantities of smoke; the formation and possible sinking of extremely viscous and dense residues; and safety concerns (ITOPF 2008).

In situ means *in place,* and the method involves controlled burning of oil at the location where spill has occurred. The ignition can be started by igniter suspended from a helicopter and the burning continues as long as the thickness of spill is more than 2 mm. After in situ burning, the amount of oil can be significantly reduced, and its adverse effects on the environment minimized. One major effect of in situ burning is to minimize its spreading. Burning on land is more difficult than burning on water. The heat generated by burning also helps some additional few percent vaporization of oil. In situ for slicks of less than 1 to 2 mm is not successful and presence of wind and current make burning more difficult (Allen 1990). In situ burning requires approval and the approval must be obtained as early as possible. For the *Exxon Valdez* spill, about 30,000 oil was burned on the second day. The largest in situ burning occurred in 1991 in Kuwait. More than one billion barrels of oil were burned over a nine-month period. In situ burning can significantly reduce storage or disposal requirements. However, the method produces large amount of smoke, with negative impact on environment (Allen and Ferek 1993).

The success of a cleanup method mainly depends on the capability and availability of well-maintained equipment and products (ships, skimmers, booms, storage, dispersants, absorbents, etc.) and—most importantly—availability of trained manpower. It also requires good information, surveillance, command and strategy or planning. Oil spill responders are dealing with several hazardous situations, such as risk of drowning, working with skimmers and debris-cutting knives, heavyweight containment booms, toxic vapors in the surrounding area, various chemicals, and explosive environment. It is very important that an appropriate safety and health plan be established before any cleanup operation begins.

Collected oil from sea or shoreline must be treated before disposal. At first, water must be removed and treated. The best option for separated oil is to process it through refining as much as possible, but this is not always possible because the oil has been weathered and is so heavy that processing may be too expensive. In addition, the oil contains seawater, debris, and other materials. The other options are landfilling in controlled sites, destruction by incineration or biological processes. However, this must be done with government and environmental regulations in the area.

8   MODELING AND SIMULATION OF OIL SPILLS BEHAVIOR

8.1   Background

As it was shown in Figure 3, once an oil spill occurs, it may go through processes such as evaporation, spreading, dissolution, degradation, emulsification, and sedimentation. State-of-the-art models include some, but not all, of these processes at varying degrees of sophistications. Field or laboratory experiments designed to calibrate or test models usually focus on only one process. For example, Spaulding (1988), Villoria et al. (1991), Riazi and Edalat (1996), Riazi and Alenzi (1999), Psaltaki and Markatos (2005), Nasr and Smith (2006), and Riazi and Roomi (2008) propose some of these models.

Perhaps the cheapest and easiest way to remove an oil spill from seawater is to vaporize and disperse it. Light petroleum fractions such as gasoline or kerosene can be completely vaporized with time; however, crude oils, which consist of some heavy compounds, may not be completely vaporized. Heavy compounds and residues tend to disperse into water or to sink into the bottom of sea. The rate at which a hydrocarbon dissolves in water is generally much lower than the rate of evaporation under the same conditions, as shown by Wheeler (1987), Green and Trett (1989), and ASCE Task Force (1996).

It is widely considered that, after volatility, the most significant property of oil components, from the point of view of their behavior in aquatic environments, is their solubility in water (Wheeler 1987). Riazi and Alenzi (1999) and Riazi and Roomi (2005) showed that the rate of oil dissolution in water is small in comparison with rate of oil evaporation, and usually the amount of oil dissolved is less than 1 percent of original mass of the spill. So the rate of oil dissolution in calculation of the overall rate of oil disappearance may be insignificant, and many numerical models developed for oil spill trajectory do not consider this process. But the dissolved concentrations of hydrocarbons in water are a concern from a toxicological viewpoint, and it is important to know the exact amount of oil dissolved in water as a result of an oil spill. Aromatic hydrocarbons, especially monoaromatics such as benzenes, are the most toxic compounds, and their amount in water determines the degree of toxicity in water. The physical process of dissolution is well understood, but the description in the case of oil spills is complicated, due to the complexity of oil composition, with hundreds of components and the necessity of describing the dissolution of a single component with component-specific parameters. The component-specific description may be necessary because toxicity is component-specific, as well. The most soluble oil components are usually the most toxic components. Even low concentrations of these toxic compounds could lead to serious effects on biological systems. New studies by the National Marine Fisheries Service show that even very low levels of weathered oil spill are toxic to fish and wildlife (*Exxon Valdez* Victims 2000).

The next section presents a mathematical model originally developed by Riazi and Al-Enzi (1999) to predict the rate of an oil spill evaporation, dissolution, and sedimentation. This model later was modified by Riazi and Roomi (2008) to predict that rate of dissolution of most toxic compounds found in an oil spill (mainly monoaromatics). They proposed mathematical relations for rates of oil evaporation and dissolution by introducing two temperature-dependent mass transfer coefficients, one for evaporation and one for the dissolution process, with variable slick thickness. They also measured some experimental data on the rate of disappearance of some petroleum products and a crude oil floating on water at different conditions. They developed a more complete model to consider the rate of sedimentation, as well as the multicomponent nature of crude oils. Such information may be used in better selection of an appropriate method for the cleanup of an oil spill.

8.2 Modeling Scheme and Method

Upon spreading, evaporation, and other dynamic processes shown in Figure 3, the thickness of the spill decreases with time. In this model, slick thickness is considered as a time variable parameter. Consider an oil spill floating on seawater surface with initial volume of V_o and initial surface area of A_o in which the subscript o indicates an initial value at time zero. The oil is assumed as a mixture of N component/pseudocomponents each specified by i. Then at any time t after the occurrence of spill, its volume (V), area (A), and the thickness (y) of the spill are given as following:

$$V = \sum_{i=1}^{N} V_i, \quad A = \sum_{i=1}^{N} A_i, \quad y = \frac{V}{A} \qquad (1)$$

For each component, the volume fraction disappeared after time t is defined as $F_{Vi} = (V_{oi} - V)/V_{oi}$. In the mathematical model, a time step of Δt is chosen and a semianalytical approach is used to formulate the model. During each time step, it is assumed that slick thickness is constant. The volume fraction of each component x_{vi} is calculated from V_i and V as $x_{vi} = V_i/V$. Volume of each component V_i can be converted to mass (m_i) through its density ρ_i ($m_i = V_i \times \rho_i$). The total mass of oil at any time is calculated from mass of individual

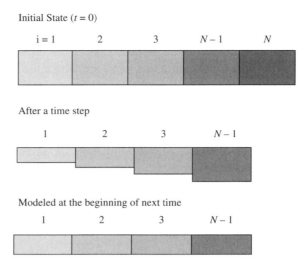

Figure 6 Modeling scheme for an oil spill

components: $m = \sum m_i$. Schematic of the model is shown in Figure 6, where the oil is assumed as a mixture of N components. Component 1 is the lightest, while N is the heaviest component in the oil. Components with specific gravity SG_i ($=\rho_i/\rho_{\text{water}}$) greater than unity sink to the bottom of the sea immediately. It is assumed that all components heavier than water sink in the first time step. Based on the remaining components, new slick thickness can be determined.

The total volume fraction of oil disappeared at time t is calculated from

$$F_V = \sum_{i=1}^{N} x_{vi} F_{Vi} \tag{2}$$

The volume fraction disappeared due to evaporation of oil (F_{Vi}^{vap}) during each time step Δt is calculated from the following relation:

$$F_{Vi}^{\text{vap}} = 1 - \exp(-Q_i^{\text{vap}} \Delta t) \quad \text{where} \quad Q_i^{\text{vap}} = \frac{K_i^{\text{vap}} Z_{\text{liq},i}^{\text{sat}}}{y_i} \tag{3}$$

In equation (3), $Z_{\text{liq},i}$ is the compressibility factor for the saturated liquid defined as ($P_i^{\text{sat}} M_i / \rho_i^{\text{sat}} RT$) in which P^{sat} is the vapor pressure of component i, R is the gas constant, and T is temperature. K_i^{vap} is the mass transfer coefficient due to evaporation given by the following relation derived specifically for Kuwait oils (Riazi and AlEnzi 1999):

$$K_i^{\text{vap}} = \frac{1.5 \times 10^{-5} T U^{0.8}}{M_i^2} \tag{4}$$

in which K_i^{vap} is in m/s, T is temperature in K, M_i is the molecular weight of component i, and U is the wind speed in m/s.

The volume fraction disappeared due to dissolution of oil (F_{Vi}^{dis}) during each time step Δt is calculated from the following relation:

$$F_{Vi}^{dis} = 1 - \exp(-Q_i^{dis}\Delta t) \quad \text{where} \quad Q_i^{dis} = \frac{K_i^{dis}C_{si}}{y_i\rho_{mi}} \qquad (5)$$

In equation (5), ρ_{mi} is liquid molar density and C_{si} is the molar solubility of component i in water at water temperature and K_i^{dis} is the mass transfer coefficient due to dissolution given by the following relation derived specifically for Kuwait oils (Riazi and AlEnzi 1999):

$$K_i^{dis} = \frac{4.18 \times 10^{-9}T^{0.67}}{V_{Ai}^{0.4}A_i^{0.1}} \qquad (6)$$

in which K_i^{dis} is in m/s, T is in K, V_{Ai} is the molar volume of component i at its normal boiling point in m³/gmol. A_i is the surface area of component i in m². Solubility of component i in water (C_{si}) is calculated from the following relation in terms of molecular weight, temperature and salt concentration in water:

$$C_{si} = \exp[(4.6 - 0.0036M_i) + (0.1 - 0.0018M_i)S_w - 4250/T] \qquad (7)$$

in which C_{si} is in mol/L, T is the absolute temperature of water in K, and S_w is the salt concentration in water in weight percent. This relation was derived from data on solubility of various oils in water at different salt concentrations and temperature. Once volume fraction of each component dissolved is calculated from equation (5), mass of each component dissolved (m_i^{dis}) may be calculated through the following relation:

$$m_i^{dis} = \rho_i F_{Vi}^{dis} V_{oi} \qquad (8)$$

Total mass of spill disappeared after time t is then calculated through sum of masses disappeared due to evaporation, dissolution, and sedimentation for all components. Total amount dissolved can be calculated as:

$$m^{dis} = \sum_{i=1}^{N} m_i^{dis}$$

Similarly, total mass of component i and total oil evaporated after time t can be calculated as discussed in detail by Riazi and Alenzi (1999). Calculation of basic parameters is discussed in the next section. After each time step, similar calculations are repeated, beginning with a new value of slick thickness calculated from previous calculations.

In the model, an oil spill is considered as a mixture of different real and pseudocomponents. These pseudocomponents are treated separately, and those components that are heavier than water with specific gravities greater than one would immediately sink into water. Although sedimentation of oil into water is a time-dependent process, in the proposed model it is considered as an instant process for the sake of mathematical simplicity. So at the beginning of the calculations, if density of component i is greater than water density at the same temperature, then the component sinks into the water and it does not enter the subsequent calculations for the rates of evaporation and dissolution.

8.3 Calculation of Model Parameters

Composition of a crude oil sample used in this modeling is given in Table 1. Besides known light hydrocarbons, heavier compounds are lumped as hexanes (C_6) and heptane-plus fraction (C_{7+}), where C_6 is a narrow-cut hydrocarbon mixture while C_{7+} is a wide hydrocarbon mixture containing hydrocarbons from heptane and heavier. Since treatment of the whole crude oil as a single component leads to significant errors in estimation of its properties, a probability density function (PDF) can be used to generate a certain number of pseudocomponents for the C_{7+} fraction, as explained by Riazi (2005).

For the C_6 and the C_{7+} pseudocomponents ($C_{7+(1)}$, $C_{7+(2)}$, $C_{7+(3)}$, ...), the PNA (paraffins, naphthenes, and various families of aromatics) composition can be estimated from the API methods, as given in ASTM Manual (Riazi 2005). According to these methods, fraction of paraffins, naphthenes, monoaromatics, and polyaromatics are determined from specific gravity, molecular weight, and refractive index of each pseudocomponent. The relation for estimation of monoaromatics is:

$$X_{MA} = -62.8245 + 59.9081 R_i - 0.0248335 m \qquad (9)$$

where X_{MA} is the fraction of monoaromatics, $R_i = n - d/2$ and $m = M(n - 1.475)$, in which n and d are refractive index and density, both at 20 C (3). Once each C_{7+} fraction group is divided into three subgroups (P, N, and A), required physical properties of these pseudocomponents can be estimated from the methods given by Riazi (2005). A summary of calculation scheme is presented in Figure 7.

8.4 Model Prediction and Results

Some experiments were conducted at 20 and 40 C, where average wind speed was 5 m/s. Initial volume of the oil spill was 500 mL (cm^3) and the initial area was 3116 cm^2. Total volume of oil dissolved in water after 174 hours was 0.76 cm^3. Salt concentration of water was 3 percent by weight, which was taken from Maseela Beach off the coast of Kuwait. Model prediction and experimental data for the mass of the crude oil spill remaining on water surface versus time at temperature of 40°C are shown in Figure 8. Specification of this crude is given in Table 1. Fractional losses (F_V) of same crude oil spill at 22 and 40°C are given in Figure 9. For the diesel oil of Table 3, the fractional loss is shown in Figure 10 at 22°C. For this oil, as it is heavier than the crude, the rate of disappearance is less than that of crude.

Predicted amounts of dissolved hydrocarbons in water for the crude oil are shown in Figures 11, 12 for temperatures of 22°C and 40°C, respectively. Amount

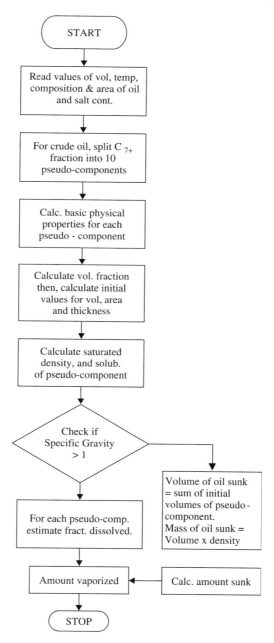

Figure 7   Summary of calculation procedure

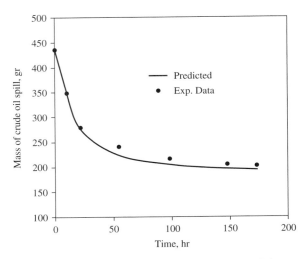

Figure 8   Model prediction for the mass of crude oil spill remaining on water surface at 40°C (Riazi and Roomi 2008).

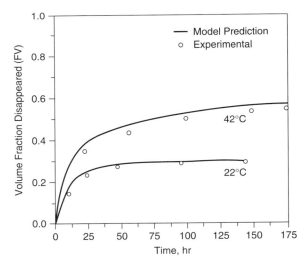

Figure 9   Rate of disappearance of Kuwaiti crude oil spill at 22°C and 42°C (Riazi and AlEnzi 1999)

of aromatics and monoaromatics dissolved in water versus time are also shown in these figures. The amount of dissolved hydrocarbons increases with increase in temperature. Calculated dissolved oil at 40°C after 174 hours is 0.8 gr versus measured value of 0.72 gr. For the slick thickness, calculated value was reduced from 1.6 to 0.75 mm, while measure values showed reduction from 1.6 to 0.78 mm, confirming validity of calculations for crude oil sample.

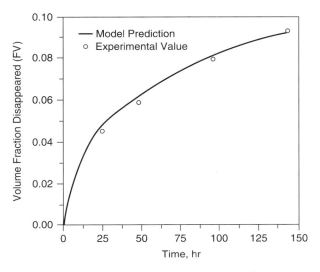

Figure 10   Rate of disappearance of a diesel fuel (Table 3) at 22°C (Riazi and AlEnzi 1999).

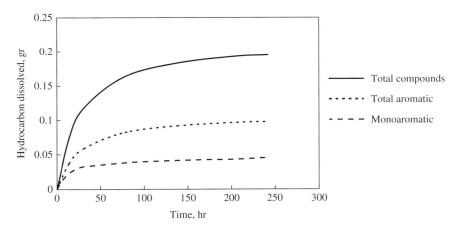

Figure 11   Model prediction for the rate of oil crude dissolution at 22°C (Riazi and Roomi 2008)

For the naphtha and kerosene samples of Table 3, the rate of dissolved aromatics and monoaromatics have been determined at 20°C at two different salt concentrations and are presented in Figures 13, 14. As shown in these figures, presence of salt causes a decrease in the amount of hydrocarbon dissolution. For light crude oil products such as naphtha, the oil spill can be completely vaporized within a few hours, and the amount of dissolved compounds remain constant after complete disappearance of oil spill due to rapid vaporization process.

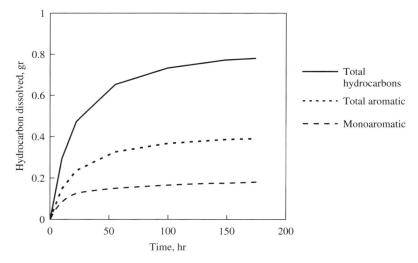

Figure 12   Model prediction for the rate of crude oil dissolution at 40°C (Riazi and Roomi 2008)

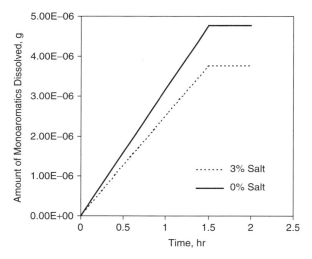

Figure 13   Rate of dissolution of monoaromatics in water at 32°C from a naphtha oil spill area of 0.32 m² (Riazi and Roomi 2008)

As shown in these figures, the amount of dissolved hydrocarbons is small because of low solubility of hydrocarbons in water, but knowledge of this concentration is important from an environmental point of view, especially the amount of aromatics in water, because of the toxic nature of these compounds. The model presented here predicts the rate of oil disappearance and concentration of different

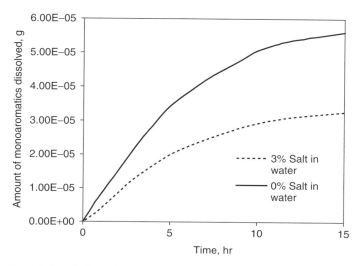

Figure 14   Rate of dissolution of monoaromatics in water at 32°C from a kerosene oil spill area of 0.32 m² (Riazi and Roomi 2008)

hydrocarbons, as well as monoaromatics in water versus time. The result could be helpful in selection of an appropriate method for clean up operations.

9   SUMMARY AND CONCLUSIONS

In this chapter, the nature of oil spills, their behavior in the marine and land, and their impact on the environment were discussed. A brief summary of major oil spills in the world, their causes, and their effects were presented. Discussions were followed by how and where to report an oil spill incident and what information is needed in the report. Some laws and regulations in the United States that are applicable to oil spill incidents were also presented. Monitoring techniques, especially use of more modern methods such as GIS, and various processes (such as evaporation, dissolution, degradation, spreading, sedimentation) that are used once an oil spill occurs in a marine environment were discussed. Then different methods of cleanup after occurrence of an oil spill, including bioremediation and biostimulation, skimming, dispersing, solidification, and burning, their advantages and disadvantages, and the equipments needed for such methods were presented. Finally, the role and importance of modeling and simulation were presented. To make any decision on the cleanup method, it is extremely helpful to predict behavior of the spill through some of existing models. This chapter described a model recently developed that can be used to predict the rate of oil spill disappearance and the rate and amount of dissolution of most toxic compounds in water.

REFERENCES

Allen, A. A. (1990), "Contained controlled burning of spilled oil during the *Exxon Valdez* oil spill," in *Proceedings of the Thirteenth Arctic and Marine Oil Spill Program Technical Seminar* (Edmonton, Alberta), June 6–8, pp. 305–313.

Allen, A. A., and Ferek, R. J. (1993), "Advantages and disadvantages of burning spilled oil," in *Proceedings of the 1993 International Oil Spill Conference,* American Petroleum Institute (Washington, DC), pp. 765–772.

AMSA (2003), *Oil Spill Monitoring Handbook*, Prepared by Wardrop Consulting and the Cawthron Institute for the Australian Maritime Safety, Canberra City, Australia.

Arab Times (1997), English Daily Newspaper, Kuwait, July 15: 3. Also see: *Arab Times*. 1998. January 17: 7.

Automatic Oil Spill Detection, ENVISYS (1999), The Norwegian Computing Center (NR). http://www2.nr.no/envisys/emergency_management/marine_oil_spill.htm.

ASCE Task Force (1996), "State-of-the-Art Review of Modeling Transport and Fate of Oil Spills," *J. Hydraulic Engineering* (November), pp. 594–609.

BBC News (2002), "Oil Spill: Consequences for Wildlife," *BBC News* (November 19), 13:29 GMT, website: http://news.bbc.co.uk/2/hi/science/nature/2491965.stm

BBC News (2006), "Court halves Exxon spill damages," *BBC News* (December 22), http://news.bbc.co.uk/2/hi/business/6204819.stm.

Breuel, A. (1981a), "How to Dispose of Oil and Hazardous Chemical Literature Review," *Arch. Hydrobiol., Adv. Limnol.* Vol. 24, pp. 1–69.

Breuel, A. (1981b), "How to Dispose of Oil and Hazardous Chemical Spill Debris," *Pollution Technol. Rev.* Vol. 87, pp. 37–71.

Breuel, A. (1981c), *Oil Spill Cleanup and Protection Techniques for Shorelines and Marshlands*, Noyes Data Corporation, Park Ridge, NJ.

Broje, Victoria (2005), *Oil Spill Publications*, University of California, School of Environmental Science and Management, Santa Barbara, http://fiesta.bren.ucsb.edu/~vbroje/index.htm.

Daniel, P., Poitevin, J., Tiercelin C., and Marchand, M. (1998), "Forecasting Accidental Marine Pollution Drift: The French Operational Plan," *Oil and Hydrocarbon Spills, Modeling, Analysis and Control, Computational Mechanics Publications*, WIT Press, pp. 43–52.

Daubert, T. E., and Danner, R.P. (1989), *API Technical Data Book–Petroleum Refining,* 5th ed., Ch. 4. Am. Pet. Inst. (API), Washington, DC, pp. 57–66.

DAWG Inc. (2005), "The Spill Control People," http://www.dawginc.com/spill-control-absorbents/oil-spill-response_ar18.php.

Draffan, G. (2005), "Major Oil Spills," http://www.endgame.org/oilspills.htm.

EHSO (2008), "Environment, Health and Safety Online," http://www.ehso.com/oilspills.php; also see the EPA webpage on oil spills: http://www.epa.gov/oilspill/.

EPA (1996), "U.S. EPA Emergency response program," Oil Spill Prevention and Response Program: http://www.rivermedia.com/consulting/er/oilspill/.

EPA Oil Spill Program Update (1998), "The U.S. EPA's Oil Program Center Report", *1* (4), U.S. Environmental Protection Agency, Washington, DC, http://www.epa.gov/oilspill/.

EPA Oil Spill Program Update (1999), "The U.S. EPA's Oil Program Center Report", *2* (2), U.S. Environmental Protection Agency, Washington, DC, http://www.epa.gov/oilspill/.

Exxon Valdez Oil Spill Trustee Council (1994), *Exxon Valdez Oil Spill Restoration, November.* Also see webpage: http://www.jomiller.com/exxonvaldez/report.html.

Exxon Valdez Victims (2000), "Remember the Valdez Spill, Save the Arctic Refuge, 1989–1999," http://www.jomiller.com/exxonvaldez/report.html.

Fleming (2008), website: http://www.oil-spill-web.com/oilspill/handbook2.htm.

Galt, J. A., Lehr, W. J., and Payton, D. L. (1991), "Fate and Transport of *Exxon Valdez* Oil Spill," *Environ. Sci. Technol.* Vol. 25, No. 2, p. 202.

Green, H. (2008), "Largest Oil Spills (The *Valdez* Doesn't Make the List)," February 28, Envirowonk, http://envirowonk.com/content/view/68/9/; For list of oil spills, see also http://www.infoplease.com/ipa/A0001451.html.

Green, J., and Trett, M. W. (1989), *The Fate and Effects of Oil in Freshwater,* Elsevier, London.

Green, J., and Trett, M. W. (1989), *The Fate and Effects of Oil in Freshwater*, Elsevier, London, 310 pp.

Huang, J. C., and Monastero, F. C. (1982), "Review of the state of the art of the oil spill simulation models," Final Report of Raytheon Ocean Systems Co.,'' Am. Pet. Inst., Washington, DC.

ITOPF (2006), "The International Tanker Owners Pollution Federation Ltd, Oil Tanker Spill Statistics: 2005," http://www.itopf.com/stats05.pdf.

ITOPF (2008), The International Tanker Owners Pollution Federation Ltd., London, UK http://www.itopf.com/spill-response/clean-up-and-response/alternative-techniques/.

Infoplease (2005), "Oil Spill Disasters," Information Please, Pearson Education. Boston, http://www.infoplease.com/ipa/A0001451.html.

Kuiper, J., and Van den Brink, W. J. (1987), *Fate and Effects of Oil in Marine Ecosystems*, Martinus Nijhoff Publishers, Boston, MA.

Leber, P. A. (1989), *Environmental Chemistry: A Case Study of the Exxon Valdez Oil Spill of 1989*, Franklin and Marshall College, Lancaster, PA, http://wulfenite.fandm.edu/exxon-valdez.htm.

Mansor, S. B., Assilzadeh, H., and Ibrahim, H. M. (2006), "Oil Spill Detection and Monitoring from Satellite Image," GIS Development 2006, http://www.gisdevelopment.net/application/miscellaneous/misc027.htm.

McAuliffe, C. D. (1986), "Organism exposure to volatile hydrocarbons from untreated and chemically dispersed crude oils in field and Laboratory," 9th Arct. Mar. Oil Prog. Tech. Semin., Edmonton, Alta., Environ. Can., Environ. Prot. Serv., pp. 497–526.

Montero, P., Blanco, J., Cabanas, J. M., Maneiro, J., Pazos, Y., Moroño, A., Pérez-Muñuzuri, V., Braunschweig, F., Fernandes, R., Leitao, P. C., and Neves, R. (2006), "Oil Spill Monitoring and Forecasting on the *Prestige-Nassau* Accident,"

Montero, P., et al. (2003) "Oil Spill Monitoring and Forecasting on the Prestige-Nassau Accident." Environment Canada's 26th Arctic and Marine Oil Spill (AMOP) Technical Seminar. Ottawa, Canada. pp. 1013–1029.

Muller, H. (1987), "Hydrocarbons in the Freshwater Environment: A Literature Review," *Arch. Hydrobiol., Adv. Limnol.* Vol. 24, pp. 1–69.

Nasr, P. E., and Smith, E. (2006), "Simulation of Oil Spills Near Environmentally Sensitive Areas in Egyptian Coastal Waters," *Water and Environment Journal*, Vol. 20, No. 1, pp. 11–18.

OPEC (2006), http://www.opec.co.uk/products_opr05.html.

Patin, S. (1999), "Environmental Impact of the Offshore Oil and Gas Industry," Trans. E. Cascio, http://www.offshore-environment.com/oil.html.

Petroleum Association of Japan (2005), "Oil Spill Response Department (PAJ –OSR): List of oil spill accidents," http://www.pcs.gr.jp/doc/incident/incidents_e.html.

Psaltaki, M. G., and Markatos, N.C. (2005), "Modeling the Behavior of an Oil Spill in the Marine Environment," *IASME Transactions*, Vol. 2, No. 9, pp. 1656–1664.

Reid, R. C., Prausnitz, J. M., and Poling, B.E. (1987), *Properties of Gases and Liquids,* 4th ed., McGraw-Hill, New York, p. 741.

Riazi, M. R. (2005), *Characterization and Properties of Petroleum Fractions*, ASTM Manual 50, Conshohocken, PA: ASTM International, http://www.astm.org/mnl50.htm.

Riazi, M. R., and Alenzi, G. (1999), "A Mathematical Model for the Rate of Oil Spill Disappearance from Seawater for Kuwaiti Crude and Its Products," *Chem. Eng. Journal*, Vol. 73, pp. 161–172.

Riazi, M. R., and Alenzi, G. (2002), "A Model for Oil Dissolution in Seawater," Presented at the IASTED International Conference on Modeling and Simulation (MS 2002), (Marina del Rey, California), May 13–15.

Riazi, M. R., and Edalat, M. (1996), "Prediction of the Rate of Oil Removal from Seawater by Evaporation and Dissolution," *J. Pet. Sci. and Eng.*, Vol. 16, pp. 291–300.

Riazi, M. R., and Roomi, Y. A. (2005), "A Predictive Model for the Rate of Dissolution of Oil Spill and its Toxic Components in Sea Water," Presented at the 230th ACS Annual Meeting, Division of Petroleum Health and Safety (Washington, DC), August 28–September 1.

Riazi, M. R., and Roomi, Y. A. (2008), "A Model to Predict Rate of Dissolution of Toxic Compounds into Seawater from an Oil Spill," *International Journal of Toxicology*, The Official Journal of the American College of Toxicology, Accepted September.

Spaulding, M. L. (1988), "A State-of-Art Review of Oil Trajectory and Fate Modeling," *Oil and Chemical Pollution*, Vol. 4, pp. 39–55.

Stiver, W., Shiu, W. Y., and MacKay, D. 1989. Evaporation times and rates of specific hydrocarbons in oil spills. Environ. Sci. Technol., 23(1): 101–105.

U.S. Department of the Interior (2008), Offshore Energy and Mineral Management, October 8, 2008 updates of the following webpage: http://www.mms.gov/tarprojectcategories/chemical.htm

U.S. Department of Interior (1987), Measuring Damages to Coastal and Marine Natural Resources, Volume 1, CERCLA 301 Project. U.S. Dep. Inter., Washington, DC. 220 pp.

Villoria, C. M., Anselmi, A. E., Intevep, S. A., and Garcia, F. R. (1991), "An Oil Spill Fate Model," *SPE23371*, pp. 445–454.

Vogt, P., and Tarchi, D. 2004. Monitoring of marine oil spills from SAR satellite data, Ispra, Italy: European Commission. http://inforest.jrc.it/documents/2004/SPIE5569-3PVogt.pdf

Wheeler, R. B. (1987), *The Fate of Petroleum in the Marine Environment*. Production Research Company Special Report, New Jersey: Exxon.

Zobell, C. (1969), Microbial Modification of Crude Oil in the Sea, API/FWPCA, American Petroleum Institute (API), Washington, DC.

CHAPTER 6

GEOLOGICAL SEQUESTRATION OF GREENHOUSE GASES

Ahmed Shafeen
Natural Resources Canada
Ottawa, Ontario, Canada

Terry Carter
Petroleum Resources Centre
Ministry of Natural Resources
London, Ontario, Canada

1   INTRODUCTION

The recent rapid rise of anthropogenic greenhouse gas content in the atmosphere has been identified to be a serious environmental concern and does not support sustainable development for the global economy. Drastic emission reduction measures are necessary in the near term to stabilize the concentration of greenhouse gas in the atmosphere to an acceptable level while longer-term non-carbon based energy options are being developed. Geological sequestration of CO_2 is a promising technology which can effect a rapid and large-scale reduction of GHG being emitted into the atmosphere and the technology is relatively mature. However, before considering CO_2 sequestration projects, a site specific in-depth techno-economic, geological and physical analysis has to be completed to assess feasibility and ensure appropriate measures are put into place to monitor system performance and guarantee public safety and environmental protection. The following section summarises the concepts of geological sequestration and the important steps to be followed for its implementation.

1.1   Concentration of Greenhouse Gases and Global Warming

The main cause for anticipated global warming is due to the continuous increase in greenhouse gas emissions from human activities. It is also well recognized that the anthropogenic emissions of greenhouse gases, especially carbon dioxide (CO_2), have the most adverse impact on climate change. The most important greenhouse gases are carbon dioxide (CO_2), methane (CH_4), nitrous oxide (N_2O), water vapor (H_2O), ozone (O_3), and chlorofluorocarbons (CFCs, such as $CFCl_3$ and CF_2Cl_2). Since the late 1950s, the atmospheric concentration of different greenhouse gases has increased significantly due to the human activity. Scientific data, such as ice core data, indicate that during the last 200 years, the carbon dioxide concentrations have increased nearly 30 percent, methane concentrations have more than doubled, and nitrous oxide concentrations have increased by about 15 percent (EPA 2003). The heat-trapping capability of the Earth's atmosphere has increased due to this enhanced concentration and subsequently contributed to the overall global warming. The contribution of different gasses to the greenhouse effect is roughly estimated as follows: CO_2 55 percent, CFCs 24 percent, CH_4 15 percent, N_2O 6 percent (Hitchon 1996). Almost half of the emitted CO_2 stays in the atmosphere; the rest is absorbed by the oceans and the biological sinks (IPCC 2001). The principal consequences of global warming, as anticipated, will be the threat of inundation of the low-lying coastal zones due to melting of the polar ice caps, regional decrease in food production, deterioration of ecosystems, and spread of diseases (Hendriks 1994). Mass human migration due to these consequences, from the affected areas to the other parts of the world, would be a great challenge to deal with in the years to come.

In order to achieve deep reductions in greenhouse gas emissions and to stabilize the emissions level, different emission reduction treaties are already in place that set the targets for different countries to reduce their greenhouse gas emission. However, implementation of these treaties would ultimately depend on the commitments and compliance from individual countries and the technoeconomic feasibility of the emission reduction projects in search for a sustainable energy future.

1.2 Options for Emission Control and Reduction

As mentioned earlier, the contribution of CO_2 is the highest among all the greenhouse gases in the atmosphere. Theoretically, any further damage to the environment can be avoided by reducing or mitigating CO_2 emission to an acceptable limit. There are methods to reduce CO_2 emission, such as development of more energy-efficient technologies, a switch to less carbon-intensive fuels, increased use of renewable energy, and increased carbon sequestration. Except for the last option, these are long-term processes and will not be able to meet the short- and medium-term emission targets set by international treaties and protocols. Deep sea or ocean sequestration has immense potential and can be achieved by direct injection of pure CO_2 into the deep ocean near the sea bed or by enhancing the net uptake by the ocean from the atmosphere by artificial means. This option is not yet suitable or practical because of the limited knowledge and uncertainties associated with the science, marine life, environmental issues, and long-term fate of the stored CO_2.

Terrestrial sequestration of carbon in the biosphere can be enhanced and maximized through methods such as reforestation, enhancement of carbon content in soils through modified agricultural practices, and reclamation of barren mining lands to establish soils and vegetation cover (Litynski et al. 2006). Again, benefits from this approach will be long-term.

There is general agreement that the best option for the near and medium-term reduction of CO_2 is through geological sequestration. The geological storage options (Hitchon 1996; IPCC 2005; Bachu 2001; Dusseault et al. 2001; and Baines and Worden 2004) include:

- Deep coal seams
- Depleted oil and gas reservoirs
- Enhanced oil recovery (EOR) in oil reservoirs

- Solution-mined caverns in salt beds and salt domes
- Deep saline aquifers

Deep coal formations can be considered for enhanced coal bed methane recovery. By this approach, a simultaneous CO_2 sequestration and methane production will be achieved from deep coal seams, where, in normal cases, methane is trapped inside the coal seam both as free gas and by adsorption onto carbon molecules. During CO_2 injection, adsorption of CO_2 by the carbon in the coal seam causes desorption of methane from the coal bed, trapping the CO_2 and releasing the methane, which is produced and sold.

CO_2 may also be injected into depleted oil and gas reservoirs, under the principle that the hydrogeological conditions that allowed the hydrocarbons to accumulate in the first place will also permit the accumulation of CO_2 in the space vacated by the produced hydrocarbons. Availability of considerable subsurface geological data makes this option attractive for CO_2 sequestration. However, the limited volume potential for long-term storage is the major disadvantage for the deep coal seams and depleted oil and gas reservoirs.

Solution-mining of salt from bedded or domal deposits of salt (halite) has created numerous manmade caverns at various locations around the world. Halite is impermeable to CO_2. Individual caverns can be filled quickly and can contain large quantities of CO_2, but these probably constitute only a local solution where other options for geological storage are not available, due to high cost and limited capacity (Dusseault et al. 2001).

CO_2 may also be injected into oil reservoirs to improve oil recovery as they near the end of their effective life in a process known as *enhanced oil recovery* (EOR). Because CO_2 is a miscible solvent with oil, the residual oil can be washed from the reservoir rock by the CO_2 solvent. The oil and CO_2 mixture is separated at the surface and the CO_2 that is returned to the surface can then be reinjected for more oil recovery. This type of EOR has been active in the United States for more than 25 years. Among all these geological sequestration options, deep saline aquifers in sedimentary basins offer the largest storage potential (Hitchon 1996, IPCC 2005). Deep saline aquifers are common subsurface fluid reservoirs and are present almost everywhere in the world. These are too deep and too saline to be used as a source of potable water or for agricultural activities. Extremely large amounts of CO_2 can be stored in these saline formations for long geological time periods. Upper estimates of worldwide storage capacities of these aquifers range from approximately 10,000 (IPCC 2005) to 50,000 Gtonnes of CO_2 (Hendriks 1994).

2   CARBON SEQUESTRATION

The terminology *carbon sequestration* is widely used for the secure storage and isolation of CO_2 once it is separated from its source gases. According to U.S. Department of Energy (DOE), carbon sequestration is defined as "the capture and secure storage of carbon that would otherwise be emitted to or remain in the atmosphere. The idea is (1) to keep carbon emissions produced by human activities from reaching the atmosphere by capturing and diverting them to secure storage, or (2) to remove carbon from the atmosphere by various means and store it" (DOE 2003).

There are three steps to carbon sequestration:

1. Separate and capture CO_2 from the flue gases and exhaust of power plants, refineries, oil sands operation and heavy oil upgrading facilities, cement plants, steel plants, ammonia plants and other chemical plants.
2. Concentrate it for transportation to and storage in distant reservoir locations.
3. Convert it into stable products by biological or chemical means or allow it to be absorbed by natural sinks such as terrestrial or ocean ecosystem.

2.1 Basic Concepts: Geology for Sequestration

Geological sequestration of CO_2 requires access to large volumes of subsurface space that will act to securely isolate the CO_2 from contact with the atmosphere for time periods of thousands to millions of years. This space is usually in the form of pore spaces between rock and mineral grains in bedrock, generally at depths greater than 800 meters to keep the CO_2 at densities approaching that of water and safely below most fresh-water aquifers. Rocks with sufficient porosity to meet these requirements generally are confined to sedimentary basins.

A sedimentary basin is a large-scale depression in the Earth's crust, within which sediments have accumulated over millions of years, compacted and lithified after burial to great depths, and which are now preserved as thick sequences of layered sedimentary rocks. A sedimentary basin may cover tens to hundreds of thousands of square kilometers and may contain thicknesses of several thousand or even tens of thousands of meters of sedimentary rocks in the basin centers, thinning towards the basin margins. Most sedimentary rocks contain pore spaces between the mineral grains. These pore spaces are occupied predominantly by saline water and less commonly by liquid and gaseous hydrocarbons and other gases (e.g., CO_2 and H_2S).

Sedimentary rocks transmit water to varying degrees, due to a property known as *permeability*. Three general types of permeable sedimentary rocks are generally recognized: aquifers, aquitards, and aquicludes. *Aquifers* are rocks with sufficient permeability to conduct water and yield significant quantities of water to springs and wells, and conversely, into which water (or CO_2) can be injected. Coarse-grained sedimentary rocks such as conglomerates, sandstone, and some carbonates have highly interconnected pore spaces and form aquifers. An *aquitard* is a rock that retards but does not prevent the movement of water and does not readily yield water to a well or spring. Examples of aquitards are fine-grained rocks such as shales, which are composed of compacted and lithified mud with poorly interconnected pore spaces. *Aquicludes* are barriers to water movement, the best example being evaporite rocks such as halite (rock salt) or anhydrite. Aquitards and aquicludes act as caprocks or seals that retard upward movement of water and any CO_2 injected into the aquifer.

Sedimentary basins have complex plumbing systems as a result of the interbedding of permeable and nonpermeable layers of sedimentary rocks of varying thicknesses and structural configurations. The flow of water and gas is lateral along bedding in the aquifers and is vertical, across the bedding, in aquitards (Gunter et al. 2004). Flow rates across aquitards are several orders of magnitude slower than along aquifers. Beds of halite and anhydrite prevent flow by self-healing of fractures and pores by salt creep and by reprecipitation of dissolved evaporite minerals in pore spaces. Intact beds of evaporites are essentially impervious to fluid flow.

To store CO_2 in sedimentary basins, it must be injected into a saline aquifer that is confined beneath an aquitard or aquiclude so that the CO_2, which is buoyant in water, does not immediately escape back to the surface. The CO_2 will rise to the top of the aquifer and will move laterally under the confining layer of rock. If the CO_2 becomes trapped in a closed structure, it can be expected to remain isolated for millions of years. Oil and gas has been trapped in this way deep in the subsurface. If the aquifer is open laterally, the CO_2 will migrate up-dip toward the surface. In most cases, fluid flow in the deep subsurface is very slow, measured in cm per year, with flow paths to the surface exceeding hundreds of kilometers. It is considered likely, however, that the CO_2 would become dissolved into the formation water in the aquifer before it could escape to the surface.

2.2 Properties of Formation Water

In simple terms, formation water is naturally occurring water that occupies the pore spaces of a rock in the subsurface. Formation waters may be fresh or saline, with salinity varying from much less than seawater to about 10 times more saline (White 1965). Deep formation waters are dominated by Cl^- anions with sodium and calcium and lesser amounts of magnesium as the dominant cations.

The solubility of CO_2 in the formation water varies, according to the degree of salinity as well as to the temperature and pressure. Generally, solubility decreases with the increase of salinity, and vice versa. Hydrodynamics of the formation water, especially the extremely slow migration of groundwater in deep aquifers, plays an important role in the sequestration process. Normally, the velocity of the formation water varies from only a few centimeters/year to a few meters/year (Hitchon 1996).

2.3 Basic Principles of Sequestration

It is preferred to carry out CO_2 sequestration at a temperature and pressure beyond the CO_2 critical state (31.1°C and 7.38 MPa). At normal atmospheric pressure, CO_2 is gaseous with a density of $1.872 \, kg/m^3$, whereas in a supercritical state, CO_2 behaves like a gas, filling all available pore space, but has a density similar to water, ranging from 150 to greater than $800 \, kg/m^3$ (Bachu 2003). This enables storage of tremendous amounts of CO_2 in a relatively small volume. Other possible ways of storage include gaseous and liquid CO_2. These, however, are not usually considered because of their limited potential, and because they can only be useful in the case of small volumes. Once the CO_2 is injected into the aquifer, it should remain in a supercritical state. According to the subsurface temperature and pressure gradients of the sedimentary basins, it is generally established that a minimum reservoir depth of 800 m is necessary for CO_2 to remain in its supercritical state (Hitchon 1996, Shafeen et al. 2004). This depth may vary according to the location of the reservoir and subsurface temperature and pressure gradient in the basin.

2.4 Mechanism of Sequestration

CO_2 can be sequestered in saline aquifers by hydrodynamic trapping of a discrete plume of CO_2, dissolution of the CO_2 in the saline water, and by mineral trapping through geochemical reactions with the water and the host rocks of the aquifer (Bachu and Adams 2003). When first injected, CO_2 displaces the saline water occupying the pore space of the aquifer. The CO_2 has a lower density than the formation water and migrates upward because of the buoyancy effect. Vertical migration stops when the CO_2 is obstructed by the caprock. Because it is contained by the caprock, the accumulated CO_2 plume starts moving laterally up-dip, preferably toward shallower depth of formation. Assuming the CO_2 remains as a separate, buoyant phase, it will continue migration until it eventually discharges at the surface. In a deep sedimentary basin, the pathway to the surface may be hundreds of kilometers long, and migration to the surface may take a million years (Gunter et al. 2004).

Once the CO_2 has moved far enough from the influence of the injection point, it can be assumed that the CO_2 plume moves at approximately the same velocity as the formation water and that CO_2 movement and dissolution of the CO_2 in the water take place simultaneously. The dissolution rate is very slow, as CO_2 is slightly soluble in water. However, over geological time, the dissolution rate will be greater than the migration rate. As a result, the CO_2 plume will diminish gradually, and thus, dissolution will be the dominant form of sequestration. This is the most desirable and practical way to sequester CO_2.

Mineral trapping is a very slow process, where the injected CO_2 reacts with both the formation water and the minerals of the formation rock and is transformed into stable minerals such as carbonate and dawsonite (Hitchon 1996). A small percentage, probably less than 2 to 4 percent, of the injected CO_2 could be fixed permanently in this way. Some mineral trapping can be expected early in the lifetime of the project. After that, the reactions are kinetically slow and limited additional reaction will occur for thousands of years (Gale 2002 and Van der Meer 2002). An in-depth site-specific analysis is required to find out the exact amount of storage by this method. However, slow process of the mineral trapping, and the unavailability of sufficient subsurface data of the target reservoir and the unknown reactions of CO_2 with the reservoir rock, leads to the significantly low contribution of mineral trapping toward the overall sequestration process.

2.5   Formation Temperature and Pressure

Formation temperature and pressure are two major parameters for sequestration. The density of CO_2 varies according to temperature and pressure, which, in turn, influences the storage capacity calculation. The temperature, pressure, and density vary according to the depth. If the temperature and pressure gradients are known, it is easy to estimate the exact values of those parameters with some simple calculations as follows (Hendriks and Blok 1993).

$$T_r = T_s + a_T \times d \tag{1}$$

$$P_r = P_s + a_P \times d \tag{2}$$

where T_r = reservoir temperature (°C), T_s = surface temperature (°C), a_T = geothermal gradient (°C/m), P_r = pressure in the reservoir (MPa), P_s = surface pressure (0.1MPa), a_P = pressure gradient (MPa/m), d = depth (m). Temperature and pressure gradients also vary according to different locations. The average geothermal gradient varies from 25–30°C/km (Schlumberger 2009). However, in some places, it might vary even from 20–60°C/km (Hendriks and Blok 1993). The normal hydrostatic pressure gradient for freshwater is 0.433 psi/ft, or 9.792 kPa/m, and 0.465 psi/ft for water with 100,000 ppm total dissolved solids (a typical Gulf Coast water), or 10.516 kPa/m (Schlumberger 2009). So it is understood that the pressure gradient depends on the chemical composition of the groundwater and normally varies between 10.5 and 12.4 kPa/m and in some cases as high as 23 kPa/m (Hendriks and Blok 1993).

In general, it is difficult to predict the actual temperature and pressure gradients for a particular formation due to the absence of necessary data related to subsurface temperatures and pressures. As an example, Figure 1 shows a typical temperature and pressure gradient for a particular area such as Michigan basin, located inside United States and Canada (Vugrinovich 1989). It is a useful tool for describing the parameters (P, T) of a certain formation. According to these gradients for the Michigan basin, a minimum depth of 865 m (corresponding to the CO_2 critical temperature in Figure 1) is essential for any sequestration activity in this area. This value may differ further if a different temperature gradient for the same basin is used (Cercone 1984, Shafeen et al. 2004). It is suggested that a new set of experimental data and core analysis explicitly meant to identify local subsurface temperatures and pressures—as well as the rock properties such as porosity and permeability—would be useful, in order to avoid any uncertainty regarding the temperature and pressure gradients.

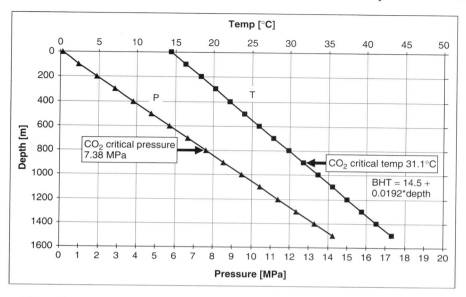

Figure 1 Subsurface temperature and pressure gradients for the Michigan basin

2.6 CO$_2$ Density

The storage capacity of a reservoir varies significantly along the depth of the reservoir due to the change in density of CO$_2$. The density of CO$_2$ in the reservoir at different combinations of temperature and pressure gradients is indicated in Figure 2. This density difference might be encountered at different locations of

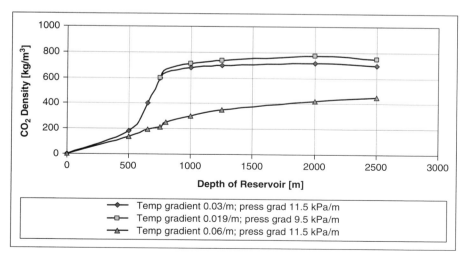

Figure 2 CO$_2$ density versus reservoir depth (adapted from Hendriks and Blok 1993, reproduced by permission of Elsevier)

the reservoirs (Hendriks and Blok 1993 and Baviere 1991). The sharp increase in density between 500 m and 1000 m is due to the fact that the temperature and pressure at that interval are close to the critical values of pure CO_2 (31.1°C and 7.38MPa). Below a depth of 1,000 m, the density becomes almost constant. No significant advantage for capacity can be achieved from density at a depth greater than 1,000 m. It is also indicative that cooler reservoirs provide more storage opportunity than warmer ones.

2.7 CO$_2$ Solubility

The solubility of carbon dioxide in formation water is an important factor for the reservoir storage capacity. Pressure has a much weaker effect on solubility (Figure 3). Solubility increases significantly from atmospheric pressure up to approximately 7MPa, and then it remains fairly constant. Usually, reservoir pressures are greater than 7 MPa, and, as a result, the pressure effect will play an insignificant role in CO_2 solubility. The solubility is also sensitive to temperature and decreases with increasing temperature. The impact of salinity, described as a total dissolved solid (TDS), is also shown clearly in the same figure (Baviere 1991). The solubility of CO_2 decreases as the salinity increases. It varies according to the degree of salinity, as well as to the temperature and pressure. The salinity of the formation water is influenced principally by sodium, calcium, and chloride ions. In the Canadian portions of the Michigan and Appalachian basins, the values of the total dissolved solids (TDS) in deep formations vary between 100,000 and 300,000 ppm, to a maximum of 390,000 ppm (Dollar et al. 1991). As an example, an average value of 200,000 ppm could be a good estimate for

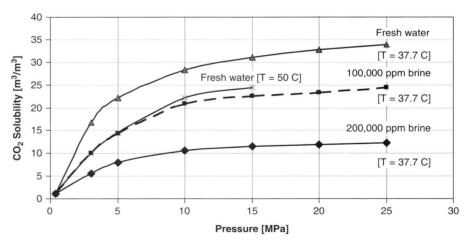

Figure 3 CO$_2$ solubility in water versus pressure and temperature (adapted from Baviere 1991, reproduced by permission of Springer)

this particular formation for any reserve capacity estimation calculation (Shafeen et al. 2004). CO_2 solubility is found to be $10m^3/m^3$ at an average salinity of 200,000 ppm and at a temperature and pressure of $37.7°$ C and 10 MPa, respectively (Baviere 1991). The effect of TDS on capacity estimation is much less than that of the porosity, sweep efficiency, and the CO_2 saturation.

3   RESERVOIR STORAGE CAPACITY

Deep subsurface aquifers can store large amounts of CO_2 in the form of compressed gas, liquid, or aqueous solutions under the influence of high-formation pressure. The storage capacity of a reservoir is determined by the size of the reservoir, the porosity, the fraction of the reservoir that can be filled with CO_2, the density of CO_2, the solubility of CO_2 in the formation water, and CO_2 formation volume factor. The formation volume factor is defined as the ratio of gas volume at reservoir conditions divided by gas volume at standard conditions. Among these parameters, the most uncertain one is the fraction of the reservoir that can be filled with CO_2 due to viscous fingering and gravity segregation, which is also known as *sweep efficiency*. Predictions of correct sweep efficiency will result in more accurate storage capacity estimation. A sweep efficiency in the range of 2 to 20 percent is found to be reasonable in storage capacity calculation (Van der Meer et al. 2002, Vugrinovich 1989, Baviere 1991 and Hendriks and Blok 1993). However, this number depends on the assumption of the presence of structural trap or any other obstruction in the reservoir. In some cases, an assumption of a lower range (5 percent) is also possible without the presence of any structural trap in the reservoir (Holloway 1996, Hendriks and Blok 1995, and Shafeen et al. 2004).

CO$_2$ can be stored either in a structural trap or hydrodynamically in the aquifer. On the one hand, the upper estimates of a reserve capacity consider that a large fraction of the pore volume of the saline aquifer could be used for storage. On the other hand, the lower estimates consider that a small fraction of the pore space that makes up the fluid traps could be used for storage. There is no universally accepted method to calculate the storage capacity of a formation. Different models are available for calculation. This section looks at three models.

3.1   Capacity Calculation: Method 1

This method can be used for a very rough estimation of the storage capacity. It suggests that one cubic meter of deep aquifer of effective porosity 20 percent may accommodate about 0.8 to 1.2 Nm^3(volume at the atmospheric pressure and at $20°C$) of carbon dioxide, which is dissolved in water. The statement is quite general and may not be suitable for any site-specific case. However, this gives a first-hand idea about the storage potential (Koide et al. 1995).

3.2 Capacity Calculation: Method 2

This is a simple method to determine the amount of CO_2 that can be stored in unexplored regions. The main assumptions for this method are as follows:

- 1 percent of the total area of sedimentary basins could be used for CO_2 storage.
- CO_2 is dissolved in the formation water.

By fixing values for other parameters required for capacity calculation, this method estimates storage capacity at approximately 492,000 tons CO_2/km^2 of the sedimentary basins. There is a large uncertainty in the assumption of 1 percent of the total area of sedimentary basins that could be used for CO_2 storage. This method is also referred to as the areal method (Koide et al. 1992).

3.3 Capacity Calculation: Method 3

In this method, CO_2 flooding and underground natural gas storage technology are applied to estimate the amount of CO_2 sequestered. The storage capacity is calculated as follows (Tanaka et al. 1995):

$$\text{Storage capacity} = \text{Displaceable volume} \\ + \text{Dissolved volume of } CO_2 \text{ in water in situ}$$

$$\text{Storage capacity} = E_f \times A \times h \times \phi \times (S_g/B_g(CO_2) \\ + (1 - S_g) \times R_s(CO_2)) \tag{3}$$

where
E_f = overall sweep efficiency, (fraction)
h = formation thickness (m)
S_g = CO_2 saturation, (fraction)
$B_g(CO_2)$ = CO_2 formation volume factor (m^3/m^3) (reservoir vol. /std. vol.)
$R_s(CO_2)$ = CO_2 solubility in water (m^3/m^3)
A = projected structural area of formation (m^2)
ϕ = porosity (fraction, dimensionless)

Depending on the availability of reservoir data, an estimate of the storage capacity can be acquired by this method. This method requires fewer assumptions than the other two.

CO_2 injection is assumed to be in a gaseous or supercritical state for equation (3). During injection, part of the formation water is displaced by CO_2 and the rest of the formation water that is not displaced plays an active role in dissolving some of the CO_2. Like other two-phase systems, the saturation of injected CO_2 in the formation depends on the relative permeabilities of CO_2 and formation water. The injected CO_2 often passes through some parts of the formation due to gravity segregation and viscous fingering phenomena that affect the sweep efficiency. The sweep efficiency depends on the reservoir thickness, vertical permeability distribution, and mobility ratio of the injected CO_2 and formation water. The mobility of a fluid in a porous medium is defined as the ratio of the effective permeability to the viscosity of that fluid (Lake 1989).

Capacity estimates based on assumed values of parameters such as porosity, permeability, CO_2 saturation, and sweep efficiency might give a rough estimate for the reservoir. A few drill-core analysis reports for the formation would give an exact value for these parameters and generate more accurate capacity estimation. Hence, all kinds of uncertainties would be eliminated in the presence of scientifically sound data collected from the core analysis. In case of porosity, a conservative estimate of 10 percent seems to be quite consistent with the available data from any core analysis. However, a horizontal permeability value of 20 to 30 md (millidarcy) could be chosen as a starting point as a number of data reported by different core analysis falls within this range. For most cases, the vertical permeability is reported as <0.01 md for different carbonate formations (Core analysis 1988, Core analysis 1968). The CO_2 formation volume factor is calculated to be 0.0029114 m^3/SCM at an average reservoir temperature and pressure similar to that of the solubility measurement (Lake 1989). CO_2 saturation and sweep efficiency are assumed to be 20 percent and 10 percent, respectively. CO_2 solubility could be assumed as 10 m^3/m^3 (CO_2 solubility in water). These numbers could be a good starting point for any reservoir capacity estimation for CO_2 sequestration. However, for accurate capacity estimation, reservoir modeling software packages such as TOUGH2, ECLLIPSE or GEM/STARS should be used (Hitchon 1996).

4 CO$_2$ TRANSPORTATION

Transportation of CO_2 from the source to the injection location is a major task for the sequestration project. A large volume of CO_2 has to be transported by a high-pressure pipeline. Several design considerations must be evaluated for CO_2 pipeline transmission systems. Extra care is needed for startup, pressurization, operation, and shutdown of the pipeline transportation because of the probability of occurrence of a two-phase mixture (Mohitpour et al. 2000).

The most practical way to transport large volumes of CO_2 is via pipeline. CO_2 can be transported as gas, liquid, or supercritical dense fluid. Transmission in gaseous phase is not economical, as it can result in two-phase flow and thus high-pressure losses, particularly in hilly terrain. The pipeline can be designed and operated similarly to a natural gas pipeline, with careful consideration given to specific differences in design and materials of construction. The desirable conditions for CO_2 transportation via pipeline are at a high pressure beyond its critical value (7.38 MPa), preferably high enough to avoid any two-phase mixtures. This ensures the liquid phase transportation and enhances the economic benefits.

The transportation temperature is mostly determined by the ambient condition (Hendriks and Blok 1995, Skovholt 1993, Birkestad 2002). The operating pressure lies between 8.62 MPa at 4°C and 15.3 MPa at 38°C. The upper and lower limits are set, respectively, by the ASME-ANSI 900# flange rating and ambient condition coupled with the phase behavior of CO_2 (Mohitpour et al. 2000, Gupta et al. 2002 and Farris 1983). The minimum temperature is influenced by the elevation, depth at burial, and offshore water temperature condition for the offshore scenario.

It is mandatory to avoid two-phase flow inside the pipeline. This could be easily achieved as long as the pressure and temperature are kept well beyond the critical values. The temperature difference between the pipeline and the reservoir has a significant impact on the density of CO_2. For example CO_2 density will drop from approximately 969 kg/m^3 (at pipeline transportation condition of 11.03 MPa and 4°C) to a -lower value depending on the actual reservoir depth and in situ pressure and temperature of the reservoir. According to equation 1 and 2, if a reservoir is at a depth of around 1000m with a pressure and temperature gradient of 10.93 MPa/km and 23°C/km, the in situ pressure and temperature will be approximately 11.03 MPa and 38°C (when the surface pressure and temperature are 0.1 MPa and 15°C). At this pressure and temperature the CO_2 density will be 682 kg/m^3 (Farris 1983). So it is a considerable decrease in density for CO_2 at the reservoir condition.

This pressure and temperature effect on density must be considered while determining the operating conditions for a pipeline. The pressure drop across the pipeline must be calculated properly in order to avoid a significant loss in pipeline pressure. The possible requirement of a booster compressor for a long pipeline could be evaluated after careful calculation of the pressure drop across the whole pipe length.

As an example, the emission from a typical 500 MW coal-fired power plant is assumed to be approximately 10,000–14,000 t/d, which is equivalent to 59–83 m^3/s at standard temperature and pressure (Gupta et al. 2002, Freund and Davidson 2002). A 356–406 mm (14–16 in.) pipeline is required to transport this amount (Farris 1983). The higher-diameter (406 mm) pipeline for this scale of project gives greater flexibility in case of additional flow of CO_2.

Impurities present in the CO_2 available from the capture process might play a significant role during transportation. It influences the vapor pressure of CO_2 and affects pipeline fracture control/propagation properties. The impurities also affect pipeline capacity and, hence, the system design. A 5 percent methane addition in a pure CO_2 stream reduces almost 10 percent of the volumetric capacity of a pipeline, and 10 percent methane addition reduces almost 16 percent of the total capacity. Similarly, 5 percent nitrogen addition with pure CO_2 reduces 13 percent of the pipeline capacity, and 5 percent hydrogen with pure CO_2 reduces 18 percent of the total volumetric capacity of CO_2 pipeline, and so on (Mohitpour et al. 2000).

5 CO$_2$ INJECTION

The practice of injecting gases and liquids into geological formations is well established. In 1915, natural gas was first injected into a natural gas reservoir in Welland County, Ontario, Canada, that had been converted for storage use (Carter and Manocha 1996). Storage of natural gas in depleted natural gas reservoirs, aquifers, or solution-mined salt caverns now occurs at numerous locations in Canada and the United States. In most cases, the gas is stored in the consuming regions where it can be utilized in the winter when the pipelines from producing regions have insufficient capacity to meet peak demands for heating of homes, businesses, and commercial establishments (Natural Gas Supply Association 2009). Injection and underground storage of liquefied hydrocarbons in solution-mined salt caverns is recorded as early as the late 1940s and 1950s in North America and Europe (Thoms and Gehle 2000).

CO$_2$ is injected into underground oil reservoirs for enhanced oil recovery (EOR) at over 70 projects worldwide (Carter et al. 2007), with no apparent negative effects. At Saskatchewan, Canada, over 1 million tonnes per year of CO$_2$ are injected into the Weyburn oil pool. This is the largest EOR program in Canada and will result in the recovery of an additional 25 million barrels of oil from the reservoir over the life of the project. At the Sleipner Vest gas field in the North Sea, the Norwegian oil company STATOIL has been injecting 1 million tonnes of CO$_2$ each year into the Utsira Formation, without any difficulties, since 1996. Injection is taking place through a single injection well into a saline reservoir in the Utsira formation, which is located 800 to 1,000 meters below the seabed (Arts et al. 2000, Zweigel et al. 2004, Statoil 2003).

CO$_2$ can be injected through a single well or multiple wells, depending on the reservoir criteria and the amount of CO$_2$. There is no significant difference in the drilling techniques of a CO$_2$ injection well or an oil and gas production well. CO$_2$ must be compressed before injection. It can be compressed either at the well site or at some intermediate location. Sometimes, pipeline pressure is sufficient for the injection and no extra compression is required. All this depends on the distance between the location of the source and the sink. In any case, a higher well-bottom pressure relative to that of the reservoir has to be maintained for injection flow and to avoid CO$_2$ blowout (Gale 2002, Shafeen et al. 2004).

In order to estimate the injection capacity, flow through the reservoir must be calculated first. Injection through a well largely depends on the degree of flowability through the reservoir. Different models are available to estimate the flow through the porous media. A few of the injection capacity calculation methods are described next.

5.1 Injection Capacity Calculation: Method 1

This estimation is based on some numerical results, generalized using regression analysis and a simple steady-state radial outflow model. It leads to the following relation for CO_2 injectivity as a function of its mobility:

$$Q_{CO_2} = 0.0208 \times \rho \times (k_h \times k_v)^{0.5} \times h \times (P_{inj} - P_{aq})/\mu_{CO_2} \qquad (4)$$

where

Q_{CO_2} = quantity of injected CO_2 (tonne/day)
k_h = horizontal permeability (md)
k_v = vertical permeability (md)
h = aquifer thickness (m)
P_{inj} = injection pressure (MPa)
P_{aq} = aquifer pressure (MPa)
μ_{CO_2} = viscosity of CO_2 at aquifer condition (mPa-s)

This equation has an average error of 9.3 percent and a maximum error of 19.9 percent. In order to determine the average CO_2 injection rate into an aquifer, it is necessary to know the depth, thickness, and absolute permeability values in both the vertical and horizontal direction. Although the equation was derived based on the experimental data from Alberta basin (in Canada), it should generate reasonably acceptable values of injection rate in other basins, too (Hitchon 1996).

5.2 Injection Capacity Calculation: Method 2

This is a widely used approach by reservoir engineers to calculate flow-through porous media. The injection capacity of the individual wells can be determined once the reservoir flow is estimated. To estimate the flow rate, a simple heuristic formula, applied by reservoir engineers, can be used. It should be noted that the following equation only gives an indication of the injection rate and not the actual one. Considerable deviation can occur due to the change in pressure at the well-bottom, reservoir equilibrium pressure, permeability, thickness of the reservoir, and CO_2 viscosity (Hendriks 1994, Dake 1978 and Lake 1989).

The equation is as follows:

$$q_s = \frac{\rho_r}{\rho_s} \times \frac{2\pi k h}{\ln(r_e/r_w)\,\mu} \times \Delta P \qquad (5)$$

where

q_s = flow rate $[M_N^3/S]$
h = thickness of the reservoir (m)
r_w = radius of the well (m)
r_e = radius of the influence sphere of the injection well (m)
ρ_r = density of the gas under reservoir conditions (kg/m^3)
ρ_s = density of the gas under standard conditions (kg/m$_N$^{3})
k = permeability of the reservoir (m^2)
μ = viscosity of CO_2 at the well bottom (Pa-s)
$\ln(r_e/r_w)$ = as a rule of thumb, it can be assumed as 7 or 8

ΔP = pressure difference between reservoir and well-bottom pressure (Pa)

where the typical values for $\rho_r = [700\,\text{kg/m}^3]$; $\rho_s = [1.95\,\text{kg/mN}^3]$; $k = [25 \times 10^{-15\text{m}2}]$; $\mu = [5 \times 10^{-5}\text{Pa s}]$; $\Delta P = [1.71 \times 10^6\text{Pa}]$. As a rule of thumb, the value of the logarithmic term can be assumed as 7.5. The approximate injection capacity per injection well normally falls within the range of 1,000 to 3,000 t/d. The flow rate calculated by this equation only gives an indication of the injection rate and deviation may occur in a practical situation (Hendriks and Blok 1993, Hendriks 1994).

6 PARAMETERS INFLUENCING THE INJECTION RATE

The flow rates predicted by equations (4) and (5) are rough estimates and, in practice, significant deviations may occur due to the influence of well-bottom pressure, reservoir thickness, porosity, and permeability, CO_2 viscosity, and reservoir homogeneity. Those parameters are discussed in more detail in the following sections.

6.1 Well-Bottom Pressure

Sufficient overpressure at the well bottom is necessary to inject CO_2 into the reservoir. It should not be too high to damage/fracture the reservoir, nor should it be too close to the reservoir pressure such that it is unable to penetrate the reservoir. It should be typically 9 to 18 percent above the reservoir in-situ pressure (Hendriks 1994, Gupta et al. 2002). Also, a much higher pressure is possible, almost as high as 90 percent of the fracture pressure (Hitchon 1996). The flow rate will increase with the higher pressure difference. The pressure at the bottom could be calculated as a function of density of CO_2 at different temperature and pressure, using compressibility factors. As an example, Figure 4 shows the change of several well-head pressures along the injection tubing at different depths for the Michigan basin, utilizing the available information from the literature (Hendriks 1994) and the pressure gradient for the Michigan basin. It is found that a well-head pressure of 8 MPa will give a rise in pressure of about 17 MPa at a depth of 1,000 m where the reservoir in situ pressure is approximately 9.5 MPa (Figure 4). This bottom-hole pressure is almost 80 percent higher than the reservoir pressure, contrary to one of the suggested values of 9 to 18 percent. As the exact difference is unknown, the value of 11.21 MPa, which is 18 percent higher than the in situ pressure, could be used in sample injection calculations for this particular basin under consideration. This uncertainty can be addressed by identifying the actual fracture pressure of a certain formation and thereby adjusting the injection pressure (Shafeen et al. 2004).

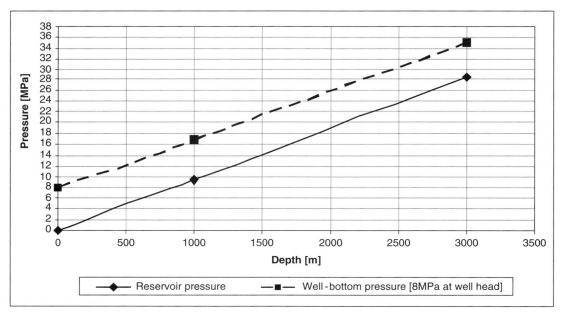

Figure 4 Well-bottom and reservoir pressure at well-head pressure of 8 MPa (adapted from Hendriks and Blok 1993, reproduced by permission of Elsevier)

6.2 Reservoir Thickness

The injection flow rate is directly proportional to the thickness of the reservoir. It also shows similar behavior as pressure difference. Clearly, thick reservoirs have the higher potential of increased flow rate, and reservoir thickness is an integral part of all storage capacity calculation methods. So, accurate prediction of the reservoir thickness, by implementing number of experimental wells and other thickness measurement tools, is an important part for the injection flow rate measurement.

6.3 Porosity and Permeability

Porosity is a measure of space in a reservoir rock that is not occupied by the solid framework of the rock. It is defined as the fraction of the total bulk volume of the rock not occupied by solids. Porosity is controlled by the depositional environment and subsequent conversion (e.g., by compaction or chemical reaction) of sediment into rock. Permeability is determined by the organization of the space, which controls the movement of fluids. Porosity is usually understood as a tortuous network of large pore bodies connected by narrower pore throats. Most of the storage of CO_2 takes place in the pore bodies, and pore throats play the major role in determining the permeability (Cole 1969, Mian 1991).

Permeability is affected by the pore space connectivity. It allows the flow of liquid or gas through the porous media. If a reservoir has high porosity and thickness but has low permeability, it is not suitable for gas injection. Permeability values may range from 1 millidarcy (md) (10^{-15} m^2) to several hundred or thousand md in a single reservoir. Permeability of a caprock should be preferably less than 0.001 md over a distance where the CO_2 plume could be reasonably expected to migrate until dissolution (Keith and Wilson 2002). Porosity has comparatively little effect on injection, other than the frontal advancement of CO_2 to a further distance from the injection well.

6.4 Viscosity of Carbon Dioxide

Viscosity plays an important role in sequestration. Low viscosity will result in poor sweep efficiency and hence a reduction in overall storage capacity. CO_2 viscosity depends on the temperature and pressure of the reservoir. At the wellhead condition, in the injection string, and at the well bottom, the viscosity is different due to the difference in pressure and temperature. CO_2 viscosity will vary between 5×10^{-5} Pa-s and 7×10^{-5} Pa-s at a temperature of 40°C and pressure of 10 to 15 MPa (Hendriks 1994, Baviere 1991).

6.5 Homogeneity of the Reservoir

Though it is often assumed that the reservoir or the formation is homogeneous, in actual practice, it is not. Geological heterogeneity, either stratigraphic (e.g., permeability) or structural, is the norm, not the exception. Stratigraphic heterogeneity may be expressed vertically as complex interlayering of rock formations of different permeability or laterally by variations in permeability that may slow, obstruct, or limit lateral migration of a plume of CO_2. The degree and nature of inhomogeneity depends on a variety of geological factors, including the sedimentary environment in which the original sediments were deposited, and subsequent burial, compaction, diagenesis, and lithification of the sediments.

The sedimentary depositional environment is the combination of physical, chemical, and biological processes associated with the deposition of the sediments, and is of particular importance, as it will determine the lateral extent and thickness of individual formations and the frequency and nature of lateral changes in the lithologic and biologic characteristics of the rock. For example, sedimentary rocks that were deposited in a deep-water marine environment in an epicontinental sea several hundreds of kilometers in diameter, such as that which occupied the Michigan Basin during the Paleozoic period of geologic time, will be laterally extensive with only gradual and relatively predictable lateral changes in thickness and lithology. In contrast, a sedimentary rock deposited in a fluvial environment will exhibit rapid lateral changes in thickness, grain size (and thus porosity), lithology and continuity and will cover a much smaller area. *Diagenesis* is the sum of all the biologic, chemical, and physical changes experienced by sediments subsequent to deposition during and after lithification into sedimentary rock.

After lithification, later tectonic uplift and deformation of the sedimentary rocks will result in structural heterogeneity. Folds and structural closures may result in accumulation of large columns of CO_2 beneath small areas of caprock. Vertical displacement of rock layers by faults may create barriers to lateral migration by juxtaposing impermeable rocks against layers of permeable rock. Faults may also create linear zones of enhanced porosity and permeability that may act as vertical or lateral pathways for rapid and focused migration of CO_2. Faults can breach reservoir caprocks and result in rapid vertical migration and premature escape of CO_2 to the atmosphere.

The existence of this nonhomogeneity leads to the spreading of CO_2 in a way that is not uniform in nature. This little deviation does not alter the calculation of capacity or the injection flow rate for the reservoir. Different values of permeability in different layers may guide us in selecting the appropriate layer for injection. Preference could be given to a layer that has higher permeability and that is covered by a low permeability layer. This may slow the upward movement and allow more lateral spreading (Hitchon 1996). Sequestration capacity can be increased by geological heterogeneity by dispersing the flow path of the migrating CO_2 plume, thus increasing the pore volume contacted by the CO_2 and increasing CO_2 dissolution and sequestration (Hovorka et al. 2004).

6.6   Caprock Integrity

Caprock integrity is a major property for site selection. The integrity and the fracture pressure of the caprock will have a great impact on sequestration activity. As the storage security depends on the confidence that can be placed in the caprock, it requires extensive surveying across the area of the aquifer. Detailed subsurface geological structure and isopach (thickness) maps, geological facies maps, regional 2-D seismic profiles, and 3-D seismic surveys of critical areas in the vicinity of the injection well(s) should be carried out to determine the nature and integrity of the caprock. It is suggested that a caprock be least 50 m in thickness with a demonstrated permeability of less than 0.001 millidarcies over a distance that the CO_2 plume could be reasonably expected to migrate until dissolution (Keith and Wilson 2002). The caprock should be tested by core analysis to determine its porosity, permeability, lithology, mineralogy and petrography, and overall integrity. The primary caprock, and the presence of any subsequent upper shales, usually increases the efficiency of the caprock and its trapping performance. Injection rate and pressure must be designed to prevent overpressuring of the reservoir and possible hydraulic fracturing of the caprock.

6.7 CO$_2$ Migration

Once CO$_2$ is injected it will tend to rise upward and will reach the top of the aquifer until it is being obstructed by the caprock. Once obstructed, the plume will spread gradually, and simultaneous dissolution into the saline formation will take place. The rate of dissolution will be controlled by the surface area of CO$_2$ in contact with the formation water. This phenomenon occurs due to the buoyancy force caused by the density difference between the supercritical CO$_2$ and the saline water present into the formation. The more it will be in contact with the formation water, the more the solubility of CO$_2$ will be, and hence the storage amount will increase (Hendriks and Blok 1993).

The CO$_2$ plume will usually move up-dip toward the shallow depth of the formation rock. Buoyancy plays a major role in case of the CO$_2$ plume movement. Migration of plume in different directions will also be influenced by the hydrodynamics of groundwater in the sedimentary basins. Related data for hydrogeology seem to be very important for the CO$_2$ plume movement. Numerous reservoir modeling studies imply that the plume migration will not be more than 25 to 50 km (King et al. 2003, Lindeberg 1997, Obdam et al. 2002 and Chadwick et al. 2002). Before exceeding this distance, the plume might dissolve completely into the aquifer, which may take up to several thousand years. Hence, the minimum distance from the injection point to the 800 m contour must be greater than 50 km for a standard reservoir in order to have all the CO$_2$ dissolved into the saline formation water within a considerable geological time period. If a reservoir is too shallow and the formation is close to 800 m depth limit, the chance of CO$_2$ becoming vapor is immense. If it vaporizes, the CO$_2$ will move vigorously upward, and depending on the amount, the potential for possible leakage will be great (Shafeen et al. 2004).

6.8 Hydrocarbon Occurrences

The presence of hydrocarbons in different geological formations will be a major issue for sequestration. The economic interest in extraction of oil and gas would deter any sequestration activity in these reservoirs. However, the presence of hydrocarbons in different formations provides evidence of the existence of successive caprocks in different layers in those regions. In these scenarios, a particular formation might be considered for sequestration with enhanced oil recovery (EOR). This simultaneous approach would enable the sequestration to be economically more viable. For EOR with sequestration, the location of the injection point would be adjusted in order to facilitate the EOR activity. A *sequestration without EOR* activity is also possible in the same formation where hydrocarbons occur if they are located just beyond the 50 km range from the injection point. It ensures any unwanted interaction between the sequestration and the hydrocarbon production activity from the same pools. However, rigorous analysis is required before proceeding to any sequestration activity in a formation where hydrocarbon pools are known to be present or may potentially be present.

7   MONITORING, RISK, AND SAFETY ISSUES

Monitoring is one of the most important aspects of the geological CO_2 sequestration. Continuous monitoring is required for the volume of injection, formation pressure, subsidence, or uplift of the surface due to formation pressure change, CO_2 plume movement, migration distance, geochemical reactions, induced seismic activity due to CO_2 injection, leakage to the atmosphere, contamination of the potable water table, contamination of the hydrocarbon resources, caprock integrity, and other related issues. Careful monitoring and handling of a sequestration project will minimize the risk and will increase the project life. Localized safety is a major concern for the sequestration site. There must be a contingency plan to handle any unforeseen safety hazards to human or to the ecosystems. Specific care is also needed for high-pressure CO_2 transportation via pipeline from the source to the injection location (Van der Meer 1992 and 1993, Holloway 1996 and Wolfgang and Rudnicki 1998).

As CO_2 exists in a reservoir at a supercritical state for thousands of years (prior to its dissolution in the saline water), there is a possibility that it can migrate out of the saline aquifer. However, as hydrocarbons are stored in various reservoirs for geological timescales, then CO_2 also has the potential to remain in the reservoir for similar timescales. The natural occurrence of CO_2 in underground reservoirs in the United States supports the evidence that CO_2 can be stored safely. The environmental consequences of any CO_2 migration out of the reservoir are very difficult to assess and requires knowledge of possible CO_2 flux rates from the source of leakage. Leakage may also occur during transportation and injection in the reservoir. Risk can be minimized by proper design and monitoring at different stages of CO_2 transportation and injection processes. A pressurized transportation network always adds concern to public safety. A slow and persistent leak from an unidentified migration pathway out of a storage reservoir could present a safety hazard if CO_2 reaches the surface. A danger of asphyxiation is also possible if the CO_2 leaks suddenly at an onshore location and accumulates in hollows in the ground surface. It could also pollute potable water above the reservoir (Hendriks and Blok 1995). Off-shore, CO_2 migration into the lake or deep sea can cause localized changes in pH that has the potential to affect marine ecosystems. Several other monitoring and safety issues are to be considered to minimize the risk during the sequestration activity (Krom et al. 1993, Sanford and Quillian 1959, USGS 2003, and NEHP 2003).

7.1   Monitoring Issues

It is necessary to have a reliable monitoring system to make the sequestration project safe and acceptable to everyone. Monitoring is required to verify the quantity of CO_2 injected, to track the CO_2 plume movement, to identify leakage of CO_2 into the atmosphere and its aftereffect on the local population and ecosystems, to protect groundwater from contamination, to prevent the contamination of the hydrocarbon resources present in the same formation, and to avoid overpressure of the formation. A large uncertainty also exists in determining these factors precisely, as the mixing and solution state of fluids within the reservoir is not accurately understood for the sequestration process.

Monitoring of CO_2 Flow Rate

In order to assess the overall performance of the sequestration project, it is important to have a reliable CO_2 flow measurement system. Measurement technology for CO_2 injection rate is available for the enhanced oil recovery projects. Similar methods could be used for CO_2 sequestration as well. A state-of-the-art measurement, monitoring, and verification (MMV) system is an integral part of a sequestration process.

Monitoring of CO_2 Plume Movement

It is important to know the extent of the CO_2 plume movement into the reservoir once the injection is started. By tracking the plume movement, it is possible to detect post-injection effects on subsurface geology and location of any leaks or preferred migration paths. Seismic surveying could be a suitable technique for monitoring CO_2 movement into the saline aquifer.

Monitoring Atmospheric Concentration of CO_2

Routine measurement of CO_2 concentration in air around the injection site and the abandoned wells is a must for early detection of any leakage of CO_2 to the atmosphere. Injection wells must be secured properly to avoid any leakage around the injection-well bore. Possibilities of leakage from the abandoned wells are discussed later. To better predict the impact of leakage to the local population and ecosystems, CO_2 flux in the leak stream should be known correctly (Holloway 1996).

Monitoring of Groundwater Contamination

Adequate groundwater monitoring wells in the upper layers of the caprock have to be present around the sequestration site. Ensuring the safety of groundwater (potable water) quality is a major concern for any sequestration project. Routine sampling of groundwater has to be done to detect CO_2 contamination. Any contamination will prove the presence of leakage path through the caprock.

Monitoring Hydrocarbon Source Contamination

The presence of hydrocarbons in different geological formations will be a major issue for sequestration. There might be some oil and gas pools located in the same formation where CO_2 is being planned to inject. Those hydrocarbon pools might still be producing, and the presence of CO_2 is not at all expected in those areas. Contamination of these hydrocarbon reserves must be avoided by planning a scientific injection system. The major interest in extraction of oil and gas would certainly deter any sequestration activity in these reservoirs. However, a strict monitoring and scientific injection process that avoids contamination of the hydrocarbon pool would permit the sequestration activity in the same area.

Monitoring Reservoir Pressure Development

CO_2 injection into the formation may lead to a pressure buildup as it displaces water. Significant increase in pressure could affect the caprock integrity, deform the land surface, or induce seismicity. A state-of-the-art reservoir pressure monitoring techniques must be in place in order to avoid any disaster. In any case, pressure must not exceed the formation fracture pressure.

7.2   Safety Issues

Safety is a major concern for the success of any sequestration project. Safe handling of the transportation and injection systems and efficient management of the subsurface geological formations are prerequisites to make the sequestration successful. Several other issues must also be addressed to consider the project acceptable from a safety point of view. Some of the safety issues related to the geological condition and design and operation of the project are discussed here.

Impacts of Faulted Zone

One of the major concerns for sequestration is leakage to the atmosphere. Careful mapping of fault locations and displacements at potential sequestrations sites is critical, because heavy leakage may occur through a faulted zone, especially when the fault affects all the geological layers from surface to the basement rock. Faulted layers are broken and displaced along fractures. Several large- and small-scale faults might develop over time to relieve the stresses within the formation rocks in a particular storage location. Some of these faults would pass through all the geological layers from surface to the basement rock, while others may pass through only a number of layers. Some of these faults are also the sites of potential hydrocarbon traps. The presence of hydrocarbon in those traps is an indication that the faults are not working as a pathway for leakage. However, some faults might prove to be a potential leakage path. Numerical modeling indicates that increased pore pressure caused by CO_2 injection changes the local stress field in the reservoir rock and caprock and cause deformation that, under some scenarios, could cause failure and leakage through faulted and fractured caprock (Rutqvist and Tsang 2002).

Impacts of Abandoned Wells

Heavy leakage of CO_2 from the reservoir may occur through abandoned wells also. Abandoned wells may act as a bypass to the atmosphere if these were not sealed properly. The status and location of previously drilled and abandoned wells may be unknown or poorly documented. Some of these might have been abandoned for the past 100 years. Other wells may have drilling records, but there are no reports available on plugging of the wells, location of plugs, plug materials, and cement type. Moreover, the quality and integrity of cement used in the early years might have severely degraded by now. The reactivity of the injected CO_2 (or mixture of gas) with this cement and its consequences must be evaluated. An in-depth analysis, based on the abandoned well data available for a particular sequestration site, unveils the identity of the wells that could be most vulnerable to leakage.

Abandoned wells are normally identified depending on their depth and availability of the plug-end date. Arbitrarily, as a first start, a minimum limit of 700 m could be set for well depths to find out the number of wells. The reason to choose 700 m as the base case is only because of their proximity to the threshold limit for CO_2 injection into the reservoir and their probable susceptibility in case of any leak from the cap rock or failure of the cement plugging. A detailed investigation is necessary to find out the real status of the abandoned wells, their ability to withstand the sequestration pressure and impact on environment in case of a failure.

Impact of Seismic Activity

Catastrophic leakage may occur due to seismic activities if the storage location is situated nearby earthquake hazard zones. Possibility of any seismic activity induced by the deep well injection of CO_2 into the target area could also be a concern for sequestration. More studies are necessary in this regard before starting any sequestration activity in a particular site.

Safeguard against Backflow into the Injection String

In order to avoid any backflow or formation damage around the injection well in case of a failure/trip of the power plant or other large point source, from which CO_2 is being sourced for sequestration, a buffer stock of CO_2 could be an option. Presence of a contingency plan for this type of unforeseen situation is essential for the injection process (Skovholt 1993).

Safe Distance from the Pipeline

A safe setback distance for development has to be maintained on either side of the onshore part of the high pressure pipeline. The usual pipeline right-of-way is the land over and around the pipeline, typically 25 feet on either side. It is also extremely important to continuously maintain the right-of-way in an environmentally sound manner and avoid general deterioration of the land due to maintenance or other line-related activities (Kinder Morgan 2009). Standard procedures are recommended to follow while laying down a pipeline for pressurized systems. Pipeline through a built-up area is not recommended, as it will increase the chance for any unforeseen accidents and the consequences of a pipeline failure.

Increased Number of Block Valves in the Pipeline

Block valves or shut-off valves are to be placed in a shorter distance, especially in the populated area. This will help isolate a leaky system quickly and reduce the damage. The advantage of the block valves will be tremendous, especially in case of a pipeline rupture. The impact of the pipeline rupture in a populated area will be severe. The normal atmospheric concentration of CO_2 is approximately 330 ppm. Although it is nontoxic in nature, a concentration of 10 to 20 percent CO_2 by volume in the air would be immediately hazardous to human life. It may cause unconsciousness, failure of respiratory muscles, and a change in the pH of the blood stream (Mohitpour et al. 2000). So the probability of a pipeline rupture and its consequence either on the populated area or on the marine life needs to be evaluated.

Impact of Water Content in CO_2 on Pipeline

Moisture content in CO_2 can cause formation of highly corrosive carbonic acid. The normal accepted level of water vapor content for CO_2 transmission is 18 to 30 lb/MMSCF (Mohitpour et al. 2000). More specifically, the industrial standard for the allowable water vapor content for the pipeline is usually not greater than 30 lb/MMSCF (Visser et al. 2008). As a result, the CO_2 available at the battery limit of the sequestration project must contain water content below the acceptable range.

CO_2 Emission during Transportation

CO_2 emission while transportation via pipeline, except the accidental release, is also an important issue. It is considered as a separate emission problem, which is common to the pipeline transportation of the pressurized gas systems. Leakage can occur through compressors, pumps, fittings, and joints. It does not depend on the length of pipeline but on the number of individual equipments and associated fittings. Extreme decompression might cause the o-rings, seals, and so on to be ineffective and might allow leakage (Mohitpour et al., 2000). If there are no booster compressors in between the pipeline, the leakage will be mainly from the equipment associated with the CO_2 capture and storage facility. It is possible to calculate the leakage emission from the number and type of equipments and equipment specific emission factors (IPCC 2006).

8   SCIENTIFIC AND METHODICAL APPROACH FOR A NEW SEQUESTRATION PROJECT

Many issues must be addressed before a sequestration project is considered. The safety of the project remains a prime concern, as is the safety of the sequestration site. Only an in-depth analysis can answer all these questions. Another obvious question is the economics of the whole project. To answer all these uncertainties, a detailed study on storage capacity, injection rate, cost estimation, and safety analysis are required. Apart from the administrative and regulatory procedures, following are some technical recommendations as a minimum requirement to undertake a new sequestration project.

8.1   Reservoir Characterization

To assess the actual storage capacity, CO_2 injection rate, and CO_2 plume migration, the first step is to identify the values of different parameters of the reservoir. In order to do that, several field and laboratory activities are required:

- *Drill experimental wells.* It is required to drill several experimental wells to find out the exact value of porosity and permeability of the formation at different locations. They might vary over a wide range for the whole area of the reservoir. It is also important because zones of low and high permeability can create preferential flow paths or channels and thus render portions of an aquifer inaccessible to injected fluids. This reduces the volume of CO_2 that can be stored. Identification of several other properties, such as CO_2 saturation, relative permeability, and capillary pressure can also be done from the rock samples.

- *Geological mapping.* Detailed subsurface geological structure and isopach (thickness) mapping, documentation of lithology and petrography from visual and microscopic examination of drill core, and regional 2-D seismic profiles and 3-D seismic surveys of both the saline aquifer and its caprock over the likely area to be affected by the CO_2 plume are essential for determining geological structure and thickness of both the saline aquifer and the caprock formations. A geological model of the depositional environment and lateral facies variations in both formations should be prepared.

- *Identify the formation water chemistry.* Formation water chemistry needs to be identified properly. CO_2 solubility largely depends on the total dissolved solids (TDS) in the saline water. As solubility varies with water chemistry, it influences the overall storage capacity of the reservoir.

- *Identify the hydrodynamics.* Hydrodynamics of the formation water must be identified as either up dip or down dip. The direction and rate of flow play an important role during sequestration, especially during CO_2 plume movement. A down-dip water flow might compel an up-dip moving plume to flow in the down-dip direction if the partially dissolved CO_2 and saline water become more concentrated than that of the formation water. Basically, a down-dip flow is preferable because of its ability of increased hydrodynamic and mineral trapping.

- *Determine sweep efficiency.* Accurate value of sweep efficiency for that particular aquifer should be determined to better predict the reserve capacity. It is a sensitive parameter, and any change in small fraction could lead to a wide change in capacity estimation.
- *Map faults and identify their characteristics and possibility of leakage.* This is critical in the site selection process. Faults can be located and mapped using a combination of seismic, high-resolution aeromagnetic surveys and structure-top geological mapping of subsurface rock formations using data from petroleum wells and stratigraphic tests.

8.2 Reservoir Modeling

Both a conceptual geological model and a numerical model should be prepared. A credible numerical model requires the actual identification and measurement of the rock properties for the particular formation. Once the properties are measured, reservoir modeling can be done for the identified zones. Different modeling software packages are available to perform reservoir modeling such as TOUGH2, ECLLIPSE, or GEM/STARS. Modeling of CO_2 injection, migration, and dissolution in the reservoir would precisely identify the exact injection location and the amount to be injected. It also would help us to find out the advantages and disadvantages of a particular injection location, as well as the whole reservoir.

8.3 Evaluation of the Pipeline Route

A detailed terrain study is necessary to assess a realistic cost for pipeline to transport CO_2 to the injection zone. Pressure drop calculation is also very important as it determines the number of booster compressors needed for the total length of the pipeline. Booster compressors contribute a significant cost to the overall operating and capital cost of the project. Several CO_2 pipeline networks, more than thousands of kilometers in length, are already in place in the United States and Canada for the transportation of CO_2 for enhanced oil recovery processes (IEAGHG/SR5 1995). So it is expected that the issues related to the pipeline routing are already resolved and well documented and could be accessed, if required, from the respective authorities.

8.4 Evaluation of the Safety of CO_2 Injection

Injection of CO_2 into a reservoir under pressure will change the stress level of the formation. It is necessary to evaluate the safety and integrity of CO_2

injection. Alteration of stress must not cause any failure within the reservoir, along the faults, or in adjacent regions outside the reservoir. Detailed study on permissible aquifer pressure associated with CO_2 injection is needed. The efficiency of injection and the safety and integrity of containment largely depend on the scientific understanding of the hydrological, mechanical, and chemical behavior of the reservoir.

8.5 Evaluation of Mineral Trapping

Several chemical reactions are expected to take place when CO_2 is injected into the formation. Though the mineral trapping might play a very insignificant role in terms of storage volume, it might influence the injection rate by changing the porosity and permeability of the formation because of mineral dissolution or precipitation. An in-depth analysis of the formation rock chemistry and their probable influence on sequestration is needed. This could be done in laboratory using available rock samples with CO_2 injection at the reservoir conditions.

8.6 Evaluation of Seismic Activity Induced by Deep-Well Injection

The proposed site must be properly investigated and monitored for induced seismic vulnerability. Deep-well injection may weaken the strength of different faults present in the target location. In the worst case, it might provide a pathway and thus loosen the containment. Even if the area is not hazardous from the seismic point of view, yet an understanding of the structural geology and its reaction to the deep-well injection is necessary. A simulation study could be an initial start to address this issue.

8.7 Evaluation of Caprock Integrity

Due to the gravity segregation, the bulk of the injected supercritical CO_2 will remain in contact with the top seal for a long period of time until complete dissolution. It should be clearly identified how the top seal will behave with the changing stress and chemical conditions. A thorough investigation is necessary to identify whether the formation can act as a good seal. Rock samples available from the experimental wells will be useful to identify their properties from different laboratory experiments.

8.8 Identification of the Abandoned Wells and Their Status

Status of the abandoned wells, if present in the storage area, and their condition of cement plugging needs to be clearly documented. Leakage might occur due to the failure of a weak cement plug and could damage the sequestration concept. There is also the possibility that the cement plugs or cement sheaths bonding casing to bedrock could degrade due to changes in down-hole conditions associated with injection of CO_2. Any possible issues with existing wells, whether plugged or not, must be resolved in advance and consideration should be given to replugging. New cement formulations may provide improved protection against cement degradation in the presence of CO_2 (Barlet-Gouedard et al. 2009).

8.9   Long-Term Monitoring and Contingency Plan

A long-term monitoring and contingency plan must be prepared for the overall success of the sequestration project. The plan must include drilling of several monitoring wells in the storage area and also into the subsequent upper layers to monitor the CO_2 movement and to detect any leakage into different layers. Solving a pollution/emission problem through geological sequestration will require considerable safety and contingency measures to make it acceptable to the public.

8.10   Addressing the Uncertainties

Uncertainties may arise from a different perspective. They may arise during the actual implementation of the project. Uncertainties in the reservoir condition during the injection process could lead to the drilling of additional wells or building of additional platforms (in case of off-shore sequestration) and may increase the unexpected workload associated with huge cost involvement. They could also arise from the federal or provincial government policy (e.g., the imposition of a temporary ban on all kinds of drilling activity in any particular area). A detailed analysis on the uncertainties must be carried out to minimize the overall cost of the project

9   CONCLUSIONS

Geological sequestration of carbon dioxide can play an important role in reducing carbon dioxide emissions to the atmosphere. Although detailed local studies to identify suitable sites have yet to be completed, potential worldwide storage capacity is huge.

Capture and storage of CO_2 is a technologically complex and expensive process. Considerable scientific research has been completed to identify the geological and engineering requirements for implementation, and many parts of the technology have already been developed and are in use in the petroleum and petrochemical industries. There do not appear to be any insurmountable technical issues.

The acceptability of the sequestration idea in a densely populated and economically rich area will be a difficult task. Establishing adequate safety measures and implementing a contingency plan in case of a blowout of an injection well could be some of the necessary preconditions for sequestration for any particular area. A state-of the-art monitoring plan for underground CO_2 movement is also necessary to enhance the confidence level. However, convincing the local community, if people theoretically reside on top of large volume of stored CO_2, or the consumers, who might not get any visible benefit in their lifetime other than paying higher electricity bills, will remain the major challenge.

The cost of building and operating a carbon capture and geological storage infrastructure is significant. Consequently, voluntary adoption of this technology is not likely, except where there is an opportunity for payback, as in the case of enhanced oil recovery projects or where tax or regulatory incentives have been established. Significant variations in cost estimation may occur, depending on the cost of individual units. Several other costs, such as cost of the monitoring wells, cost of the right-of-way, or any other unforeseen cost might alter the number for capital investment significantly.

There remain some uncertainties about cost and reliability, and in most countries there is still no legal, policy, or regulatory framework in place to guide or encourage implementation. In time, these issues and the remaining technical issues may be resolved.

REFERENCES

Arts, R., Brevik, I., Eiken, O., Sollie, R., Causse, E., and van der Meer, B. (2000), "Geophysical Methods for Monitoring Marine Aquifer CO_2 Storage—Sleipner Experiences," *5th International Conference on Greenhouse Gas Control Technologies* (Cairns, Australia), August.

Bachu, S. (2001), "Geological Sequestration of Anthropogenic Carbon Dioxide: Applicability and Current Issues," *In* L. Gerhard, W.E. Harrison, and B. M. Hanson (eds.), *Geological Perspectives of Global Climate Change*, American Association of Petroleum Geologists, AAPG Studies in Geology No. 47, pp. 285–304.

Bachu, S., and Adams, J. J. (2003), "Sequestration of CO_2 in Geological Media in Response to Climate Change: Capacity of Deep Saline Aquifers to Sequester CO_2 in Solution," *Energy Conversion and Management*, Vol. 44, pp. 3151–3175.

Bachu, S. (2003), "Screening and Ranking of Sedimentary Basins for Sequestration of CO_2 in Geological Media in Response to Climate Change," *Environmental Geology*, Vol. 44, pp. 277–289.

Baines, S. J., and Worden, R. H. (2004), "Geological Storage of Carbon Dioxide," *In* S. J. Baines, and R. H. Worden (eds)., *Geological Storage of Carbon Dioxide*, Geological Society, London, Special Publications, 233, pp.1–6.

Barlet-Gouedard, V., Rimmele, G., Porcherie, O., Quisel, N., and Desroches, J. (2009), "A Solution Against Well Cement Degradation under CO_2 Geological Storage Environment," *International Journal of Greenhouse Gas Control*, Vol. 3, pp. 206–216.

Baviere, M. (1991), *Basic Concepts in Enhanced Oil Recovery Processes*, Elsevier Applied Science, London and New York.

Birkestad, H. (2002), "Separation and Compression of CO_2 in an O_2/CO_2-fired Power Plant", T01-262, M.Sc. Thesis, Dept. of Energy Conversion, Chalmers University of Technology, Sweden.

Carter, T. R., Gunter, W., Lazorek, M., and Craig, R. (2007), "Geological Sequestration of Carbon Dioxide: A Technology Review and Analysis of Opportunities in Ontario,"

Ontario Ministry of Natural Resources, Applied Research and Development Branch, Climate Change Research Report CCRR-07, 24 p.

Carter, T. R., and Manocha, J. S. (1996), "Underground Hydrocarbon Storage in Ontario," *Northeastern Geology and Environmental Sciences*, Vol. 18, Nos. 1/2, pp. 73–83.

Carbon Sequestration-Research and Development (1999), U.S. Department of Energy, http://www.fe.doe.gov/coal_power/sequestration/reports/rd/chap1.pdf. Accessed September 2003.

Cercone, K. R. (1984), "Thermal History of the Michigan Basin," *American Association of Petroleum Geologists Bulletin*, Vol. 68, No. 2, pp. 130–136.

Chadwick, R. A., Zweigel, P., Gregersen, U., Kirby, G. A., Holloway, S., and Johannessen, P.N. (2002), "Geological Characterization of CO_2 Storage Sites: Lessons from Sleipner, Northern North Sea," *Proceedings of the 6th International Conference on Greenhouse Gas Control Technologies* (Kyoto, Japan), Vol. 1, pp. 321–326.

Cole, F. W. (1969), Reservoir Engineering Manual, Gulf Publishing Company, Houston, Texas.

Core Analysis Report (1968), Atlas No.1 Dunwich 1-23 IV, Willey Field, Cambrian Formation, CoreLaboratories-Canada Ltd.

Core Analysis Report (1988), Ram BP #5 Raleigh 1-17, 1-17-XIII, Cambrian Formation, GEOTECHnical resources Ltd., Calgary, Canada, 1988.

Dake, L. P. (1978), "Fundamentals of Reservoir Engineering," Elsevier Scientific Publishing Company, Amsterdam, the Netherlands.

DOE (2003), http://www.fe.doe.gov/coal_power/sequestration/reports/rd/chap1.pdf. Accessed September 2003.

Dollar, P., Frape, S. K., and McNutt, R. H. (1991), "Geochemistry of Formation Waters, Southwestern Ontario, Canada and Southern Michigan, USA—Implications for Origin and Evolution," Ontario Geoscience Research Grant Program, Grant No. 249, Ontario Geological Survey, Open File Report 5743, 72 p.

Dusseault, M., Bachu, S., and Davidson, B. (2001), Carbon Dioxide Sequestration Potential in Salt Solution Caverns in Alberta, Canada. Solution Mining Research Institute, Clarks Summit, PA, U.S.A. Fall 2001 Technical Meeting, October 8–10, 2001, Albuquerque, New Mexico.

Farris, C. B. (1983), Unusual Design Factors for Supercritical CO_2 Pipeline, *Energy Progress*, Vol. 3, No. 3, pp. 150–158.

Freund, P., and Davidson, J. (2002), "General Overview of Cost, IEA Greenhouse Gas R&D Programme," *IPCC Workshop on Carbon Capture and Storage* (Regina, Canada).

Gale, J. (2002), "Geological Storage of CO_2: What's Known, Where Are the Gaps and What More Needs to Be Done," *Proceedings of the 6th International Conference on Greenhouse Gas Control Technologies* (Kyoto, Japan), Vol. 1, pp. 207–212.

EPA (2003). "Global Warming—Climate," U.S. Environmental Protection Agency, http://yosemite.epa.gov/oar/globalwarming.nsf/content/Climate.html. Accessed September 2003.

Gunter, W. D., Bachu, S., and Benson, S. (2004), "The Role of Hydrogeological and Geochemical Trapping in Sedimentary Basins for Secure Geological Storage of Carbon Dioxide." *In* S. J. Baines and R. H. Worden (eds)., *Geological Storage of Carbon Dioxide*, Geological Society, London, Special Publications, Vol. 233, pp.129–145.

Gupta, N., Smith, L. A., Sass, B. M., Bubenik, T. A., Byrer, C., and Bergman P. (2002), "Engineering and Economic Assessment of Carbon Dioxide Sequestration in Saline Formations," *Journal of Energy and Environmental Research*, Vol. 2, No. 1, pp. 5–22.

Hendriks, C. (1994), *Carbon Dioxide Removal from Coal-Fired Power Plants*, Kluwer Academic Publishers, Dordrecht.

Hendriks, C. A., and Blok, K. (1993), "Underground Storage of Carbon Dioxide," *Energy Conversion and Management*, Vol. 34, No. 9–11, pp. 949–957.

Hendriks, C. A., and Blok, K. (1995), "Underground Storage of Carbon Dioxide," *Energy Conversion and Management*, Vol. 36, Nos. 6–9, pp. 539–542.

Hitchon, B. (ed.), *Aquifer Disposal of Carbon Dioxide: Hydrodynamic and Mineral Trapping-Proof of Concept*, Geoscience Publishing Ltd., Sherwood Park, Alberta, Canada, 1996; p. 165.

Holloway, S. (1996), "An Overview of the Joule II Project: 'The Underground Disposal of Carbon Dioxide,' " *Energy Conversion and Management*, Vol. 37, Nos. 6–8, pp. 611–618 and 1149–1154.

Hovorka, S. D., Doughty, C., Benson, S. M., Pruess, K., and Knox, P. R. (2004), "The Impact of Geological Heterogeneity on CO2 Storage in Brine Formations: A Case Study from the Texas Gulf Coast," *In* S. J. Baines and R. H. Worden (eds.), *Geological Storage of Carbon Dioxide*, Geological Society, London, Special Publications, 233, pp.147–163.

IEAGHG/SR5 (1995), "Transport and Environmental Aspects of Carbon Dioxide Sequestration," *IEA Greenhouse Gas R&D Programme*, Cheltenham, UK.

IPCC (2001), "IPCC Third Assessment Report—Climate Change 2001," http://www.ipcc.ch/. Accessed September 2003.

IPCC (2005), in B. Metz, O. Davidson, H. de Coninck, M. Loos, and L. Meyer (eds.), *Special Report on Carbon Dioxide Capture and Storage*, Cambridge University Press, London, Chapter 5, pp.195–276.

IPCC Guidelines for National Greenhouse Gas Inventories (2006), http://www.ipcc-nggip.iges.or.jp/public/2006gl/pdf/2_Volume2/V2_5_Ch5_CCS.pdf. Accessed February 24, 2009.

Keith, D., and Wilson, M. (2002), "Developing Recommendations for the Management Of Geological Storage of CO_2 in Canada," A report prepared for Environment Canada, Saskatchewan Industry and Resources, Alberta Environment, and British Columbia Energy and Mines.

Kinder Morgan (2009), "KINDER MORGAN CO_2 -Transportation," www.kindermorgan.com/business/co2. Accessed February 25, 2009.

King J. E., Gibson-Poole, C. M., Lang, S. C., and Paterson, L. (2003), "Long-term Numerical Simulation of Geological Storage of CO_2 in the Petrel Sub-Basin, North West Australia," Australian Petroleum Cooperative Research Centre. http://www.apcrc.com.au/GreenhouseFrameset.htm. Accessed September 2003.

Koide, H., Takahashi, M., Tsukamoto, H., and Shindo, Y. (1995), "Self-trapping Mechanism of Carbon Dioxide in the Aquifer Disposal," *Energy Conversion and Management*, Vol. 36, No. 6–9, pp. 505–508.

Koide, H., Tazaki, Y., Noguchi, Y., Nakayama, S., Iijima, M., Ito, K., and Shindo, Y. (1992), "Subterranean Containment and Long-Term Storage of Carbon Dioxide in Unused Aquifers and in Depleted Natural Gas Reservoirs," *Energy Conversion and Management*, Vol. 33, No. 5–8, pp 619–626.

Krom, T. D., Jacobsen F. L., and Ipsen, K. H. (1993), "Aquifer-Based Carbon Dioxide Disposal in Denmark: Capacities, Feasibility, Implications and State of Readiness," *Energy Conversion and Management*, Vol. 34, Nos. 9–11, pp. 933–940.

Lake, L.W. (1989), *Enhanced Oil Recovery*, Prentice Hall, Upper Saddle River, NJ.

Lindeberg, E. (1997), "Escape of CO_2 from Aquifers," *Energy Conversion and Management*, Vol. 38, Suppl., pp. 235–240.

Litynski, J. T., Klara, S. M., McIlvried, H. G., and Srivastava, R. D. (2006), "An Overview of Terrestrial Sequestration of Carbon Dioxide," United States Department of Energy Fossil Fuel Energy R&D Program; Climatic Change, 74, pp 81–95.

Mian, M. A. (1991), *Petroleum Engineering Handbook for the Practicing Engineer*, Vols. 1, 2, Penwell Publishing Company, Tulsa, Oklahoma.

Mohitpour, M., Golshan, H., and Murray, A. (2000), *Pipeline Design and Construction: A Practical Approach*, American Society of Mechanical Engineers Press, New York.

National Earthquake Hazards Program (NEHP) (1985), Seismic Zoning Map of Canada, http://www.seismo.nrcan.gc.ca/hazards/zoning/seismiczonea_e.php. Accessed September 2003.

Natural Gas Supply Association (2009), "Natural Gas Storage," http://www.naturalgas.org/naturalgas/storage.asp. Accessed January15, 2009.

Obdam, A, Van der Meer, L, May, F., Kervevan, C, Bech, N., and Wildenborg, A. (2002), "Effective CO_2 storage capacity in aquifers, gas fields, oil fields and coal fields," *Proceedings of the 6th International Conference on Greenhouse Gas Control Technologies*, Kyoto, Japan, Vol. 1, pp. 339–346.

Rutqvist, J., and Tsang, C. (2002), "A Study of Caprock Hydromechanical Changes Associated with CO_2-Injection into a Brine Formation," *Environmental Geology*, 42, pp. 296–305.

Sanford, B. V., and Quillian, R. G. (1959), "Subsurface Stratigraphy of Upper Cambrian Rocks in Southwestern Ontario," Geological Survey of Canada, Ottawa.

Schlumberger (2009), "The Oilfield Glossary: Where the Oil Field Meets the Dictionary," http://www.glossary.oilfield.slb.com/. Accessed January 25, 2009.

Shafeen, A., Douglas, P., Croiset, E., and Chatzis, I. (2004), "CO_2 Sequestration in Ontario, Canada. Part-1: Storage Evaluation of Potential Reservoirs," *Energy Conversion and Management*, Vol. 45, pp. 2645–2659.

Shafeen, A., Douglas, P, Croiset, E., and Chatzis, I. (2004), "CO_2 Sequestration in Ontario, Canada. Part-II: Cost Estimation," *Energy Conversion and Management*, Vol. 45, pp. 3207–3217.

Shafeen, A., Douglas, P., Croiset, E., and Chatzis, I. (2004). "Assessment of Geological CO_2 Sequestration in Ontario," M.A.Sc. thesis, Department of Chemical Engineering, University of Waterloo, Waterloo, Ontario, Canada.

Statoil (2003), http://www.statoil.com/. Accessed September 2003.

Skovholt, O. (1993), "CO_2 Transportation System," *Energy Conversion and Management*, Vol. 34, Nos. 9–11, pp. 1095–1103.

Tanaka, S., Koide, H., and Sasagawa, A. (1995), "Possibility of Underground CO_2 Sequestration in Japan," *Energy Conversion and Management*, Vol. 36, Nos. 6–9, pp. 527–530.

Thom, R. L., and Gehle, R. M. (2000), "A Brief History of Salt Cavern Use," in *Eighth World Salt Symposium*, Elsevier, The Hague, Netherlands, 1400 p.

USGS National Seismic Hazard Mapping Project (2003), U.S. Geological Survey, http://geohazards.cr.usgs.gov/eq/index.html. Accessed September 2003.

Van der Meer, L. G. H. (1992), "Investigations regarding the storage of carbon dioxide in aquifers in the Netherlands," *Energy Conversion and Management*, Vol. 33, Nos. 5–8, pp. 611–618.

Van der Meer, L. G. H. (1993), "The Conditions Limiting CO_2 Storage in Aquifers," *Energy Conversion and Management*, Vol. 34, Nos. 9–11, pp. 959–966.

Van der Meer, L. G. H. (2002), "CO_2 Storage in the Subsurface," *Proceedings of the 6th International Conference on Greenhouse Gas Control Technologies* (Kyoto, Japan), Vol. 1, pp. 201–206.

Visser, E., Hendriks, C., Barro M., Molnvik, M.J., Koeijer, G., Liljemark, S., and Gallo, Y. (2008), "Dynamis CO_2 quality recommendations," Vol. 2, pp. 478-484.

Vugrinovich, R. (1989), "Subsurface temperatures and surface heat flow in the Michigan Basin and their relationships to regional subsurface fluid movement," *Marine and Petroleum Geology*, Vol. 6, No. 1, pp.60–70.

White, D. E. (1965), "Saline Waters of Sedimentary Rocks," *American Association of Petroleum Geologists*, Memoir Vol. 4, pp. 342–366.

Wolfgang, R.W., and Rudnicki, J. W. (ed). (1998), "Terrestrial sequestration of CO_2—An assessment of research needs," www.science.doe.gov/bes/geo/Publications/CO_2report.pdf. Accessed September 2003.

Zweigel, P., Arts, R., Lothe, A.E., and Lindeberg, E. (2004), "Reservoir Geology of the Utsira Formation in the First Industrial-Scale Underground CO_2 Storage Site (Sleipner Area, North Sea)," *In* S. J. Baines and R. H. Worden (eds.), *Geological Storage of Carbon Dioxide*, Geological Society, London, Special Publication, Vol. 233, pp.165–180.

CHAPTER **7**

CLEAN-COAL TECHNOLOGY: GASIFICATION PATHWAY

J. Guillermo Ordorica-Garcia, Ali Elkamel, Peter L. Douglas, and Eric Croiset
Department of Chemical Engineering
University of Waterloo
Waterloo, Ontario Canada

1   INTRODUCTION

Coal is a very vital fuel and will continue to remain important in the future as well. Around 23 percent of the primary energy is produced by coal, in addition to the generation of about 39 percent of electricity. Furthermore, 70 percent of the world's steel production is dependent on coal feedstock. According to the International Energy Agency, there is an expected increase of 43 percent in the use of coal from 2000 to 2020 (World Nuclear Association 2008). Coal is also more advantageous than other sources of energy since it very abundant on the earth as well as significantly cheaper. It has been estimated that there are over 250 billion tons of recoverable coal in the United States, which is sufficient to supply 250 years of the current consumption. This implies that since coal is available in large amounts, the market costs for coal will also be low compared to the other fossil fuels. Coal costs approximately $0.50 to $1.00 per gigajoule, whereas natural gas, for example, costs $2 to $2 per gigajoule. This lower cost of coal enables one to budget a certain amount of money for carbon dioxide disposal during sequestration. According to the World Nuclear Association, every dollar difference between coal and natural gas results in an increase of the cost of carbon dioxide capture by $22/ton of carbon dioxide. Coal plants give out more carbon dioxide than natural gas plants. Thus, in the end coal is more cost-effective as more carbon dioxide will be removed for a cheaper price.

The biggest challenge is to be able to burn coal without increasing the global carbon dioxide levels. In 2006, the World Nuclear Association estimated that burning coal releases about 9 million tonnes of carbon dioxide into the atmosphere, out of which 70 percent is generated by power plants. Thus, zero-emission coal technologies have been introduced so that the abundant coal resources available can be utilized in the future without contributing to global warming.

In this chapter, relevant information concerning power plant technology and carbon dioxide capture techniques will be presented. *Zero-emission* coal technologies are being developed in order to eliminate the amount of carbon dioxide being emitted from coal-fired power plants. In this chapter, different clean-coal technologies are discussed with an emphasis on gasification technologies and CO_2 capture. Other clean coal technologies are discussed elsewhere in other chapter (e.g., CO_2 transportation and storage, combustion technologies, etc.).

Section 2 of this chapter covers current gasification techniques, including commercial gasification processes. Section 3 corresponds to IGCC power plants. Their advantages, stage of development, and selected commercial IGCC projects are presented. Section 4 introduces the integrated coal gasification fuel cell combined cycle (IGFC). Finally, Section 5 describes all major CO_2 capture technologies. Selected examples of commercial carbon dioxide projects are also included.

2   COAL GASIFICATION

Coal gasification generally refers to the reaction of coal with oxygen and steam, to yield a gaseous product for use directly as a fuel in a power plant or as a feed to synthesis of other gaseous or liquid chemicals.

Combined cycle power plants that employ synthetic gas from coal as a fuel are called integrated gasification combined cycle (IGCC) power plants. In the current section, the details concerning the production of synthetic gas from coal will be reviewed. The description and performance of IGCC plants will be discussed in Section 3.

Gasification is, in essence, an incomplete combustion. There are similarities between coal combustion and gasification, but the pollutant formation processes are different in gasification. The latter is classified as a clean-coal technology, because under reducing conditions, sulfur from coal is converted mostly to H_2S rather than SO_2, and nitrogen is converted mostly to NH_3, and almost no NO_X is formed.

Depending on the type of gasifier and the operating conditions, coal gasification can yield gases for different uses. Low-heating-value gas is produced by gasification with air and steam, for use as an industrial fuel and as a power generation fuel. A medium-heating-value gas results from gasification with oxygen and steam, and can be used as a fuel gas or as a chemical synthesis feedstock. High-heating-value gases are generated by shift conversion or hydrogasification and can be used as a substitute natural gas (SNG).

In gasification, coal reacts with oxygen or air and steam. The organic material is converted into various gases, such as CO, CO_2, H_2, CH_4, and traces of sulfur-bearing or nitrogen-bearing gases, while ash is a residue in the form of dry powder or molten slag. The fundamental reactions involved in such a process are shown as follows:

Devolatilization

$$4C_nH_m \rightarrow mCH_4 + (4n - m)C$$

Char Combustion

$$2C + O_2 \rightarrow 2CO$$
$$C + O_2 \rightarrow CO_2$$

Char gasification

$$C + H_2O \rightarrow CO + H_2$$
$$C + CO_2 \rightarrow 2CO$$
$$C + 2H_2 \rightarrow CH_4$$

Gas-phase reactions

$$2CO + O_2 \rightarrow 2CO_2$$
$$2H_2 + O_2 \rightarrow 2H_2O$$
$$H_2O + CO \rightarrow H_2 + CO_2 \text{ (water-gas shift)}$$
$$CO + 3H_2 \rightarrow CH_4 + H_2O \text{ (methanation)}$$

The reaction mechanism of gasification is complex. It has been confirmed that through the carbon structure there exist active carbon-free sites, due to lattice imperfection or discontinuities. These sites provide unpaired electrons, required for chemisorption of reacting gas constituents to form surface complexes. The rates of formation and removal of these surface complexes, as well as the number and extent of coverage of the active carbon-free sites, determine the rate and the order of each reaction.

Like all chemical reactions, the gasification reactions tend to state of equilibrium, which may be one-sided. This means that after a sufficient reaction time, depending on the temperature and pressure, a fixed relation exists between the amounts of starting materials and end products, which is expressed by the equilibrium constants.

The oxygen-consuming reactions proceed rapidly to completion at normal gasification conditions. By contrast, the reactions of carbon with steam and carbon dioxide never reach equilibrium. For the water-gas shift and the methane forming reactions, raising the temperature displaces the equilibrium composition toward the starting materials. An increase of pressure, for the methane-forming reactions, favors the formation of the end products, whereas the water-gas shift reaction is almost pressure independent.

In the light of these relations, it is clear that, on the one hand, the formation of carbon monoxide and hydrogen, by gasification, will occur to a large extent only at high temperatures and normal pressure. On the other hand, the formation of methane is favored by a high hydrogen pressure and a relatively low reaction temperature. However, the actual composition of the product gas, from the gasification of coal, is critically dependent on the rate at which equilibrium is attained, as described by the reaction kinetics.

2.1   Gasification Techniques

Coal gasification processes can be classified in a variety of ways. These classifications are based on the method used to generate heat for the gasification reactions, on the method of contacting reactants, and on the physical state of the residue removed.

On the basis of the technique of heat transfer chosen, the gasification processes may be autothermic or allothermic. In autothermic processes, the heat required is generated in the reactor itself, by using oxygen and steam as the gasifying medium. The great advantage of them is that any losses associated with heat transfer are avoided. All the gasification processes that are currently operated on industrial scale are autothermic. In *allothermic* processes, the heat required is generated either by gaseous, liquid (molten slag, salts or metals), or solid heat carriers, which are brought into direct contact with the coal, or indirectly via heat exchange surfaces.

On the basis of the mode of conveyance of the coal and the gasifying medium, the gasification processes can be classified as fixed-bed, fluidized-bed, or entrained-bed. Apart from the flow pattern of the reactants and products, these processes differ from each other with respect of the type and rank of coal that can be used, the size distribution of the coal, the residence time, and the reaction temperature and pressure.

Finally, on the basis of the physical state of the residue removed, the gasification processes can be classified as slagging and nonslagging.

The choice of a gasifier will generally depend on nine factors:

1. The production rate of energy
2. The turndown requirements
3. The desired heating value of the syngas (low HV $= 3.8–7.6\,MJ/m^3$, medium HV $= 10.5–16\,MJ/m^3$, high HV $> 21\,MJ/m^3$)

4. The operating temperature and pressure
5. The allowed gas purity (sulfur, carbon gas)
6. The allowed gas cleanliness (tars, soot, ash)
7. Coal availability, type and cost
8. The gasifier end-use locations and interactions
9. Plant size constraints

Fixed Bed

Fixed bed in gasification technology actually refers to a bed of lump coal maintained at a constant depth between two fixed extremities (Figure 1). The coal is supplied at the upper extremity, while the residue is removed at the lower one, and this end provides a support for the bed. The coals moves countercurrent to the rising gas stream, through five definable, but not distinctly separate, zones:

1. Coal drying and preheating zone
2. Devolatilization zone
3. Gasification zone
4. Oxidation zone
5. Ash zone

As the coal slowly descends, the hot gases produced in the combustion and gasification zones exchange energy with the colder solid. After drying, the coal is pyrolized in the devolatilization zone, releasing methane, higher hydrocarbons, and tar. In the gasification zone, the char reacts with steam and carbon dioxide, resulting in absorption of heat and a consequent progressive fall in temperature. When coal reaches the combustion zone, it is finally combusted, providing

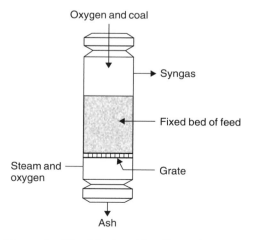

Figure 1 Fixed-bed gasifier (Vamvuka 1999)

heat for the process and leaving an incombustible residue, the ash. The ash is discharged from the base, either through a grate, if it is in dry form, or through some form of slag tap (Figure 1).

In general, fixed-bed gasification systems are simple and reliable, offering a high thermal and coal efficiency, because of low heat losses and minimal carbon fine-ash carryover in the gas stream. They are suitable for lump coals (6–50 mm) and have a limited capability to accept materials smaller than 3 mm in size. Caking coals cannot be used without pretreatment. The solid residence time may be of the order of several hours, while the gas residence time is on the order of seconds.

Fluidized Bed

In fluidized bed (Figure 2), the coal is supplied as a powder, usually less than 3 mm in size, which is then maintained in a suspended state of continuous random motion by the gasifying agent. No reaction zones are present in this system, as they are in fixed-bed gasification. There is a very intensive internal mixing of the solid particles. The fluidized state results in a very homogeneous temperature distribution throughout the whole volume of the turbulent layer, so that gasification reactions take place more uniformly. The composition of gas varies along the bed, since combustion processes predominate in the lower part of the bed and gasification processes predominate in the upper part.

The characteristics that are advantageous to coal gasification are the high efficiency of heat transfer, the uniform and moderate temperature, high specific gasification rate, the high product uniformity and moderate temperature, tolerance to a wide range of feedstock compositions, and the capability of accepting coal fines.

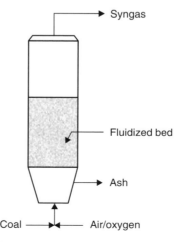

Figure 2 Fluidized-bed gasifier (Vamvuka 1999)

Fluidized-bed coal gasification also implies disadvantages: a substantial amount of fine particles is present in the raw gas, and this can cause erosion, fouling, or plugging of downstream equipment. Caking coals may agglomerate, altering the hydrodynamic behavior of the system and lowering the rate of gasification.

Low-rank coals display advantages to higher-rank coals for gasification in fluidized-bed systems, due to their higher reactivity and noncaking character.

Entrained Bed

In the entrained-bed gasifier (Figure 3), pulverized coal (particle size less than 0.12 mm) is entrained with the gasifying agent to react in concurrent flow, in a high temperature atmosphere, from which part of the ash may separate as a liquid slag.

Virtually all types of coal can be gasified by this technique. Swelling and caking characteristics do not affect the operability of the unit, since each discrete particle is separate from others in the flowing gas stream. The full entrainment of the particles requires relatively high gas velocities, and this will lead to high throughputs and residence times of only a few seconds.

Entrained-bed gasifiers are simple in design and have few internal components. Their dimensions are determined by the particle size and the reactivity of the fuel, the reaction temperature, and the gas velocity. The main disadvantages come from the low concentration of fuel in the gasifying medium and the concurrent flow of the reactants. This arrangement lowers the gasification rates at the reactor outlet and leads to unfavorable conditions for heat transfer, resulting in high gas outlet temperatures, compared with fixed and fluidized-bed processes. Another disadvantage is that due to the high temperature requirements, the oxygen consumption in this gasifier is relatively high, compared with the other types.

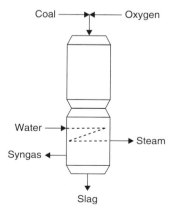

Figure 3 Entrained-bed gasifier (Vamvuka 1999)

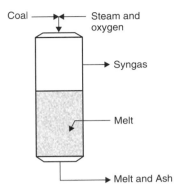

Figure 4   Molten bath gasifier (Vamvuka 1999)

Molten Bath

Most of the molten-bath processes involve the gasification of coal in direct contact with steam and either air or oxygen, in a molten slag reservoir (Figure 4). Development of such systems has been encouraged by their potential to overcome the disadvantages of gas-solid reaction systems.

High temperatures are normally required to maintain the bath molten, and this favors high reaction rates—hence, high throughputs. The rapid release of volatile matter causes the coal to disintegrate, resulting in a large increase in surface area, which, in turn, also increases the reaction rates. Unlike fixed, fluidized, and entrained-bed gasifiers, the majority of molten bath gasifiers utilize coarsely ground coal, including fines.

Advantages of such a process are that strongly caking coals, even with high ash or sulfur contents, can be gasified, and retention of sulfur in the melt can result in the production of an essentially sulfur-free gas stream. The main disadvantages are the high oxygen requirements, the high heat losses, and the problems associated with severe corrosion caused by the high-temperature molten salts and metals.

2.2   Commercial Gasification Processes

This section gives a detailed description of two well known existing commercial gasification processes. Process flow diagrams of both installations are explained and the details of the operation of the units and the ranges of operating conditions are given.

Shell Gasification Process

This gasification process is a high-pressure version of the Koppers-Totzek gasifier, developed in 1952. The technology is owned by Shell International Petroleum and Krupp-Koppers (Vamvuka 1999). A schematic of the Shell gasification process is shown in Figure 5.

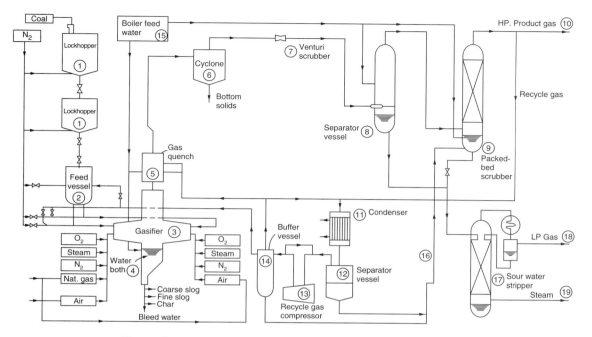

Figure 5   Shell entrained-flow gasification pilot plant (Lee 1996)

Dried, pulverized coal is fed to opposed burners through lock hoppers while oxygen and steam enter the gasification chamber. Coal particles are gasified as they move along the gasifier. The reaction temperature at the burner reaches about $1990°C$, and the hot raw syngas leaves the unit at about $1480°C$. High burner temperatures are required to ensure slagging operation. Most of the ash is converted to slag, but some fly ash is present in the gaseous stream exiting the gasifier. Syngas is cooled by product gas recycle quench and by waste heat boilers.

The Shell process was demonstrated commercially at the Shell Coal Gasification Plant in Houston, Texas, which started operations in 1987 (Vamvuka 1999), and also at a facility in Deer Park, Texas. Some of the most relevant performance data of the latter are shown in Table 1.

The advantages of this gasifier are that no tars, phenols, or liquid hydrocarbons are produced due to the high gasification temperature. The gasifier can handle any solid fuel, including high-ash bituminous coals. High coal conversions and gas efficiencies can be achieved by this gasifier.

Studies in the Netherlands have shown that a 250 MW unit using one Shell coal gasifier of 2,000 tonnes per day combined with ABB's type 13E gas turbine is an ideal "building block" for capacity additions in plant size ranging from 250 to 1000 MW. Its prominent features are ultra-low emissions, nonleachable slag and sulfur byproducts, maximum fuel flexibility, option for cogeneration, short

Table 1   Operating Results for Shell Gasification of Sandow Lignite (Vamvuka 1999)

	High-Ash Lignite	Low-Ash Lignite
Lignite fed to plant t/d	334	248
Oxygen/lignite ratio	0.8777	0.865
Syngas composition, dry, vol percent		
Carbon monoxide	60.59	61.82
Hydrogen	28.20	28.21
Carbon dioxide	5.38	4.47
H_2S+COS	0.71	0.80
$N_2+Ar+CH_4$	5.08	4.83
Sweet syngas production		
Mass basis kg/h	12,060	10,380
HHV GJ/h	155	131
Sulfur removal from syngas (%)	99.1	99.8
Carbon conversion (%)	99.7	99.4
Cold gas efficiency (% HHV)	78.8	80.3
Slag production, kg/h	2,210	
Flyslag production, kg/h	410	
Acid gas, kg/h	1,690	

lead times, low cooling-water consumption, high thermal efficiency (43 percent), and low capital cost (Luthi et al. 1989).

Texaco Gasification Process

Texaco has more than 50 years of experience in the field of partial oxidation of petroleum residual oils. With a special division devoted solely to gasification and more than 70 commercial gasification plants in operation around the world, Texaco has consolidated in the present as one the most recognized international providers of gasification equipment (Falsetti et al. 2001). In the adaptation of the partial oil-residual oxidation process to coal gasification, Texaco developed a special unit to slurry finely ground coal. The gasifier, which is a one-stage vertically mounted cylindrical pressure vessel, is shown with auxiliary units in Figure 6.

The feed coal is pulverized and slurried in a wet rod mill. The slurry water consists of recycled condensate and make-up water. The coal slurry is oxidized with oxygen or air in the refractory-lined gasifier chamber on the top. A typical reactor is 2.7 m in diameter and 4.6 m in height, operating at pressures of about 40 bar. The high temperature produced from the exothermic reaction of coal and oxygen evaporates the water, which, in turn, reacts with the coal. Gasification temperatures range from 1,260°C to 1,490°C. Under these conditions, coal is converted primarily to CO, H_2, CO_2 and small amounts of methane. No tars or oils are generated. Some fuel-bound nitrogen is converted to ammonia, while sulfur is reduced to hydrogen sulphide and trace amounts of COS.

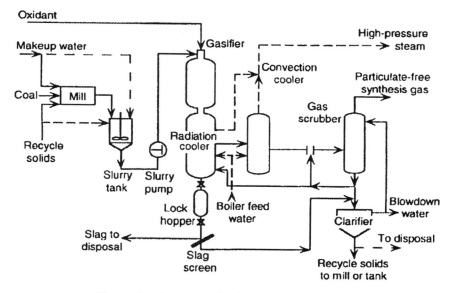

Figure 6 Texaco gasification process (Lee 1996)

The produced syngas is cooled in radiative and convective heat exchangers, before wet-particulate scrubbing. The gas has a medium heating value and is suitable for use in power production or as a chemical feedstock. The ash leaves the gasifier as molten slag through lock hoppers located at the bottom of the gasifier.

The advantages of this gasifier are its flexibility to accept any kind of coal, very high throughputs (especially when compared to fluidized-bed and fixed-bed gasifiers), and short residence time. Among the disadvantages, high oxygen consumption is the main concern, followed by the production of more CO_2 compared to dry-feed injection (Vamvuka 1999).

The best-known Texaco-based power plant is the 120 MW "cool water" plant, located in the Mojave Desert in California. It started operations in 1989, converting approximately 900 tonnes/day of coal to syngas. The SO_2, NO_X, and particulate emissions from this plant are only one tenth of the United States New Source Performance standards (NSPS) for coal-fired plants (Vamvuka 1999).

3 INTEGRATED GASIFICATION COMBINED CYCLE

Coal is a relatively abundant fuel all over the world. It could provide energy for several more decades if an efficient and clean technology could be developed to use coal for electric power generation. Such a technology of integrated coal gasification and steam cycle for power generation, now known as IGCC, was conceived during the 1970s and demonstrated in many projects during the 1980s. IGCC technology has developed over the past decade and is now ready for commercial application. IGCC is expected to be the key evolving power technology of the twenty-first century due to its high efficiency and "cleaner and greener" characteristics, thus minimizing the impact on the environment.

3.1 Process Description

A diagram of a basic IGCC plant is shown in Figure 7. The gasifier receives coal from the coal handling and preparation equipment and oxidant from an air treatment plant, and gasifies the coal through a partial oxidation/reduction process involving coal, oxygen, and steam. The fuel output from the gasifier is in the temperature range of 537 to 1426°C, and is treated for removal of particulates and sulfur. This gas cleanup can involve cooling (cold gas cleanup) or no cooling (hot gas cleanup). A sulfur scrubber removes sulfur from the fuel gas and processes it to a salable product. The gas cooler produces steam, which is utilized by the steam turbine and sometimes by the gasifier, too. The clean gas is combusted in a turbine to generate electricity, and the heat from the exhaust gas in recovered and used to generate steam. This steam is sent to a steam turbine, and more electricity is generated.

Solid waste from the gasifier includes ashes and other byproducts. The particulate scrubber system removes particulates, trace alkali metals, and fuel-bound nitrogen converted to ammonia in the gasifier. Hydrogen sulphide is removed from the fuel gas, and the turbine feed is virtually free of pollutant precursors.

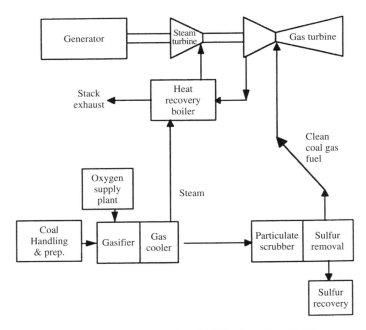

Figure 7   Schematic of an IGCC plant (Lee 1996)

Typical designs for IGCC systems use cold gas cleanup (CGCU), including low-temperature removal of H_2S and particulates from the coal syngas, sulfur byproduct recovery, and syngas moisturization to reduce NO_X formation in the gas turbine. Dry physical and chemical hot gas cleanup (HGCU) techniques reduce the efficiency penalty associated with syngas cooling.

Cold Gas Cleanup

The design basis for the CGCU system is the use of fluidized-bed gasifiers. Coal is partially combusted with oxygen and gasified with steam in a reducing atmosphere to yield a gas containing CO and H_2 as main constituents. Oxygen for the gasifier is provided by an air separation plant. Steam is supplied from the plant steam cycle. The hot fuel gas is cooled in a steam generator, and then enters a low-temperature cooling section. As part of the low-temperature cooling, nearly all of the particulate matter and ammonia in the fuel gas are selectively removed by wet scrubbing. The fuel gas, at a temperature of about 40°C, then enters an absorption-based acid removal unit, where H_2S (the primary sulfur species in the fuel gas) is selectively removed. The clean fuel gas is then combusted in a combined cycle system.

Removal of ammonia in the fuel gas in wet scrubbing systems substantially reduces the amount of fuel-bound nitrogen in the fuel gas. In conventional gas turbine combustors, most of the fuel-bound nitrogen is converted to NO_X. Thermal NO_X emissions are controlled either by moistening the fuel gas or by steam injection to the gas turbine combustor to reduce the flame temperature.

In IGCC systems with CGCU, there are two main waste streams: liquid condensates from the high temperature gas and blowdown from the fuel gas wet scrubber. Solid waste includes gasifier bottom ash and particulate cake from the scrubber system.

Hot Gas Cleanup

The HGCU systems reduce or eliminate the need for syngas cooling prior to particulate removal and desulfurization. This improves the plant efficiency and reduces or eliminates the need for heat exchangers and process condensate treatment. Thus, HGCU offers a more highly integrated system in which a major wastewater stream is eliminated.

The primary features of this design, compared to the IGCC system with CGCU, are as follows:

- Elimination of oxygen plant by using air, instead of oxygen, as the gasifier oxidant
- In situ gasifier desulfurization with limestone
- External (not in the gasifier) desulfurization using a high-temperature removal process
- High-efficiency cyclones for particulate removal

- Elimination of heat exchangers for fuel gas cooling at the gasifier exit
- Elimination of sulfur recovery and tail gas treating
- Addition of a circulating fluidized-bed boiler for sulphation of spent lime-stone (to produce an environmentally acceptable waste) and conversion of carbon remaining in the gasifier ash.

In-bed desulfurization is expected to result in 90 percent sulfur capture within the gasifier. The external desulfurization process for zinc ferrite, an iron and zinc compound with a spinel-type crystal structure, is expected to reduce the sulfur content of the syngas to 10 ppm, resulting in low SO_2 emissions from the gas turbine combustor. Upon regeneration, the sulfur captured by the zinc ferrite sorbent is evolved in an offgas containing SO_2, which is recycled to the gasifier for capture by the calcium-based sorbent.

Thermal NO_X emissions are expected to be quite low for air-blown IGCC. However, the HGCU system assumed here does not remove fuel-bound nitrogen from the fuel gas. The IGCC/HGCU system is not expected to have any liquid discharges other than those normally associated with the steam cycle and plant utilities. The solid waste streams include bottom ash, fines collected in the secondary cyclones, and spent zinc ferrite sorbent.

HGCU technology has not reached a commercial status compatible with commercial electricity generation yet. One of the latest applications of HGCU was in the Puertollano (ELCOGAS) plant in Spain. As of November 1999, serious fouling problems linked to the use of limestone for sulfur recovery in the gasifier, as well as increasing syngas pressure drop in the ceramic candle filters, were reported (Granatstein et al. 2001). The Tampa Electric IGCC plant (1996) is another project where HGCU was to be applied, but after initial test work, support for this option was discontinued. The main factor for such a decision was the high maintenance and low availability derived from HGCU implementation (USDOE/TECO 2000).

3.2 Advantages

IGCC systems represent a promising approach for the cleaner and more efficient use of coal for power generation, offering low levels of SO_2 and NO_X emissions along with salable solid byproducts, and zero or low wastewater discharges.

IGCC systems combine several desired attributes and are becoming an increasingly attractive option among the emerging technologies. First, IGCC plants provide high-efficiency energy conversion (up to 44 percent), with the prospect of even higher efficiencies if higher temperature turbines or viable hot gas cleanup systems are employed. Second, very low emission levels for sulfur and nitrogen species have been demonstrated at many facilities around the world (Texaco P&G 2001). Third, IGCC plants produce flue gas streams with concentrated CO_2, as well as high levels of carbon monoxide that can be easily converted to CO_2. Recovery of CO_2 in IGCC systems is potentially less expensive than in the conventional combustion systems. Moreover, CO_2 recovery can be done in conjunction with hydrogen sulphide removal by using several commercial technologies (Granatstein et al. 1999).

In IGCC systems, environmental control is required not just to meet environmental regulations but also for proper plant operation. In particular, contaminants such as sulfur, particulates, and alkali must be removed prior to fuel gas combustion to protect the gas turbine components from erosion, corrosion, and deposition.

The approach to emissions control in an IGCC plant is fundamentally different from that in a pulverized-coal-fired power plant. Emission control strategies focus on the fuel gas, which is pressurized and has a substantially lower volumetric flow rate than the flue gas, which flows near atmospheric pressure of coal combustion power plants. Furthermore, sulfur in the fuel gas is in reduced form (mostly H_2S), which can be removed by a variety of commercially available processes.

Another attractive feature of IGCC technology is that it can be progressively implemented in new or existing plants. This technology is modular in nature and offers the possibilities of phased construction. A schematic representation illustrating one of many possible phased-construction scenarios for an IGCC plant is presented in Figure 8.

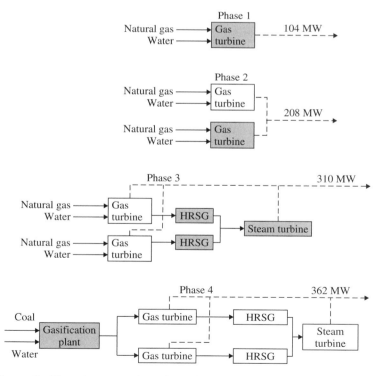

Figure 8   Phased construction of an IGCC power plant (Joshi et al. 1996)

The simple scheme in Figure 8 is based in two gas turbines. In the first phase, a single gas turbine is installed, to be fired by natural gas. A second gas turbine is added in the second phase, increasing the capacity of the plant, which continues to be operated as a peaking unit. A steam cycle is added next, increasing the efficiency of the plant and enabling it to operate as a base load combined plant fired by natural gas. In the fourth and final phase, a coal gasification plant is constructed and the entire facility is operated as a coal-fired IGCC power plant. (Joshi et al. 1996).

3.3 State of Development

In the past few years, IGCC has matured as a technology, and its niche marketplace has become clear. Texaco dominates the market, with 12 commercial projects that began operation from mid-1999 to late-2000. Refinery bottoms/petroleum coke are the preferred fuels. Currently there are 160 commercial gasification plants in operation, under construction, or in the planning design stages in 28 countries. Capacity of these plants is approaching 430 million m^3 of syngas per day, which are equivalent to almost 800,000 bbl of oil (Granatstein et al. 2001).

The cumulative worldwide gasification capacity of 60,833 MW_{th} of synthesis gas is steadily increasing (proceedings of the 1999 Gasification technologies conference). The current growth in gasification is mostly in electrical power generation. Much of the current surge in gasification capacity is in industrial polygeneration, which is defined as the production of steam and power, plus synthesis gas for hydrogen or chemicals.

Worldwide IGCC Status

The following information was obtained from a study by Simbeck presented at the 1999 Gasification Technologies Conference in California. SFA Pacific has developed a database for gasification technologies based on capacity and growth, location, technology, application, and primary feedstocks. The study was based on information and gasification technologies/projects available to SFA Pacific from nonproprietary public sources. The complete database has records for 329 gasification projects, representing a total of 754 gasifiers (including many small pilot plants and shutdown/dismantled units). To assess the role of gasification technology in current world markets, it is more useful to review only data for the commercial-scale real and actively planned gasification projects.

Figure 9 presents the worldwide gasification growth.

Gasification by Location

Western Europe is the leading region for gasification, based on MW_{th} of syngas output of real and planned projects. Asia and Australia are the second-largest region, with China dominating the region's gasification market. North America

MW$_{th}$ syngas

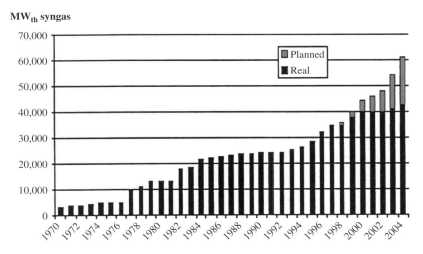

Figure 9   Cumulative worldwide gasification capacity and growth

is the third region of the world where there is extensive use of gasification. Most of the newest U.S. gasification activity is for large electric power generation projects (IGCC and oil refinery polygeneration), with a smaller emphasis on synthesis gas chemical production.

Gasification by Application

Synthesis gas for chemicals continues to be the dominant application or product of gasification. This application includes 89 real projects accounting for 18,361 MW$_{th}$ of synthesis gas output. Power generation however, is gaining ground quickly and represents most of the recent and planned additions, with 20 new projects planned to generate 10,758 MW$_{th}$ of synthesis gas output.

Gasification by Primary Feedstocks

Coal and petroleum (mainly heavy residues) are the dominant feedstocks for gasification projects, as shown in Figure 10. Coal is used in 29 real projects; petroleum feedstocks fuel 56 real projects, while petcoke is the choice fuel of only 5 projects. Low-quality residual oil and petcoke currently have price advantages over coal.

Gasification by Technology

The dominant commercial technologies are listed in Table 2. Texaco currently enjoys a well-established position.

IGCC Status in Canada: The success of the Cool Water IGCC demonstration plant in California in the mid-1990s caused widespread interest in Canada in

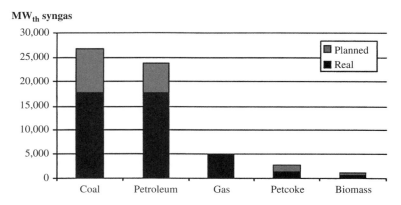

Figure 10   Worldwide gasification by primary feedstock

Table 2   Dominant Gasification Technologies/Licensors

Technology/Licensor	Total Capacity (MW_{th})	Percentage of Total	Number of Projects
Texaco	16,483	40	63
Shell	8,967	21	28
Lurgi Dry ash	11,842	28	7

the last decade. IGCC is under active consideration in New Brunswick, Ontario, Nova Scotia, Saskatchewan, and Alberta. However, there are no IGCC projects currently under operation or construction in Canada yet.

The CANMET Energy Technology Center has conducted the bulk of the IGCC-related research in the country. They have developed detailed ASPEN Plus simulation models of IGCC power plants burning specific types of Canadian coals and of biomass gasification processes for Canadian sites (Granatstein et al. 1999).

A major evaluation of IGCC for three sites in Canada—Nova Scotia, Ontario and Alberta—was completed in 1991 by Bechtel Canada Inc. The work was sponsored by the Canadian Electrical Association and was funded by the Government of Ontario, Ontario Hydro, Environment Canada and members of the Canadian Coal Gasification Technical Committee. Conceptual engineering design, plant performance, the selection of gasification processes, emission projections, and capital cost estimation were carried out as part of the work undertaken by Bechtel (MacRae 1991).

Additionally, the Coal Association of Canada proposed in the mid-1990s the design, construction, and operation of an IGCC demonstration project under conditions unique to Canada. The proposed study would address CO_2 removal (to be potentially used for enhanced oil production in Western Canada), environmental emissions, site selection, process selection, and capital and operating costs.

This would result in a project description and cost estimate in sufficient detail to evaluate the economics of IGCC in specific Canadian applications.

The province of Saskatchewan is also supporting joint government-industry assessment of the technical and economic feasibility of using IGCC for post-2000 power generation. Saskatchewan is currently considering the testing of a lignite coal in a pilot gasification unit to determine the gasification characteristics of the coal.

3.4 Commercial IGCC Projects

In this section two commercial IGCC installations are described. The process operations of both installations are described as well as their generation capacity.

Tampa Electric

This 250-MW IGCC is located in Mulberry, Polk County, Florida. It started operations in 1996. It utilizes commercially available oxygen-blown entrained-flow coal gasification technology licensed by Texaco Development Corporation (Texaco). Coal is ground with water to the desired concentration (60 to 70 percent solids) in rod mills. The gasifier is designed to utilize about 2,200 metric tons per day of coal (dry basis).

An air separation unit (ASU) separates ambient air into 95 percent pure oxygen for use in the gasification system and sulfuric acid plant, and nitrogen is sent to the advanced GE MS 7001F gas turbine (GT). The addition of nitrogen in the GT combustion chamber has dual benefits. First, this additional mass flow has the advantage of producing higher GT power output. Second, the nitrogen acts to control potential NO_X emissions by reducing the combustor flame temperature, which, in turn, reduces the formation of thermal NO_X in the fuel combustion process.

The coal/water slurry and the oxygen are then mixed in the gasifier process feed injector producing coal syngas. The gasifier is designed to achieve greater than 95 percent carbon conversion in a single pass. The gasifier is a single vessel feeding into one radiant syngas cooler (RSC), which was designed to reduce the gas temperature to 760°C while producing 113 bar saturated steam.

After the RSC, the gas is split into two parallel convective syngas cooler boilers (CSC), where the temperature is further reduced to less than 426°C and additional high-pressure steam is produced. The syngas is then further cooled in the gas to gas exchangers, where the heat in the raw ash–loaded syngas is exchanged with either clean, particulate-free syngas or nitrogen.

Next, the particulates and hydrogen chloride are removed from the syngas by intimate contact in the syngas scrubbers. Most of the remaining sensible heat of the syngas is then recovered in low-temperature gas cooling by preheating clean syngas and heating steam turbine condensate. A final trim cooler reduces the syngas temperature to about 40°C for the cold gas cleanup (CGCU) system.

The CGCU system is a traditional amine scrubber type that removes most of the sulfur from the syngas. Sulfur is recovered in the form of sulfuric acid, which has a ready market in the phosphate industry in the central Florida area.

Most of the ungasified material in the coal exits the bottom of the RSC into the slag lock hopper, where it is mixed with water. These solids generally consist of slag and uncombusted coal products. As they exit the slag lock hopper, these nonleachable products are salable for blasting grit, roofing tiles, and construction building products.

Since Polk Power Station's first gasifier run in July 1996, the gasifier has operated over 29,500 hours, and the combustion turbine has operated over 28,400 hours on syngas, producing over 8.5 million MWh of electricity. Polk Power Station's IGCC plant produced a record 2 million MWh on syngas in 2000. This is an increase of over 250,000 MWh over 1999. Polk Power Station's gasifier reached its project goal of 80 percent on-line availability during 2000. The project completed it operations period under the Clean Coal Program in October 2001.

Tests have included evaluation of various coal types on system performance. Kentucky #11, Illinois #6, and three Pittsburgh #8 coals were tested for their: (1) ability to be processed into a high concentration slurry, (2) carbon conversion efficiency, (3) aggressiveness of the slag in regard to refractory wear, and (4) tendency toward fouling of the syngas coolers. Kentucky #11 coal proved to have the best overall characteristics.

As a result of the Polk Power Station demonstration, Texaco-based IGCC can be considered commercially and environmentally suitable for electric power generation utilizing a wide variety of feedstocks. Sulfur capture for the project is greater than 98 percent, while NO_X emissions are 90 percent less than that of a conventional pulverized coal-fired power plant. The integration and control approaches utilized at Polk can also be applied in IGCC projects using different gasification technologies (Clean Coal Technology Program).

Sarlux

At 550 MW, Sarlux is the world's largest IGCC power plant. It is owned by Saras Raffienerie S.P.A. of Milan and Enron Corp. of Houston. It was recently constructed by a consortium including Snamprogetti S.p.A. of Milan and GE Power Systems of Schenectady on the Italian island of Sardinia. The IGCC plant is located at the Saras Oil Refinery in Sarroch, the second largest European refinery. The plant has been running on syngas since August 2000, and produces 551 megawatts of electricity, 285 metric tons of process steam for the refinery, as well as 20 million standard cubic feet a day of hydrogen feedstock (Valenti 2000).

The Sarlux IGCC plant gasifies 150 metric tons/h of tar residue produced by vacuum visbreaking at the nearby Sarroch refinery. The visbreaking tar is pumped to three Texaco gasifier units, and the resulting raw syngas is scrubbed with water and cooled before being sent to the COS Hydrolysis Unit. The hydrolyzed syngas is cooled and fed to a Selexol Sulfur Removal Unit. Acid gas is treated in a Claus unit and the tail gas is compressed and recycled back to the Selexol unit.

The clean syngas is later sent to the hydrogen removal and recovery units. These units include a membrane section and a pressure swing absorption (PSA) section that produce pure hydrogen, available to refinery battery limits.

The syngas is saturated with water (35 to 40 percent mol) and fed to the gas turbines of the combined cycle unit, at about 20 bar and 200°C. The combined cycle unit consists of three GE STAG 109E, single-shaft combined cycle systems built by GE and its subsidiary, Nuovo Pignone of Florence. Each GE STAG (steam and gas) system consists of a GE MS9001E gas turbine, a GE 109E condensing steam turbine, a double-end generator, and a heat recovery steam generator.

The turbines are started up by distillate oil, injected with steam to control nitrogen oxide formation, and then switched over to syngas. Distillate oil also serves as the backup fuel for the Sarlux turbines.

Each Sarlux turbine produces up to 186 MW of electricity while meeting Italian emission levels of 30 parts per million for nitrogen oxides and sulfur oxides. GE adds the 40 percent moisture to the fuel to reduce NO_X formation.

By July 2001, the plant passed all the reliability and guaranteed performance tests. The average power output was 516 MWe, with 90 percent hydrogen and steam export. All the consumption figures were below the guaranteed ones; specifically, the fuel consumption was 7 percent below the designed value. All the emissions were below the maximum allowed values (Collodi 2001).

The Sarlux IGCC plant will generate about 4 billion kilowatt-hours of electricity annually that will be sold to ENEL. This energy will be distributed throughout Sardinia's electrical grid.

4 INTEGRATED COAL GASIFICATION FUEL CELL COMBINED CYCLE (IGFC)

IGFC is a triple-combined power-generation system that combines the IGCC process with fuel cells. There are two types of fuels cells: low-temperature fuel cells and high-temperature fuel cells. For the IGFC process, high-temperature fuel cells are used as they are able to achieve high-efficiency power generation. The IGFC process is the most efficient amongst all coal technologies, as it reduces carbon dioxide most effectively. In addition, this process also creates minimal environmental impact and is considered as the "ultimate" coal-based power generation technology since it is able to achieve 50 to 55 percent efficiency in power generation, along with reducing emissions effectively. The characteristics of IGFC are compared with the IGCC and conventional coal-fired power generation in Figure 11.

Besides allowing high-efficiency power generation, the IGFC is also convenient because a wide variety of coals can be used in this process. Furthermore, the

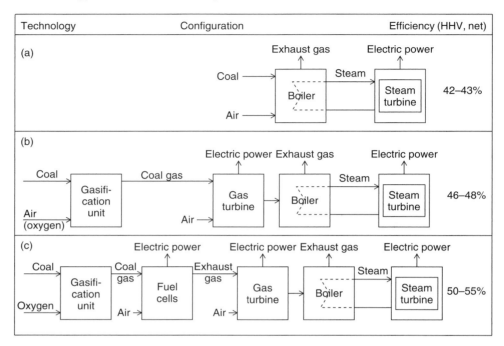

Figure 11 Comparison of characteristics of IGFC with IGCC and conventional coal-fired plant

environmental pollutants can be easily eliminated with the help of a gas cleanup unit in the downstream gasification process. The high carbon dioxide concentrations in the exhaust gas enable this process to get rid of the carbon dioxide at a relatively low cost. Near-zero emissions can be easily achieved through this process by transporting the captured carbon dioxide with the help of pipelines or tankers, as well as sequestrating and storing it underground.

5   CARBON DIOXIDE CAPTURE AND REMOVAL

Coal has traditionally been used in a significant portion of electricity generation in North America. There have been recent initiatives to reduce the levels of CO_2 emissions from power plants in the near future. Natural gas–based power plants produce less CO_2 per kW of power output compared to a coal-based power plant of the same output. This is due to two main factors:

- Natural gas has a lower carbon-to-hydrogen ratio compared to that of coal for the same level of thermal input.
- Natural gas–based systems have higher power generating efficiencies compared to coal-based systems, utilizing the same or similar power generation equipment.

Carbon dioxide removal methods vary, depending on type of power plant. In conventional gas and pulverized coal-fired units, CO_2 can be removed from the exhaust gas in an absorber/stripper system. The partial pressure of CO_2 in the exhaust gas in these units is fairly low, due to the combined effect of the low concentration of CO_2 in the gaseous stream and its near-atmospheric pressure. Low CO_2 partial pressures mean large and costly removal equipment. However, coal-based technologies such as IGCC produce more concentrated and higher-pressure CO_2 streams, which offer opportunities for low-cost CO_2 removal. High-partial pressure carbon dioxide may be removed from the syngas prior to combustion in oxygen-blown IGCC plants. This results in relatively cheaper separation due to the increased driving force. Hence, IGCC systems may be the most cost-effective coal-based power plants, if CO_2 removal is required.

5.1 Methods

In the present, there are a variety of methods for CO_2 capture. Many of them are still under development, and only a few have seen commercial application. Thus, in this paper, only commercially proven technologies will be reviewed.

Göttlicher and Pruschek (1997) have classified CO_2 removal systems into five process families:

- Process family I comprises processes with CO shift or steam reforming before CO_2 removal.
- Process family II covers processes where fuel is combusted in an atmosphere of oxygen mixed with recycled CO_2 or steam.
- Process family III includes all kind of fossil fuel–fired power plants in which CO_2 is removed from the flue gas after combustion at the exhaust end of the plant.
- Process family IV includes the Hydrocarb processes, whereby carbon is separated from the fuel prior to combustion
- Process family V deals with CO_2 separation in fuel cells suitable for the use of fossil-fuel-derived gasses

The selection of a capture method for a particular process family depends largely on a number of interacting factors, such as the total pressure of the feed gas, partial CO_2 pressure, partial pressure of the impurities, inlet feed gas temperature, utilities available, economic and energy considerations, required plant size, life, and location. Additionally, the possibility of integrating the CO_2 capture unit into the overall plant (utilizing low grade heat for regeneration or refrigeration in case of physical solvents) should be contemplated. In the final analysis, there is no substitute for commercial operating experience as the final deciding factor.

Figure 12 provides a good reference chart for the selection of a specific commercial CO_2 capture method based on the partial pressures of the undesirable

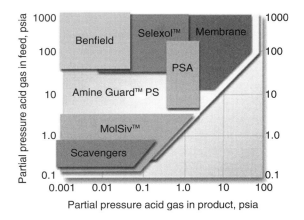

Figure 12 CO_2 removal solution selection chart (UOP 2001)

gases in the feed and product streams. Regardless of the trade names, most commercial capture methods can be grouped into three basic categories:

1. *Liquid scrubbing (absorption).* Methods are based on counter current contacting of a gas in either a packed or trayed column with a liquid solvent that absorbs the CO_2 either chemically or physically in a closed-loop system. The CO_2 is regenerated either thermally or by pressure reduction. The solvents used are of three types:

   a. *Chemical solvents.* CO_2 reacts with the solvent by forming a weakly bonded intermediate compound that can be thermally destroyed, thus regenerating the original solvent and yielding a concentrated CO_2 stream. Typical solvents are MEA, MDEA, DEA, and hot potassium carbonate.

   b. *Physical solvents.* CO_2 is physically absorbed in the solvent according to Henry's law and subsequently regenerated using either or both heat addition or pressure reduction to produce a concentrated CO_2 stream. Typical solvents are Selexol®, Purisol, and Rectisol.

   c. *Hybrid solvents.* They combine the best characteristics of both chemical and physical solvents and are usually composed of a number of complementary solvents. Unfortunately, these solvents have currently not reached a fully commercial development stage yet.

2. *Membranes.* Membranes are thin, semipermeable barriers that selectively separate some compounds from others. They are, in most instances, made of polymeric materials. In the case of carbon dioxide removal, two membrane operations seem to be relevant: gas separation and gas absorption.

3. *Others.* Adsorption processes and cryogenic technologies for CO_2 capture have been frequently mentioned in the literature (Meisen and Shuai 1997; Göttlicher and Pruschek 1997; Reimer 1993), but their large-scale

commercial application is limited and will not be discussed in detail in this chapter.

Physical Absorption

Today, the commercially proven physical solvent processes and their solvents for acid gas removal (CO_2 is considered an acid gas, as it becomes highly corrosive in the presence of water) are as follows:

- Estasolvan—tributyl phosphate or TBP
- Fluor Solvent—propylene carbonate or PC
- Purisol—normal methyl pyrrolidone or NMP
- Rectisol—methanol
- Selexol—dimethyl ether of polyethylene glycol or Selexol
- Sepasolv-MPE—mixture of polyethylene glycol dialkyl ethers or Sepasolv

Most of the equilibrium data are proprietary to the process licensors. Therefore, complete information about solvent performance cannot be published without violating existing secrecy agreements.

All of the solvents are noncorrosive and nontoxic and require only carbon steel construction for a simple cycle process scheme. They operate in a temperature range of –20°C to 30°C in most commercial operations.

According to Newman (1985, p. 131), the most widely used technology for CO_2 removal in synthesis gas applications is Selexol, while the Fluor Solvent seems to be the preferred option for natural gas applications. "Selexol has a clear experience advantage over all other solvents in all applications involving H_2S and CO_2 removal in hydrocarbon systems. Fluor Solvent and Selexol both enjoy a clear experience advantage over the other processes in applications for CO_2 removal only."

Selexol

The Selexol process was introduced over 30 years ago. As of January 2000, over 55 Selexol units had been put into commercial service. The Selexol process has been the dominant acid-gas removal system selection in gasification project awards within the past three years (UOP 2000).

The Selexol process is well suited for the selective removal of H_2S and other sulfur compounds, or for the bulk removal of CO_2. It concurrently removes COS, mercaptans, ammonia, HCN, and metal carbonyls. The Selexol process uses Union Carbide's Selexol solvent, a physical solvent made of a dimethyl ether of polyethylene glycol. Selexol is chemically inert and not subject to degradation.

A schematic of the process is shown in Figure 13.

Acid gas partial pressure is the key driving force for the Selexol process. Typical feed conditions range between 300 and 2000 psia, with acid gas composition (CO_2 + H_2S) from 5 percent to more than 60 percent by volume. The product

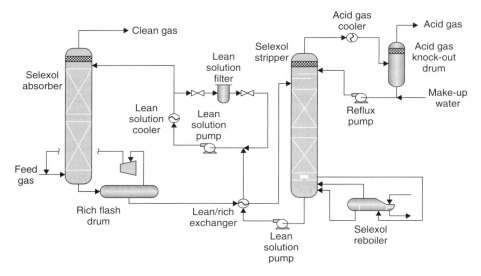

Figure 13 The Selexol process (UOP 2000)

specifications achievable depend on the application and can be anywhere from ppm up to percent levels of acid gas.

The Selexol process is primarily used in the following applications and markets:

- Selective removal of H_2S and COS in integrated gasification combined cycle (IGCC), with high CO_2 rejection to product gas (85 percent and above) and high sulfur (25 to 80 percent) feed to the Claus unit
- Selective removal of H_2S/COS plus bulk removal of CO_2 in gasification for high-purity H_2 generation for refinery or fertilizer use
- Natural gas treating to achieve either LNG or pipeline specification with dew-point reduction

There are a number of configurations available, depending on the specific application. The Selexol process with selective H_2S removal is illustrated in Figure 13. Examples of industrial plants using the process are presented in Section 5.2.

A study by Jacobs Engineering Consultancy in cooperation with General Electric and sponsored by Texaco demonstrated a 900 MW IGCC power plant with Selexol CO_2 capture designed to capture 75 percent of the feed carbon as CO_2 precombustion (Texaco P&G 2001). The plant uses Texaco quench gasifiers, followed by a sour shift system, a physical absorption acid gas removal, a sulfur recovery system, and a combined cycle consisting of two GE 9FA gas turbines, and a single gas turbine. A unique feature of this design is the possibility to operate the process in two different modes: without CO_2 capture and with CO_2 capture, with minor modifications to the plant.

An industrial application of the Selexol process that is relevant to the present work is Farmland's coke gasification to ammonia project. The plant uses Texaco quench gasifiers and a Selexol acid gas removal (AGR) unit, which together with a pressure swing absorption unit (PSA) yield a high purity hydrogen stream for ammonia production. The CO_2 capture is above 90 percent, and the carbon dioxide stream has a purity of 95 percent (Breckenridge et al. 2000).

Chemical Absorption: Chemical absorption systems are continuous scrubbing systems where CO_2 reacts with the solvent, thus forming a weakly bonded intermediate compound, which is then thermally broken down, regenerating the solvent and releasing the absorbed CO_2. Typical solvents are amine or carbonate based, such as MEA, MDEA, DEA, ammonia, and hot potassium carbonate.

These processes are recommended for low CO_2 partial pressures in the syngas stream. Additionally, SO_2, O_2, hydrocarbons, and particulates must be removed from the gaseous feed stream entering the chemical absorption unit. SO_2 and O_2 create problems because they react with the chemical solvent to form stable salts that cannot be easily regenerated. Hydrocarbons and particulates cause operating problems in the absorber. The two absorption processes most commonly applied to remove CO_2 from flue gases are the MEA (monoethanolamine) process and the activated potassium carbonate process (Riemer 1993).

MEA

Monoethanolamine (MEA) was developed about 60 years ago. MEA is the most commonly used technology for power plant flue gas scrubbing applications, and the process is well documented. A simplified reaction of CO_2 with an amine is represented by the following general equation.

$$[CO_2]_{aq} + 2[AMINE] \underset{k2}{\overset{k1}{\longleftrightarrow}} [AMINE.H]^+ + [AMINE.COO]^-$$

The forward reaction is the absorption process, while the opposite reverse reaction is the regeneration step. Under high pressure and low temperature, the forward reaction (absorption) is favored, while under low pressure and high temperature, the reverse (regeneration) reaction is favored.

The main advantages in using MEA/water solutions are its high CO_2 reactivity, its high limit load (0.5 mole of CO_2 per mole MEA), and its low molecular weight. Its main drawbacks are the stability of the carbamate ion that makes the regeneration more heat demanding and the fact that SO_2, NO_x, and O_2 have degrading effects on amine solvents. The amines form heat-stable salts with SO_2 and NO_x. The presence of oxygen causes the oxidation of the absorbent, which causes metal corrosion (Reimer 1993).

There are only two major industrial installations in which CO_2 is captured from flue gases by means of MEA scrubbing (Meisen and Shuai 1997). They are the ABB Lummus Crest (1995) designs based on aqueous monoethanolamine solutions at Trona, California, and Shady Point, Oklahoma. No IGCC power plants with MEA-based CO_2 capture have been built yet, mainly due to the fact that the MEA process is economically viable only for gas streams with a low-CO_2 partial pressure (Leci 1996), as opposed to high CO_2 partial pressures, which are the case for gasification power plants.

Membranes

Membranes have become a widely established technology for CO_2 removal since their first use in this application in 1981. UOP reported that more than 130 Separex® membrane systems had been put into operation around the world (UOP 2007). The advantages of this technology are that they are simple, steady-state operation, have no moving parts (lower maintenance and operating costs), have modular construction and are adaptable to changing feed flows and compositions. There are two main types of membranes: gas separation membranes and gas absortion membrances.

Gas Separation Membranes

These membranes are solids and operate on the principle that their porous structure permits the preferential permeation of mixture constituents. The main concerns in the design and operation of gas separation membranes are in terms of selectivity and permeability. In most petroleum, natural gas, and chemical applications, the CO_2-bearing gas mixture is fed to the membrane separator, which typically consists of a large number of hollow cylindrical membranes arranged in parallel. The CO_2 passes preferentially through the membranes and is recovered at reduced pressure on the shell side of the separator.

Gas separation membranes thus far have not been widely explored for CO_2 capture from flue gases due to the comparatively high mixture flows and the need for flue gas pressurization. The largest flow currently supported by the Separex membranes is 250 MM SCFD (UOP 1999), which is insufficient for utility companies' requirements. Feron et al. (1992) reported that the cost of a membrane system was double those of conventional amine separation processes. Riemer (1993) reports that the use of membranes in an IGCC plant design for the IEA caused an efficiency drop of 16.1 percent, while increasing the cost of electricity by a factor of 2.28 in relation to the same plant without CO_2 capture.

Gas Absorption Membranes

Gas absorption membranes consist of microporous solid membranes in contact with a liquid absorbent. The gas component to be separated must diffuse through the solid membrane, enter the liquid absorbent stream and thus be separated from the gas feed mixture. This arrangement results in independent control of gas and liquid flows and minimization of entrainment, flooding, channeling, and foaming. The equipment also tends to be more compact than conventional membrane separators.

Gas absorption membranes hold considerable promise, provided their physical and chemical stability can be assured, since they have higher selectivities and permeabilities than separation membranes. Various investigators (Meisen and Shuai 1997; Riemer 1993) have suggested that more research must be concerted on developing absorption liquids for use with gas absorption membranes that have lower energy requirements and better stability. Another interesting research area is the development of a gas absorption membrane system for H_2/CO_2 separation operating at high pressures (Riemer 1993).

Adsorption and Cryogenic

The principle behind adsorption processes is the significant intermolecular forces that exist between gases (CO_2 included) and the surfaces of certain solid materials. Some of the adsorbents that may be applicable to removal of CO_2 in power generation applications are alumina, zeolite, or activated carbon. Conditions such as temperature, partial pressures, surface forces, and adsorbent pore sizes determine if single or multiple layers of gases will be adsorbed, and if the adsorption will be selective. The adsorbents are normally arranged as packed beds of spherical particles. There are two major adsorption technologies: pressure swing adsorption (PSA) and temperature swing adsorption (TSA).

In PSA, the gas mixture flows through the packed beds at elevated pressures and low temperatures until the adsorption of the desired constituent approaches equilibrium conditions at the bed exit. The beds are then regenerated by stopping the flow of the feed mixture and reducing the pressure of the saturated bed until the trapped gases escape the bed. Technologies requiring a vacuum to regenerate are called VPSA units. The regeneration cycles are relatively short and are measured in seconds.

Thermal (or temperature) swing adsorption employs high-temperature regeneration gas to drive off trapped gases. The regeneration cycles are considerably long, and are measured in hours. The long regeneration cycle means larger quantities of adsorbent are required than with PSA systems.

The inherent advantage of adsorption processes is their relatively simple, albeit unsteady, state operation. However, adsorption is not yet a highly attractive approach for CO_2 removal in a large electrical utility scale treatment of gases because the capacity and CO_2 selectivity of available adsorbents are low. There is, however, the possibility that adsorption may become attractive when combined with another capture technology (Meisen and Shuai 1997).

Cryogenic separation of gas mixtures is done by compressing and cooling the gas mixtures in several stages to induce phase changes in CO_2 and sometimes other mixture components. CO_2 may be collected as a solid or liquid, together with other components from which it may be distilled. Water vapor in the CO_2 feed mixture leads to the formation of solids and ice, which can cause major plugging problems in the separation system.

The basic advantage of cryogenic processes is that, provided the CO_2 feed is properly conditioned, high recovery of CO_2 and other feed constituents is possible. This may also facilitate the final use or sequestering of CO_2. The largest disadvantage of cryogenic separation processes is their inherent energy intensive nature (Meisen and Shuai 1997).

5.2 Commercial Application of CO_2 Capture Processes

Three commercial units for CO_2 capture are described in this section. Two in the US and one overseas.

Selexol—Farmland Industries Coke to Ammonia Gasification Project, Kansas
Farmland Industries added a petroleum coke gasification unit adjacent to its existing refinery in Coffeyville, Kansas. The new gasification unit would produce a synthesis gas for the production of ammonia. Approximately 1,100 metric tons/day of petroleum coke from the refinery were to be gasified to produce 1,100 metric tons/day of ammonia. A portion of the ammonia is subsequently upgraded to urea-ammonium nitrate (UAN) solution. Figure 14 presents a block flow diagram of the new gasification process at Coffeyville.

Petroleum coke from the refinery coker is crushed and slurried before feeding the gasifier. The gasifiers (one operating and one spare) have been relocated from Texaco's Cool Water Demonstration facility in Daggett, California. This

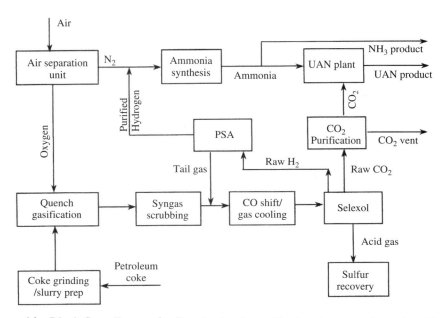

Figure 14 Block flow diagram for Farmland coke gasification to ammonia project (Breckenridge et al. 2000)

plant successfully demonstrated the Texaco gasification process by converting coal to power from 1984 to 1989. Both gasifiers use Texaco's quench gasification Technology.

The coke slurry, along with oxygen purchased across the fence from an adjacent air separation unit, are the feeds to the gasifier. The raw syngas exiting the gasifier is scrubbed of carbon fines and subsequently shifted to hydrogen. The quench section of the gasifier provides sufficient water for high conversion in the two-stage shift reactor. Hot gas from the shift reactor is cooled by generating high-pressure steam and medium-pressure steam, followed by boiler-feed water preheat and air and water cooling. The cooled syngas is fed to the Selexol unit for acid gas removal.

The Selexol process removes H_2S and most of the CO_2 in the syngas stream. The final hydrogen purification is accomplished in a downstream POLYBED™ PSA unit, which removes the remainder CO_2 along with CO, methane, part of the argon, and water. The tail gas from the PSA unit is recycled upstream of the Selexol unit to capture the hydrogen lost in the PSA unit. The purified hydrogen is combined with nitrogen from the ASU and converted to 1,100 metric tons/day of ammonia in the ammonia synthesis loop. Six hundred t/d of ammonia, along with a purified CO_2 stream from the Selexol process, are upgraded to 1,500 metric tons/day of UAN solution in the UAN plant. A unique CO_2 purification scheme was developed to remove traces of H_2S and COS from the product CO_2 feeding the UAN plant.

An optimized Selexol process configuration was developed for the Coffeyville facility, which produces an acid gas stream containing above 40 percent H_2S while using 100 percent flash solvent regeneration in the CO_2 removal section of the plant.

The syngas feed to the Selexol unit contains 50 percent hydrogen, 40 percent CO_2, 0.5 percent COS, and H_2S (mole basis), and it is saturated with water. The feed conditions are greater than 35 bar and 40°C. The hydrogen product has a purity of 99.3 percent mole and 5 ppb sulfur. The CO_2 product stream has a purity of 99 percent and 1 ppm sulfur content.

MEA—ABB Lummus Crest Power Plant, California

One of the successful integration of MEA technology for CO_2 capture is in a coal power plant in Oklahoma (Barchas 1992). The facility recovers CO_2 as a food-grade liquid with a capacity of 200 ton/day based on a conventional absorption/stripping system using a 15 to 20 percent wt MEA solution. It was jointly designed and implemented by Kerr-McGee Chemical Corp and ABB Lummus Crest Inc. The process has the flexibility to operate with both coal-fired plants (16 percent vol. CO_2) and natural gas turbines (3 percent vol. CO_2), both operating with 3 percent excess oxygen.

Sodium hydroxide is the solvent used to remove SO_2 and is recovered as a sodium sulphate salt. The amount of SO_2 is reduced from 100 ppm to less than 10 ppm. However, the removal of SO_2 levels below 100 ppm may not be necessary, because when SO_2 reacts to form MEA-sulphate, that reaction can be further converted to sodium sulphate, thus liberating MEA. Obviously, there are economic trade-offs by using this method. The study notes that below 50 ppm SO_2, removal is not economically justified. The corrosion inhibitors are not mentioned in the report specifically, as they are proprietary, but it was noted that stainless steel was used in the construction where applicable. The process recovers 90 percent of the CO_2 with a purity of 99.99 percent volume (dry basis). Future units will be modified to reduce CO_2 production costs by 10 to 30 percent. The modification entails flashing the CO_2-rich stream before it is fed into the stripper. This results in higher CO_2 concentration in the stripper feed steam and therefore less steam used in the stripping and regeneration cycle.

Membrane—Separex in Qadirpur, Pakistan

The two largest CO_2 removal membrane systems in the world are the Separex™ units installed in Qadirpur and Kadanwari, in Pakistan (Dortmund et al 1998). The Separex membrane system in Qadirpur is the largest membrane-based natural gas plant in the world. It is designed to process 265 MM SCFD of natural gas at 59 bar. The CO_2 content is reduced from 6.5 percent to less than 2 percent. The unit was designed to also provide gas dehydration to pipeline specifications. The owner has plans to expand the system to approximately 400 MM SCFD of processing capacity in the near future.

The Qadirpur membrane system was designed and constructed in two 50 percent membrane trains. Each membrane train consists of a conventional pretreatment section and a membrane section. The pre-treatment section has filter coalescers, guard vessels, and particle filters. Membrane feed heaters are also included in this design to maintain stable membrane process conditions.

6   SUMMARY

Coal as an energy source has a high environmental impact because it releases high emissions of carbon dioxide and other impurities such as sulfur into the atmosphere. The coal technologies discussed in this chapter such as the integrated gasification combined cycle (IGCC), integrated coal gasification fuel cell combined cycle (IGFC), and carbon capture and removal are effective in helping reduce emissions from coal power plants. In this chapter, an overview was provided on such technologies The processes involved in each technology were described, and the emission reduction potential were discussed. A number of process flow diagrams and statistical tables were provided to better illustrate the different processes and the future trends in the use of coal as an energy source in various countries.

REFERENCES

Barchas, R. (1992). "The Kerr–McGee/ABB Lummus Crest Technology for Recovery of CO_2 from Stack Gases," *Energy Conversion Management*, Vol. 33, pp. 333–340.

Brdar, R. D., and Jones, R. M. (2000), "GE IGCC Technology and Experience with Advanced Gas Turbines," GE Power Systems, Schenectady, New York.

Breckendridge, W., Holiday, A., Ong, J. O. Y., and Sharp, C. (2000), "Use of Selexol © Process in Coke Gasification to Ammonia Project," presented at the Laurance Reid Gas Conditioning Conference (Norman, Oklahoma), February 27–March 1.

Clean Coal Technology Program, http://www.fossil.energy.gov/programs/powersystems/cleancoal/

Collodi, G. (2001). "Commercial Operation of ISAB Energy and Sarlux IGCC Projects." Gasification Technologies, 2001, San Francisco, CA.

Dortmund, D. and Doshi, K. (1998). "Recent Developments in CO_2 Removal Membrane Technology." 14th EGPC Petroleum Conference, Cairo, Egypt, Oct. 1998.

Falsetti, James S., and Preston, W. E. (2001), "Gasification Offers Clean Power from Coal," *Power*, Vol. 145 (March/April), p. 50.

Feron, P., Jansen, A., and Klaassen., R. (1992). "Membrane Technology in Technology in Carbon Dioxide Removal," *Energy Conversion Management*, Vol. 33, p. 421.

Göttlicher, G., and Pruschek, R. (1997), "Comparison of CO_2 Removal Systems for Fossil-Fueled Power Plant Processes," *Energy Conversion and Management*, Vol. 38, No, 1, pt. Supplement, p. 173.

Granatstein, D. L., Talbot, R. E., and Anthony E. J. (2001), "Application of IGCC Technology in Canada—Phase IV," NR Canada-CANMET Energy Technology Centre. CETC01-01, June.

Joshi, Medha, Lee, M., and Sunggyu , (1996), "Integrated Gasification Combined Cycle—A Review of IGCC Technology," *Energy Sources*, Vol. 18, p 537.

Lee, Sunggyu (1996). *Alternative Fuels*. Taylor & Francis, Washington, United States.

Luthi, H. K., and Hope, T. (1989), "A Study for the Application of Shell's proven Coal Gasification Technology Integrated with the ASEA BROWN BOVERI State of the Art Gas Turbine Type BE into a 250 MW Gasification Combined Cycle (GCC) Module," *Proceedings of the Intersociety Energy Conversion Engineering Conference*. Vol. 1, p. 411.

MacRae, K. M. (1991), "COAL. New Coal Technology and Electric Power Development," Canadian Energy Research Institute, Calgary, Alberta.

Matchak, R., Rao, A. D., Ramanathan, V., and Sander, M. T. (1984), "Cost and Performance for Commercial Applications of Texaco-Based Gasification Combined-Cycle Plants". EPRI Report No. AP-3486, Prepared by FLUOR Engineers, Inc for Electric Power Research Institute, Palo Alto, California.

McDaniel, J. E. (2002), "Tampa Electric Polk Power Station Integrated Gasification Combined Cycle Project Final Technical Report," Tampa Electric Company. August, Tampa, Florida.

Meisen, A., and Shuai, X. (1997), "Research and Development Issues in CO_2 capture," Energy Conversion and Management, Vol. 38, No. 1 pt. Supplement, p. 37.

Natural Resources Canada (1997), "Canada's Energy Outlook 1996–2020," Natural Resources Canada, April.

Newman, Stephen A. (1985), "Acid and Sour Gas Treating Processes," Gulf Professional Publishing Company. Houston, Texas.

Nowacki, Perry (1981), "Coal Gasification Processes," Noyes Data Corporation, Park Ridge, NJ.

Riemer, P. (1993), "The Capture of Carbon Dioxide from Fossil Fuel Fired Power Stations," IEA Green House Gas Report IEAGHG/SR2.

Siemens-Westinghouse Power Generation (2002a), http://www.siemenswestinghouse.com/en/plantrating/index.cfm. Accessed August 2002.

Siemens-Westinghouse Power Generation (2002b), http://www.siemenswestinghouse.com/img/pool/turbine_1.jpg. Accessed August 2002.

Simbeck, Dale (2001), "Worldwide Gasification Survey," Paper presented at the 18th Annual International Pittsburgh Coal Conference (Newcastle, Australia), December 4.

Tampa Electric Company (1996), "Final Public Design Report: Technical Progress Report. July 1996," Report prepared for U.S. Department of Energy, Office of Fossil Energy. Federal Energy Technology Center. Morgantown, West Virginia.

Texaco Power and Gasification (2001), "Coal Gasification for Power Generation. A new Role for Coal in Meeting America's Clean Energy Challenge," Texaco Inc, New York.

UOP (2007), CO_2 Removal / Separex. http://www.uop.com/objects/Separex_Membrane_Sys.pdf

UOP (2000), Selexol Process. UOP Gas Processing. Illinois, United States. http://www.uop.com/gasprocessing/TechSheets/Selexol.pdf.

UOP (2001), http://www.uop.com/framesets/gas_processing.html.

USDOE/Tampa Electric Company. (2000), "Clean Coal Technology: the Tampa Electric Integrated Gasification Combined Cycle Project," Topical Report No. 19, July.

Vamvuka, D. (1999), "Gasification of Coal," *Energy Exploration and Exploitation*, Vol. 17, No. 6, p. 515.

World Nuclear Association (Feb 2008), "Clean Coal Technologies," http://www.world-nuclear.org/info/inf83.html. Retrieved Aug 20, 2008.

CHAPTER 8

AN INTEGRATED APPROACH FOR CARBON MITIGATION IN THE ELECTRIC POWER GENERATION SECTOR

Ali Elkamel, Haslenda Hashim, Peter L. Douglas, and Eric Croiset
Department of Chemical Engineering
University of Waterloo
Waterloo, Ontario Canada

1 INTRODUCTION

CO_2 is the main greenhouse gas and is suspected to be the principal gas responsible for global warming and climate change. Fossil fuel power generation plants are being challenged to comply with the Kyoto Protocol developed by the United

Nations Framework Convention on Climate Change (UNFCC). There are several possible strategies to reduce the amount of CO_2 emitted from fossil fuel power plants. Potential approaches include increasing power plant efficiency, load balancing (optimal adjustment of the operation of existing generating stations to reduce CO_2 emissions without making structural changes to the fleet), and fuel switching (switching from carbon-intensive fuel to less carbon-intensive fuels, such as from coal to natural gas). Although fuel switching may be a likely possible option, greater reduction of CO_2 from power plant flue gas in a short term is expected to be technically possible using carbon capture and storage (CCS). However, CCS is energy intensive and requires large amounts of supplemental energy. For this reason, new power plants to supply electricity to the grid based on growth rate demand, as well as to eventually supply supplemental energy for the CO_2 capture processes, are often considered. These include subcritical pulverized coal-fired (PC), PC with carbon capture (PC+CCS), integrated coal gasification combined cycles (IGCC), IGCC with carbon capture (IGCC+CCS), natural gas combined cycles (NGCC), and NGCC with carbon capture (NGCC+CCS). The overall objective is in general to determine the best strategies to reduce emissions at a certain target and to satisfy substantial growth in demand at the most economical way.

This chapter will present a systematic modeling approach that can determine the optimal structure necessary to meet a given CO_2 reduction target while maintaining or enhancing power to the grid for a fleetwide network of electricity generating stations. The approach incorporates power generation and CO_2 emissions from a fleet of generating stations using a variety of fuels and focuses on coal power plants. The approach can be applied on an existing fleet, as well as recommend new additional generating stations as well as CCS retrofit on existing generating stations to meet a specified CO_2 reduction target and electricity demand at the minimum overall cost. One can then determine the best mix of fuel, capacity for existing and new plants and technologies for capturing CO_2 as well as the location of carbon capture plants to put online.

Numerous researchers have developed energy models for assessing conventional and advanced technologies (i.e., PC, IGCC, and NGCC) and the role of carbon capture and sequestration technologies on CO_2 emissions abatement for the electric sector. Rubin et al. (2004), for instance, developed the integrated environmental control model (IECM) to provide plant level performance, emissions, and cost estimates for a variety of environmental control options for coal-fired power plants. The model is built in a modular fashion that allows new technologies to be easily incorporated into the overall framework. A user can then configure and evaluate a particular environmental control system design. Current environmental control options include a variety of conventional and advanced systems for controlling SO_2, NO_x, particulate, and mercury emissions for both new and retrofit applications. The IECM framework now is being expanded to incorporate a broader array of power-generating systems and carbon management options (multipollutant). Johnson and Keith (2004) proposed a linear programming (LP) programming model to illustrate the importance of considering competition between new plants with and without capture, and the economic plant dispatch in analyzing mitigation costs. Model results illustrate how both carbon capture and sequestration technologies and the dispatch of generating units vary with the price of carbon emissions, and thereby determine the relationship between mitigation cost and emissions reduction.

Wise and Dooley (2004) developed the battle carbon management electricity model (CMEM), a new electricity generation and dispatch optimization model to explore the effect of carbon taxes and constraints on investment and operating decisions for new generating capacity, as well as on the operation and market value of existing plants in a specific region of the United States and across three carbon dioxide scenarios. In computing the least cost decision to meet the constraints, the model considered the economic trade-offs among several factors, including capital costs for new plants, capacities of existing plants, efficiencies, operating and maintenance costs, availabilities for new and existing plants, CCS technologies for new plants and CCS retrofit for existing units, natural gas and coal prices, carbon prices resulting from the imposed CO_2-emissions constraints, and the hourly load profile of electricity demand.

Gielen and Taylor (2007) analyzed the potential role of CO_2 capture and storage for 15 world regions, using the IEA Secretariat's Energy Technology Perspectives (ETP) model. This linear programming model determined the least-cost energy system for the period 2000 to 2050. Emission reduction strategies accounted for in this model include renewable energy, energy efficiency, as well as CCS on existing and new plants.

A number of studies examined the prospects of incorporating new PC, IGCC, and NGCC with and without CCS in the electricity generation sector. Narula et al. (2002) considered replacing existing coal plants with new plants (i.e., NGCC, IGCC, and PC with and without CO_2 capture) and studied the impact of the incremental cost of CO_2 reduction on cost of electricity (COE) by implementing different technology options and by comparing (COE) without and with CO_2 removal/avoidance. Rao and Rubin (2002) conducted more advanced plant-level analyses of CCS by incorporating uncertainty and variability for about 30 independent model parameters using the *integrated environmental control model* (IECM). Such an analysis considered the following effects: energy efficiency improvement, changing fuel prices, response of electricity demand on price changes, effect of electricity demand by incorporating advanced power generation and CCS technologies, fuel switching, and plant retirement.

Vassos and Vlachou 1997 developed a methodology to investigate strategies to reduce CO_2 emissions from the electricity sector. The model enabled investigation of optimal strategies for satisfying a limit on CO_2 emissions from the electricity sector. The model can also estimate the cost of CO_2 emission limit and the optimal tax required to achieve the optimal strategy for reducing emissions to the desired level. In selecting the optimal strategy, the methodology considered commercial technologies and other new types of technology (i.e., new coal or lignite, natural gas technologies with CO_2 capture). John and Kelly (2004) reviewed the technologies that could be used to capture CO_2 from use of fossil fuels. They described three main overall methods of capturing CO_2 in power plants; postcombustion capture, oxyfuel combustion, and precombustion capture. They also discussed the impacts of different CO_2 capture technologies on the thermal efficiencies and costs of power plants based on recent studies carried out by process technology developers and plant engineering contractors.

Genchi et al. (2002) developed a prototype model for designing regional energy supply systems. Their model calculates a regional energy demand and then recommends a most effective combination of 11 different power supply systems to meet required CO_2 emission targets with minimum cost. The new energy system to be installed included co-generation systems, a photo voltaic cell system, unused energy in sewage and garbage incineration, and solar energy water supply. Linares and Romero (2002) proposed a group decision multiobjective programming model for electricity planning in Spain based on goal programming (GP). The objective was to minimize the total cost of the electricity generation, CO_2 emission, SO_2, NO_x, and radioactive waste. The model is capable of estimating the capacity to be installed for the year 2020 under four different social groups: regulators, academic, electric utilities, and environmentalists. The preferences by the groups were expressed as weights in the model that affect the different main criteria in the objective function.

Mavrotas et al. (1999) developed a mixed 0–1 Multiple Objective Linear Programming (MOLP) model and applied it to the Greek electricity generation sector for identifying the number and output of each type of power unit needed to satisfy an expected electricity demand. The first objective was to minimize the annual electricity production cost and the second objective dealt with the minimization of the total amount of SO_2 emissions. However, the model did not consider CO_2 mitigation and sequestration.

Bai and Wei (1996) developed a linear programming model to evaluate the effectiveness of possible CO_2 mitigation options for the electricity sector in Taiwan. The strategies they considered included fuel alternatives, reduced peak load, energy conservation, improving power generation efficiency, and CO_2 capture. They found that the combination of reduced peak production and increased power plant efficiency with CO_2 conservation was an effective strategy to meet significant CO_2 emission reductions.

Climaco et al. (1995) developed new techniques that incorporate multiple objective linear programming and demand-side management (DSM). These techniques are able to determine the minimum expansion cost by changing the levels and forms of the electricity use by the consumers and generating alternatives from the supply side. The model also considered the emissions caused by the electricity production.

Noonan and Giglio (1977) studied and developed an optimization program for planning investments in electricity generating systems. The optimization program determined the mix of plant types, sizes of the individual plants to be installed, and the allocation of installed capacity to minimize total discounted cost while meeting the system's forecasted demand for electricity. This problem is referred to as the *generation planning problem* (GPP). In order to comply with the variation in electricity demand, the electricity demand to be met was described by the load duration curve (LDC). The model was applied to New England Generation Planning Task Force.

Loulou and co-workers (1996, 1999a, 1999b) developed a mathematical programming model to examine least-cost strategies to reduce GHG emission using MARKAL. MARKAL is a comprehensive energy-economy optimization model. This model has been successfully applied on a multiregional basis in Canada to evaluate trading of emission permits and electricity exchanges between provinces (i.e., Ontario, Quebec, and Alberta). The MARKAL model is extended to examine the energy system in the United States and India to evaluate mitigation strategies for the industrial sector, commercial sector, residential, power plants, and transportation. The model accounted for a detailed description of a region's Reference Energy/Environmental System (RES). RES is a set of input and output of energy systems and economic parameters, such as annual fixed costs, variable costs, bounds on market shares, and so on. In this manner, all aspects of the transformation of energy to supply end-use demands through a series of technologies by representing technology components and demand activities are described with detailed information. Process technologies included refinery, power plants, and transportation of energy carriers within the region.

2   OVERVIEW OF CO_2 CAPTURE IN THE ELECTRICITY SECTOR

Electricity generation is one of the major sources of carbon dioxide emissions, particularly coal-fired power plants. Capture and sequestration significantly reduce CO_2 emissions from the power plant. Retrofitting the existing fleets of coal-fired plants with carbon capture is of interest, given the recent increase in natural gas prices. Retrofitting CCS on existing natural gas plants would also be possible. CCS can reduce CO_2 emissions by 85 to 95 percent compared to the same processes without CCS, but it is a relatively costly emission reduction strategy. Furthermore, CCS requires a large amount of energy, especially for regeneration. This reduces the overall energy efficiency of power generation typically by 10 percent, which is a large energy penalty and results in a substantial increase in the cost of power generation by 40 percent above the current levels (Costa-Pierce 1996).

There are two main options for coal-fired power plants with CO_2 capture: flue gas scrubbing with amine solvent and oxyfuel combustion. Amine scrubbing is generally considered to be the most technically proven option, but also the most expensive due to the large energy input required for solvent regeneration. Other processes like membrane separation, cryogenic fractionation, and adsorption technologies are also possible to separate the carbon from the flue gases, but "they are even less energy efficient and more expensive than chemical absorption" (Herzog et al. 1997). Typical chemical solvents are amine or carbonate based, such as monoethanolamine (MEA), diethanolamine (DEA), ammonia and potassium carbonate but MEA is the most commonly used for flue gas application. The physical absorption process typically uses a solvent such as dimethylether of polyethylene glycol (Selexol) and cold methanol (Rectisol®). Because the existing coal power plants have a low concentration of CO_2 (13 to 15 percent wet basis) in the flue gas, amine based solvents have been viewed as the potential solution to this problem (Singh et al. 2003). By contrast, Selexol is a better CO_2 capture technology for IGCC due to the high-pressure synthesis gas. Chemical absorption imposes an energy penalty of about 15 percent to 30 percent for natural gas and 30 percent to 60 percent for coal plants (Herzog et al. 1997, David and Herzog 2002). Regeneration of physical solvents is not as energy intensive as for chemical absorption, and energy penalties for IGCC plants are about 15 percent (David and Herzog 2002). Thus, the incremental cost of applying capture is lower for IGCC plants than for conventional natural gas and coal plants. The CO_2 capture system is energy intensive, energy requirements consume about 22 percent of gross plant capacity, mostly for sorbent regeneration (54 percent) and CO_2 product compression (36 percent). Sorbent circulation and fan power account for the remaining share (10 percent) of the total energy consumption of CO_2 capture unit (Rubin et al. 2004).

3 SUPERSTRUCTURE REPRESENTATION

In order to represent all mitigation options for a network of power plants in an easy to depict graphical form, a superstructure approach is adopted. Such a superstructure (Figure 1) can represent a power-generating fleet and CO_2 emission reduction techniques such as fuel balancing, fuel switching, and CCS. The following notation is adopted throughout this chapter: C, NG, O, N, H and A represent a set of coal, natural gas, oil, nuclear, hydroelectric and alternative energy power plants, respectively. C_i, NG_i, O_i, N_i, H_i and A_i represent existing coal, natural gas, oil, hydroelectric and alternative energy power plants, respectively. New power plants are represented by N_i^{new}, H_i^{new}, A_i^{new}, PC_i^{new}, NG_i^{new}, IG_i^{new}, for nuclear, hydroelectric, alternative energy, supercritical pulverized coal (PC), natural gas combined cycle (NGCC) and integrated gasification combined cycle (IGCC) power plants. $C - C_i$, $C - NG_i$, $C - O_i$, represent CO_2 capture on

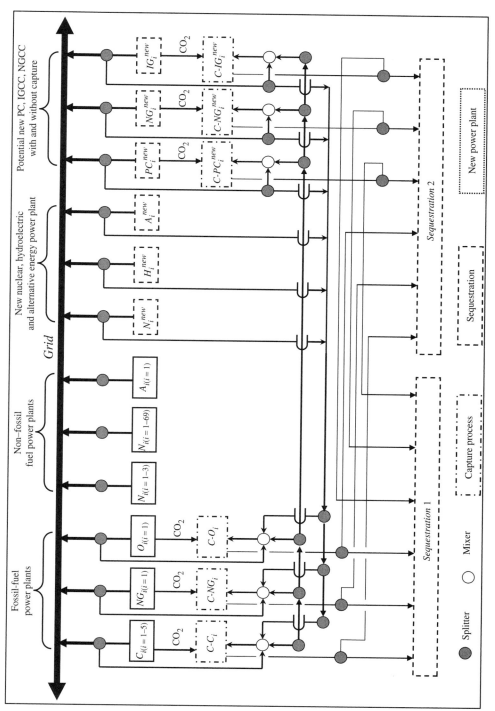

Figure 1 Generic superstructure

existing plants, whereas $C - PC_i^{new}$, $C - NG_i^{new}$, $C - IG_i^{new}$ represent CO_2 capture on new fossil plants. *Sequestration 1* and *sequestration 2* represent potential locations for CO_2 sequestration.

Figure 2 represents a representative configuration of a network of power stations. The mix of all electricity generated is injected into the electricity grid. It is assumed at this stage that there is no CO_2 capture at any existing power generating unit. Figure 3 illustrates the fuel-balancing technique to decrease CO_2

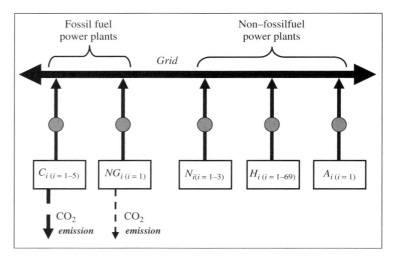

Figure 2   Current OPG fleet situation

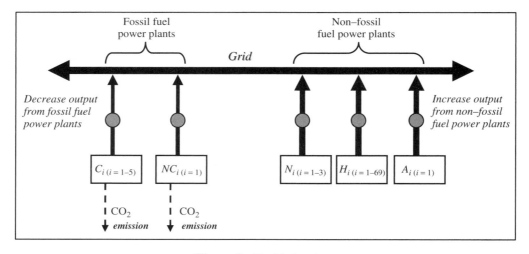

Figure 3   Fuel balancing

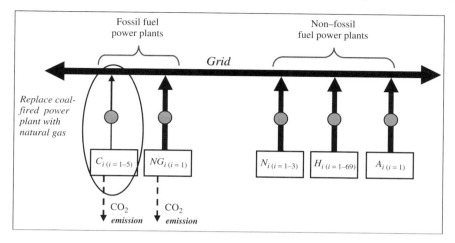

Figure 4   Fuel switching

emissions by adjusting the operation of the fleet of existing electric generating stations (e.g., increasing the load on existing non–fossil fuel power plants and decreasing the load on existing fossil fuel power plants).

In this case, it is desired to determine the optimal distribution load for all power stations to maintain electricity to the grid while reducing CO_2 emission to a certain target. Figure 4 illustrates another technique—the so-called fuel switching technique, which represents switching from carbon-intensive fuels to lower carbon intensive fuels (e.g., switch from coal to natural gas). If significant amounts of CO_2 emissions from fossil fuel power plants need to be reduced, it will be necessary to employ CO_2 capture and sequestration technologies. Figure 5 represents the superstructure for cases of CCS on existing and potential new power-generating stations. This superstructure also illustrates the supply of energy for CO_2 capture processes on given coal plants, from the plants themselves, which leads to a decrease in net plant output. Therefore, new power plants often need to be constructed to make up for this energy shortfall. Two potential locations for CO_2 sequestration are also illustrated in Figure 5.

In order to determine what is the best generating plant load distribution (existing and new), mix of fuels, and CO_2 capture processes to meet electricity growth in demand for a given CO_2 reduction target, two main groups of continuous variables need to be defined:

1. E_{ij} represents electricity generated/load distribution from the ith fossil fuel boilers using fuel j (e.g., $j = 1$ for coal and $j = 2$ for natural gas).
2. E_i represents electricity generated/load distribution from the ith existing non–fossil power plants (e.g., nuclear, hydroelectric, alternative energy

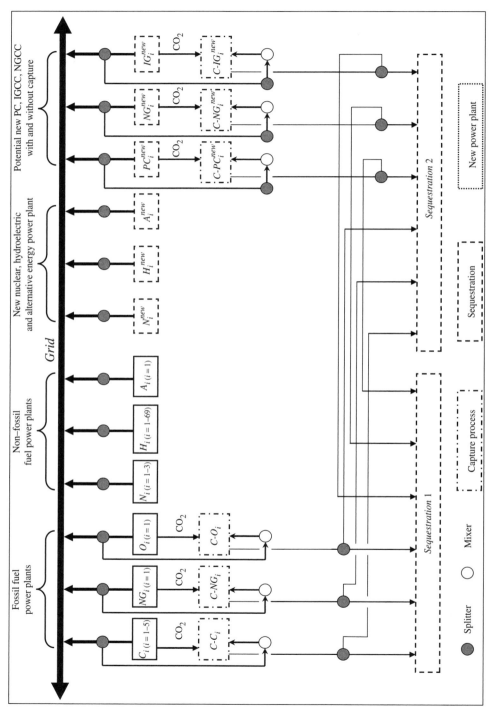

Figure 5 Superstructure including potential CO_2 capture on fossil stations

power plants) and new candidate power plants (e.g., PC, IGCC, NGCC), with and without capture. The fuel type or source of energy does need to be specified here, since the nature of the plant dictates what source of energy to be used.

Four sets of binary variables should also be defined:

1. Fuel switching (coal to natural gas) for the ith coal-fired boiler is represented by the binary variable, X_{ij}

$$X_{ij} = \begin{cases} 1 & \text{if } j\text{th fuel is used in } i\text{th coal-fired boiler} \\ 0 & \text{otherwise} \end{cases}$$

2. The existence/nonexistence of ith potential new boiler with and without capture is defined by the binary variable, y_i

$$y_i = \begin{cases} 1 & \text{if } i\text{th plant is selected} \\ 0 & \text{otherwise} \end{cases}$$

3. The binary variable, z_{ik} is introduced to model the selection of kth capture process on ith existing coal-fired boiler

$$z_{ik} = \begin{cases} 1 & \text{if } i\text{th capture process is put online} \\ 0 & \text{otherwise} \end{cases}$$

4. Selection of sth potential location for CO_2 sequestration from ith power plant is indicated by the binary variable, w_{is}.

$$w_{is} = \begin{cases} 1 & \text{if } s\text{th location of sequestration is selected} \\ 0 & \text{otherwise} \end{cases}$$

The formulation of the model is presented next.

4 COST CALCULATION

The objective is to synthesize an existing power generating fleet, incorporating carbon capture retrofit (CCS) on existing coal-fired power plants and constructing new power plants with or without capture to meet a given CO_2 reduction target while maintaining or enhancing electricity to the grid at the minimum overall cost. The cost function represents the total system cost that includes: operating cost of electricity generation for the fleet of generating stations, retrofitting cost associated with fuel switching from coal to natural gas, retrofitting cost for carbon capture retrofit on existing coal-fired power plants, and capital and operational cost for hypothetical new power plants and sequestration cost. Note that in the case of a fossil fuel already operating on gas (i.e., $j = 1$), there is no retrofitting cost involved (i.e., $R_{i1} = 0$).

The objective function is given by:

$$\text{OF} = \underbrace{\sum_{i \in F} \sum_{j} C_{ij} E_{ij} + \sum_{i \in \text{NF}} C_i^{\text{NF}} E_i}_{\text{operational cost for existing plants}} + \underbrace{\sum_{i \in F^c} \sum_{j} R_{ij} X_{ij}}_{\text{retrofit cost}} +$$

$$\underbrace{\sum_{i \in P^{\text{new}}} S_i^{\text{new}} E_i^{\text{max}} y_i + \sum_{i \in P^{\text{new}}} C_i^{\text{new}} E_i^{\text{new}}}_{\text{capital and operationl cost for new additional stations}} +$$

$$\underbrace{\sum_{i \in F^c} \sum_{k} S_{ik}^c z_{ik} + \sum_{i \in F^c} \sum_{k} C_k^c E_{ik}}_{\text{capital and operational cost of carbon capture retrofit}} + \underbrace{\sum_{i \in F^c \cup P^{new}} \sum_{s} C_{is}^{\text{seq}} \alpha_i \varepsilon_i w_{is}}_{\text{sequestration cost}}$$

(1)

where C_{ij} is the electricity generation cost per MWh if jth fuel is used in ith fossil fuel boiler; $C_i (C_i^{\text{new}})$ is the electricity generation cost per MWh for ith non–fossil fuel power plant (ith new hypothetical boiler); *C_k^c is operational cost for kth CO_2 capture process ($/MWh);* C_{is}^{seq} is sequestration cost from ith boiler to sth storage location ($/tonne CO_2 captured); $E_i (E_i^{\text{new}})$ is the electricity generated (MWh/year) from ith non–fossil fuel power plant *(ith new boilers)*; E_{ij} is the electricity generated (MWh/year) from ith fossil fuel boilers when using jth fuel*;* E_{ijk} is electricity required for kth CO_2 capture process (MWh/year) incorporated in ith coal-fired boiler that is running with the jth fuel; R_{ij} is the retrofitting cost for switching ith coal-fired boiler to jth fuel expressed in US$/year; S_i^{new} is capital cost for new power plant in $/KW; S_{ik}^c is annualized capital cost for kth capture process in $/year; ε_{ik} is the fractional use of CO_2 capture process; X_{ij} is a binary variable that indicates fuel selection or whether the plant should be shut down for ith fossil fuel boiler; i is the set of existing fossil fuel boilers, non–fossil fuel power plants and new boilers with and without capture; j is the set of jth fuel selection that consists of coal or natural gas; F is the set of fossil fuel boilers including coal (F^c) and natural gas (F^{ng}); NF is the set of non–fossil fuel power stations, including nuclear, hydroelectric, and wind turbine sources; and P^{new} is the set of hypothetical new boilers that include pulverized coal, $P^{PC}(P^{PC\text{cap}})$, integrated gas combined cycle, $P^{\text{IGCC}}(P^{\text{IGCCcap}})$ and natural gas combined cycle, $P^{\text{NGCC}}(P^{\text{NGCCcap}})$ with and (without capture). Note that in the case of non–fossil fuel plants, no associated binary variables are defined, as the fuel type for these plants is known a priori.

5   CONSTRAINTS

The minimization of the objective functions represented by equations (1) is subjected to the following constraints.

5.1 Energy Balance/Demand Satisfaction

The total electricity injected to the grid comes from existing non–fossil power, from new boilers (E_i^{NF}/E_i^{new}) and from fossil generating stations operating with jth fuel, E_{ij}. The supplemental energy required for CO_2 capture processes, however, results in an electricity reduction/power *derate*. As a generalization, the total energy required for capture process on existing boiler can be supplied from: existing non–fossil power generation, $E_i^{NF'}$, potential new boilers, $E_i^{new'}$, from fossil fuel boiler itself operating with jth fuel, E_{ij}' or energy from the grid, G_k. The net electricity generation for the whole fleet must be equal to or greater than the desired total electricity demand:

$$\left[\sum_{i \in NF} E_i^{NF} + \sum_{i \in P^{new}} E_i^{new} + \sum_{i \in F} \sum_j E_{ij} \right] - \sum_i \sum_k E_{ik} \geq \text{Demand} \quad (2)$$

5.2 Energy Balance on Capture Process

CO_2 capture processes are energy intensive, the supplemental energy required for the kth capture process, E_{ik} is supplied from ith existing non–fossil power generation, ith potential new boilers, ith fossil fuel boilers itself from the grid, G_k.

$$E_{ik} = \sum_{i \in NF} E_i^{NF'} + \sum_{i \in P^{new}} E_i^{new'} + \sum_{i \in F^c} \sum_j E_{ij}' + G_k \quad \forall i \in F^c, \forall k \quad (3)$$

In this chapter, the energy for CO_2 capture process is consider to be extracted from the steam cycle of the existing fossil power plant itself; therefore equation (3) can be simplified to:

$$E_{ik} = \sum_{i \in F^c} \sum_j E_{ij}' \quad \forall i \in F^c, \forall k \quad (4)$$

5.3 Capacity Constraint on Capture Process

$$E_{ik} \leq z_{ik} E_k^{max} \quad \forall i \in F^c, \forall k \quad (5)$$

The parameter E_k^{max} represents the maximum supplemental energy required for the kth capture technology. It also ensures that the energy required for any kth capture process is zero when no capture process is assigned to the ith coal-fired boiler. Note that capture process considered to be incorporated only on coal-fired boilers because natural gas is a less-carbon-intensive fuel and emits less CO_2 emission. z_{ik} is the binary variable that represents the selection of the kth retrofit carbon capture on the ith existing coal-fired boiler.

5.4 Fuel Selection and Plant Shutdown

For ith fossil fuel boilers, either the process is operating with one chosen fuel or is shut down. This constraint is represented by introducing the binary variable, X_{ij}, that represents the fuel selection (coal or natural gas) or plant shut down:

$$\sum_j X_{ij} \leq 1 \qquad \forall i \in F \tag{6}$$

5.5 Plant Capacity Constraints

Existing Fossil Fuel Boilers

$$E_{ij} \leq MX_{ij} \qquad \forall i \in F, \forall j \tag{7a}$$

Non–Fossil Fuel Power Plants

$$E_i \leq M \qquad \forall i \in \text{NF} \tag{7b}$$

New Power Plants

$$E_i \leq My_i \qquad \forall i \in P^{\text{new}} \tag{7c}$$

These constraints set upper bounds on energy produced from the different electricity generating stations. They also ensure that the energy production from fossil fuel plants ($i \in F$) is zero when no fuel is assigned to the plant and a decision of plant shutdown has been made. The parameter M is any large number and represents an upper bound on energy production for ith non–fossil fuel power plant/new boiler. M can be chosen to be the maximum installation capacity, E_{ij}^{max}, for fossil fuel boilers and E_i^{max} for non–fossil fuel plants and for new hypothetical boilers.

5.6 Upper Bound on Operational Changes

The electricity generated from the ith unit cannot exceed the current electricity generation for the unit by r_i (the maximum increase in the base load, E_i^{current} due to operational constraints). E_i^{max} is the maximum installed capacity of the ith potential new boiler.

Existing Fossil Fuel Boilers

$$E_{ij} \leq (1 + r_i) E_i^{\text{current}} X_{ij} \qquad \forall i \in F, \forall j \tag{8a}$$

Non–Fossil Fuel Power Plants

$$E_i^{\text{NF}} \leq (1 + r_i) E_i^{\text{current}} \qquad \forall i \in NF \tag{8b}$$

New Power Plants

$$E_i^{\text{new}} \leq E_i^{\text{max}} y_i \qquad \forall\, i \in P^{\text{new}} \tag{8c}$$

Comparing constraints (7b) and (8b), it is clear that both represent an upper bound on E_i. Because constraint (8b) is tighter, constraint (7b) is redundant and does not have to be included in the model. Constraints (8a and 8c), by contrast, include binary decision variables that are essential in the model implementation—especially in the case of plant shutdowns and to indicate existence/nonexistence of new hypothetical boilers.

5.7   Lower Bound on Operational Constraints

The annual capacity factor for each power plant must be greater than some minimum; otherwise, the plants will be shut down.

Existing Fossil Fuel Boilers

$$f_{ij} \geq l_{ij} \times X_{ij} \qquad \forall\, i \in F, \forall\, j \tag{9a}$$

Non−Fossil Fuel Power Plants

$$f_i \geq l_i \qquad \forall\, i \in \text{NF} \tag{9b}$$

New Power Plants

$$f_i \geq l_i y_i \qquad \forall\, i \in \text{NF} \tag{9c}$$

where $l_{ij}(l_i)$ is the minimum annual capacity factor for the ith fossil fuel boiler (non−fossil fuel plant and one new boiler); and f_{ij} (f_i) is the corresponding annual capacity factor. The relationship between the annual capacity factor and electricity generation is given as follows:

Existing Fossil Fuel Boilers

$$E_{ij} = f_{ij} E_{ij}^{\text{max}} \qquad \forall\, i \in F, \forall\, j \tag{10a}$$

Non−Fossil Fuel Power Plants

$$E_i = f_i E_i^{\text{max}} \qquad \forall\, i \in \text{NF} \tag{10b}$$

New Power Plants

$$E_i = f_i E_i^{\text{max}} y_i \qquad \forall i \in P^{\text{new}} \tag{10c}$$

where E_{ij}^{max} (E_i^{max}) is the installed capacity of the ith fossil (non–fossil fuel or new boiler).

5.8   Selection of CO_2 Capture Process to Be Installed in Power Plant i

The location for CO_2 capture to put online on an ith existing coal-fired boiler must be determined. This constraint only applies to the case of carbon capture retrofit on existing coal-fired power plants.

$$\sum_k z_{ik} \leq 1 \qquad \forall i \in F^c \tag{11}$$

If the existing coal-fired boilers shut down, no capture process will be put online.

$$z_{ik} \leq \sum_j X_{ij} \qquad \forall i \in F^c \tag{12}$$

This constraint is associated to constraints (7a), which indicates that the ith fossil fuel boilers will be shut down if the binary variable (fuel selection variable), X_{ij}, is equal to 0. This constraint is to ensure that no capture process will be put online if the coal-fired boilers are shut down.

5.9   Selection of New Power Plants

This chapter has considered six types of technologies to supply supplemental energy for capture processes, as well as to meet growth rate demand. These technologies are supercritical pulverized coal (PC), supercritical pulverized coal with capture (PC + CCS), Integrated gas combined cycle (IGCC), Integrated gas combined cycle with capture (IGCC + CCS), Natural gas combined cycle (NGCC) and natural gas combined cycle with capture (NGCC + CCS). Different boiler performance and sizes for each technology can be incorporated in the model.

$$\sum_i y_i \leq b \qquad \forall i \in P^{\text{new}} \tag{13}$$

where y_i represents the existence/nonexistence of the ith hypothetical new boiler for every technologies to supply supplemental energy for capture process and injected electricity to the grid, whereas b is the number of boilers of new technologies that can be incorporated.

5.10 Selection of CO_2 Sequestration Location

For Identical Coal-Fired Boilers

This constraint is to ensure that only one sequestration site will be selected for one identical coal-fired boiler. Only one sequestration location will be selected for one coal-fired stations and new power plants with capture that consist of several identical boilers:

$$\sum_s w_{is} \leq 1 \qquad \forall i \in F^c \cup P^{PCcap} \cup P^{IGcap} \cup P^{NGcap} \tag{14a}$$

$$w_{is} + w_{i's'} \leq 1 \qquad \forall i \in F^g \cup P^{new}, \forall s, i' \neq i, s' \neq s \tag{14b}$$

where w_{is} is a binary variable that indicates the selection of an sth potential location for CO_2 sequestration and F^g represents coal-fired power plants that consist of several boilers.

5.11 CO_2 Sequestration Must Be Determined Once Capture Process Is Put Online

Once carbon capture retrofit is implemented on an ith existing coal-fired boiler or on a new boiler, an sth potential location for CO_2 sequestration is determined in order to store the CO_2 captured securely and permanently.

$$\sum_s w_{is} = \sum_k z_{ik} \qquad \forall i \in F^c \cup P^{PCcap} \cup P^{IGCCcap} \cup P^{NGCCcap} \tag{15}$$

5.12 Emission Constraint/CO_2 Balance

CO_2 emissions from all existing coal-fired boilers and new potential boilers, α_i (million tonne/yr) are defined as follows:

Existing Fossil Fuel Boilers

$$\alpha_i = \sum_j CO_{2ij} \, E_{ij} \qquad \forall i \in F \tag{16a}$$

New Power Plants

$$\alpha_i = CO_{2i} E_i \qquad \forall i \in P^{new} \tag{16b}$$

where CO_{2ij} is the CO_2 emission from the ith existing fossil fuel boiler using the jth fuel per electricity generated and CO_{2i} is CO_2 emission from new boilers (tonne CO_2/MWh). CO_{2ij} and CO_{2i} are calculated using basic chemical equations that relate the production of CO_2 emission to the quantity and quality of fuel burned.

Existing Fossil Fuel Boilers

$$CO_{2ij} = 0.03667(EF)_{ij} \qquad i \in F \qquad (17a)$$

New Power Plants

$$CO_{2i} = 0.03667(EF)_i \qquad \forall i \in P^{new} \qquad (17b)$$

where 0.03667 is the conversion factor from coal to CO_2, EF_{ij} is the CO_2 emission factor of the ith fossil fuel station using jth fuel, and EF_i is the CO_2 emission factor of the ith potential new boilers as represented next:

Existing Fossil Fuel Boilers

$$EF_{ij} = \frac{1}{\eta_{ij}} \left(\frac{\%C}{HHV_{ij}} \right) \qquad \forall i \in F, \forall j \qquad (18a)$$

New Power Plants

$$EF_i = \frac{1 - \varepsilon_{ik}}{\eta_i} \left(\frac{\%C}{HHV_{ij}} \right) \qquad \forall i \in P^{new} \qquad (18b)$$

where η_{ij} is the efficiency of the ith fossil fuel boiler while operating on jth fuel and η_i is the efficiency of potential new boilers. $\%C$ represents the percentage of carbon content and HHV_{ij} is the fuel higher heating value.

In constraints (16), CO_2 emissions from fossil power plants, α_i will be captured in kth capture process also can be defined as $\alpha_i^{seq} + \alpha_i^{released}$, where α_i^{seq} is CO_2 captured by kth capture process and $\alpha_i^{released}$ is CO_2 emitted to the atmosphere. For the case of CO_2 capture with kth capture process, $z_{ik} = 1$ and ε_{ik} fraction of CO_2 captured,

$$\alpha_i^{released} = \alpha_i \left(1 - \sum_k \varepsilon_{ik} z_{ik} \right) \qquad \forall i \in F^c \qquad (19a)$$

$$\alpha_i^{seq} = \alpha_i \sum_k \varepsilon_{ik} z_{ik} \qquad \forall i \in F^c \qquad (19b)$$

If no CO_2 capture exists, $z_{ik} = 0$, then all CO_2 emitted from the ith fossil fuel boilers, α_i, will be released to the atmosphere:

$$\alpha_i^{released} = \alpha_i \qquad \forall i \in F^c \qquad (20a)$$

$$\alpha_i^{seq} = 0 \qquad \forall i \in F^c \qquad (20b)$$

Note that constraints (19a, 19b) and (20a, 20b) only apply to existing coal-fired boilers. This case is also called *carbon capture retrofit*, and the binary variable, z_{ik}, indicates the existence/nonexistence of CO_2 capture process on ith existing coal-fired boilers. Besides CO_2 emissions from existing fossil fuel boilers, the new boilers (e.g., PC, IGCC, and NGCC) with and without capture also contribute to the total CO_2 emissions. Thus, the total CO_2 emission (million tonne per year) from existing fossil fuel boilers and new hypothetical boilers can be written as follows:

$$\sum \alpha_i^{\text{released}} = \sum_{i \in F} \alpha_i \left(1 - \sum_k \varepsilon_{ik} z_{ik} \right) + \sum_{i \in P^{\text{new}}} CO_{2_i} E_i \qquad (21)$$

Substitution of equation (16a) into equation (21) results in:

$$\sum \alpha_i^{\text{released}} = \sum_{i \in F} \left[\left(\sum_j CO_{2_{ij}} E_{ij} \right) \left(1 - \sum_k \varepsilon_{ik} z_{ik} \right) \right] + \sum_{i \in P^{\text{new}}} CO_{2_i} E_i \quad (22)$$

Finally, the annual total CO_2 emissions from all existing fossil fuel boilers and potential new boilers must satisfy a specific CO_2 reduction target, $\%CO_2$.

$$\sum_{i \in F} \left[\left(\sum_j CO_{2_{ij}} E_{ij} \right) \left(1 - \sum_k \varepsilon_{ik} z_{ik} \right) \right] + \sum_{i \in P^{\text{new}}} CO_{2_i} E_i \leq (1 - \%CO_2) CO_2$$

$$(23)$$

5.13   Logical Constraints

Other constraints can be imposed on the model. These include:

- Geographical/management constraint—certain power plants are assigned to supply certain locations, while other plants cannot supply some specific locations.
- Fuel or resource constraint—the fuel supply (i.e., natural gas) is limited by pipeline capacity.

6   ILLUSTRATIVE CASE STUDY

In this section, the integrated approach described earlier is applied to the case of the Ontario Power Generation company (OPG). This company operates 79 electricity generating stations, 5 coal fired, C ($i = 1 - 5$), 1 natural gas, NG($i = 6$), 3 nuclear, N ($i = 7 - 9$), 69 hydroelectric, H ($i = 10 - 78$), and 1 wind turbine, A ($i = 79$). At nominal levels, OPG generates about 115.8 TWh and injects it into the grid. No CO_2 capture process currently exists at any OPG power plant; about 36.7 million tonnes of CO_2 were emitted in 2002, mainly from fossil fuel power plants. There are 27 fossil fuel boilers at the six fossil fuel stations: 4 boilers at Lambton (L1–L4), 8 boilers at Nanticoke (N1–N8), 8 boilers at Lakeview (LV1–LV8), 1 boiler at Atikokan (A1), 4 boilers at Lennox (L1–L4) and Thunder Bay has 2 boilers (TB1–TB2). Currently, 4 (out of 27) boilers operated by Lennox are running on natural gas and the other 4 boilers are running

Table 1 Ontario Power Generation fossil fuel power stations

Station	Fuel	Heat rate (GJ/MWh)	Installed Capacity (MW)	Number of units	Annual capacity factor	Operational cost ($/MWh)	CO_2 emission (tonne/MWh)
Nanticoke 1 (N1)	Coal	9.88	500	2	0.75	20	0.93
Nanticoke2 (N2)	Coal	9.88	500	6	0.61	20	0.93
Lambton1 (L1)	Coal	9.84	500	2	0.5	22	0.94
Lambton2 (L2)	Coal	9.84	500	2	0.75	17	0.94
Lakeview (LV)	Coal	10.8	142	8	0.25	23	0.98
Lennox (LN)	NG	7.82	535	4	0.15	47	0.65
Thunder Bay (TB)	Coal	11.7	155	2	0.55	20	1.03
Atikokan (A)	Coal	9.82	215	1	0.44	20	1.03

on coal. Table 1 summarizes OPG's current fossil fuel generating stations. The operating costs for nuclear, hydroelectric, and wind turbine were estimated to be US$21/MWh, US$3.30/MWh, and US$2.70/MWh, respectively. Note that, currently, natural gas is the most expensive fuel used by OPG, which is US$47/MWh (OPG 2002). Since the main objective here is to study CO_2 emission reduction through fuel balancing, fuel switching, and CO_2 capture, no attempt is made to study the effect of improved technology. An improvement in boiler technology will, in principle, lead to an efficiency higher than the assumed efficiency of 35 percent. The index i ($i = 1 - 79$) represents all OPG's power plants. The index j ($j = 1 - 2$) represents the fuel selection, $j = 1$(coal), 2(natural gas). The retrofit cost was estimated to be US$20 million/1000 MW with a 30-year lifetime and 15 percent annual interest rate. In order to translate CO_2 emissions into cost, $\beta = 0.03$ US$/Kg CO_2 emission was assumed from the literature (ESCAP-UN 1995). The reserve margin, r_i for load distribution for all OPG's fleet power plants is set in this case study to be 1 percent higher than the current operational level due to the design constraints and the lower bound was set to be 10 percent. In other words, the plants have to operate at least with 10 percent annual capacity factor (ACF); otherwise, the plants will be shut down.

The current estimated cost of retrofitting an existing pulverized coal plants with MEA capture is about $190/tC (Simbeck 2002). David and Herzog (2002) have proposed an effective way to compare the mitigation costs of different plants. The mitigation cost in $/tonne CO_2 avoided with the net power generated by the two plants remaining the same output can be calculated as follows:

$$\text{Cost of } CO_2 \text{ avoided (\$/tonne } CO_2 \text{ avoided)} = \frac{COE^{cap} - COE^{ref}}{CO_2^{cap} - CO_2^{ref}} \quad (24)$$

where:
COE^{cap} = Cost of electricity for plant with carbon capture ($/MWh)
COE^{ref} = Cost of electricity for reference plant ($/MWh)
CO_2^{cap} = CO_2 emission from plant with carbon capture (tonne/MWh)
CO_2^{ref} = CO_2 emission from reference plant (tonne/MWh)

COE consists of three main components; COE due to capital cost, COE due to operation and maintenance (O&M), which includes variable cost, VOM (e.g., fuel, chemicals, utilities, waste disposal) and fixed cost, FOM (e.g., rental, interest), and COE due to capture cost. The latter component (COE^{cap}) is applicable only to carbon capture retrofit on existing coal-fired stations and potential new power plants with capture. In this chapter, capital costs of all existing fossil fuel power stations were assumed to be paid off; hence, COE for an existing fossil fuel power plants (COE^{ref}) is then determined only by O&M cost of the base plant. Note that VOM cost is dominated by fuel prices.

According to Rubin et al. (2004), the energy penalty associated with CO_2 capture process can be calculated using the following equation:

$$EP = 1 - \frac{\eta_{CCS}}{\eta_{ref}} \tag{25}$$

where EP is the energy penalty (fraction reduction in output), η_{CCS} and η_{ref} are the net efficiencies of the capture plant and reference plant, respectively. This energy penalty also can be defined as the reduction in plant output for a constant fuel input and is called *plant derating*. For the case of CCS retrofit on existing coal-fired power plant, the energy penalty is assumed to be 22 percent, as suggested by Rubin et al. (2004). The details of existing coal-fired stations for the case study with and without capture characteristics are shown in Table 2. *Ref* in the table represents existing coal-fired power plant characteristics, whereas *Cap* represents the plant performance once CCS is incorporated. Cost estimation of retrofit carbon capture on existing coal-fired power plant is also estimated based on the calculation performed by Rubin et al. (2004) and is presented in Table 3.

New state-of-the-art PC, IGCC, and NGCC with and without capture cost estimation for different sizes required a standard methodology and the same economic assumption. In this chapter, the performances of power plants were

Table 2   Parameters of Retrofit CO_2 Capture on Existing Coal-Fired Power Plants

Parameters	N1		N2		L1		L2		LV		TB		A	
	ref	cap	ref	cap	ref	cap	ref	cap	ref	cap	ref	cap	ref	cap
Gross capacity (MW)	512	512	512	512	515	515	515	515	150	150	163	163	230	230
Net power gen. (MW)	490	382	490	382	493	385	493	385	142	111	155	120	215	168
Heat rate, HHV (GJ/MWh)	9.88	12.7	9.88	12.7	9.84	12.6	9.84	13.1	10.8	13.8	9.82	12.5	9.82	12.5
Capacity factor, ACF	0.75	0.75	0.61	0.61	0.5	0.5	0.75	0.75	0.25	0.25	0.55	0.55	0.44	0.44
CO_2 capture (%)	—	90	—	90	—	90	—	90	—	90	—	90	—	90
CO_2 emission (tonne/MWh)	0.93	0.09	0.93	0.09	0.94	0.09	0.94	0.09	0.98	0.09	1.02	0.11	1.02	0.11

Table 3 Cost Comparison of Retrofit CO_2 Capture on Existing Coal-Fired Power Plants

Parameters	N1		N2		L1		L2		LV		TB		A	
	Ref	Cap	Ref	Cap	Ref	Cap	Ref	Cap	Ref	Cap	Ref	Cap	Ref	Cap
CO_2 capital cost (M\$)[a]	—	236	—	236	—	238	—	270	—	290	—	86	—	121
O&M cost (\$/MWh)[a]	20	13.9	20	13.9	22	13.9	17	13.9	23	13.9	20.0	13.9	20.0	13.9
COE (¢/KWh)[a]	2.0	4.01	2.0	4.27	2.2	4.55	1.7	4.16	2.3	8.0	2.0	5.11	2.0	5.75
Capture cost (\$/tonne CO_2 avoided)[a]	—	24.0	—	27.1	—	26.6	—	27.1	—	64.5	—	33.8	—	40.7

obtained from the literature (Rubin et al. 2004; Ordorica et al. 2004; and McDaniel 2002). Data for new supercritical PC power plants without capture at two different capacities (500 MW, 575 MW) and with capture at three different capacities (500 MW, 670 MW, 710 MW with 90 percent CO_2 capture) were obtained from Rubin et al. (2004), reported in 2001\$US. Plant performance and cost (mid-2001 US\$) for a 250 MW IGCC were gathered from real plant data (McDaniel 2002). Two different plant designs and cost estimation of IGCC with capture (500 MW with 80 percent CO_2 capture and 513 MW IGCC with 60 percent CO_2 capture) were based on a study by Ordorica et al. (2004). Plant performance for NGCC without capture of two different plant sizes (326 MW, 395 MW) were obtained from Parson (2000) and for 517 MW from Rubin et al. (2004). Plant characteristics for two different sizes of NGCC with capture (517 MW and 750 MW with 90 percent CO_2 capture) were obtained from Rubin (2004). Finally, the cost of all proposed new power plants should be adjusted to the same economic assumption (e.g., 2004 US\$ for this case study) using *Chemical Engineering Plant Cost Indexes* with coal price of US\$1.2/GJ and natural gas price of US\$4/GJ. Tables 4–9 summarize the plant performance and cost comparison for all new power plants with and without capture.

For CO_2 capture on existing coal-fired boilers, absorption with MEA is the only option considered in this case study. For new power plants with capture, two possible CO_2 recovery techniques are considered: MEA for PC and NGCC. Selexol-based CO_2 capture was chosen for IGCC with capture. According to Herzog (2002), NGCC have the highest energy requirement for CO_2 capture process (0.354 KWh/kg of CO_2 processed) due to the low content of CO_2 in the flue gas compared to PC plants (0.317 KWh/kg of CO_2 processed). The model in this case study excludes the addition of new nuclear, renewable energy, and hydroelectric capacity.

Table 4 Plant Performance for PC with and without Capture

Plant Performance	PC without Capture		PC with Capture		
	PC1[a]	PC2[b]	PC+CCS1[a]	PC+CCS2[a]	PC+CCS3[b]
Gross plant size (MW)	500	575	500	670	710
Net power generation (MW)	458	524	341	458	492
Heat rate, HHV(GJ/MWh)	9.12	9.16	12.19	12.17	12.1
Plant efficiency, HHV	39.4	39.3	29.6	29.5	29.9
Capacity factor, ACF	75	75	75	75	75
CO_2 emission (tonne/MWh)	0.919	0.89	0.122	0.122	0.118
CO_2 capture (%)	—	—	90	90	90
CCS energy penalty (%)	—	—	24.9	24.9	23.9

[a]PC1 is a supercritical boiler equipped with SCR, ESP, and FGD. PC+CCS1 is a plant that used the same coal input as PC1 results to lower net power generation due to supplemental energy for MEA capture process. PC+CCS is a bigger plant size constructed so that the CO_2 capture plant produces the same net power generated as PC1 (Rubin 2002).
[b]PC2 and PC+CCS3 are a supercritical boiler equipped with SCR, ESP and FGD followed by MEA system for CO_2 capture respectively (Rubin 2004).

Table 5 Economic Evaluation of PC with and without Capture

Cost Parameters	PC without Capture		PC with Capture		
	PC1[b]	PC2[c]	PC+CCS1[b]	PC+CCS1[b]	PC+CCS2[c]
Plant capital cost ($/KW)[a]	1588	1413	2613	2468	2271
Cost of electricity ($/MWh)[a]					
Capital	36.0	32.3	59.7	56.3	51.8
O&M	2.5	2.47	18.0	18.1	18.2
Fuel	10.9	11.0	14.6	14.6	14.5
Capture cost ($/tonne CO_2 avoided)[a]	—	—	59.3	55.3	50.2

[a]Cost is adjusted according to standard methodology and reported in 2004 US$ using Chemical Engineering Plant Cost Indexes (Chemical Engineering Magazine 1990–2004) with coal price is US$1.2/GJ.
[b]PC1 is the plant reference for PC+CCS1 and PC+CCS2 (Rubin 2002).
[c]PC2 is the plant reference for PC+CCS3 (Rubin 2004).

7 EFFECT OF DIFFERENT MITIGATION STRATEGIES

This section will discussed two main results related to the case study of the previous section: (1) the effect of cost of electricity (COE) by adjusting the CO_2 emission reduction by fuel balancing, fuel switching, or capturing and (2) the effect of CO_2 reduction on the distribution of electricity generation for existing

Table 6   Plant Performance for IGCC with and without Capture

Plant Performance	IGCC without Capture			IGCC with Capture	
	IGCC1[a]	IGCC2[b]	IGCC3[c]	IG+CCS1[b]	IG+CCS2[b]
Gross plant size (MW)	315	400	675	650	653
Net power generation (MW)	250	350	583	489	513
Heat rate, HHV(GJ/MWh)	7.37	7.9	8.78	10.46	9.97
Plant efficiency, HHV	48.9	45.4	41	34.4	36.1
Capacity factor, ACF	85	85	85	85	85
CO_2 emission (tonne/MWh)	0.7	0.75	0.81	0.176	0.39
CO_2 capture (%)	—	—	—	80	60
CCS energy penalty (%)	—	—	—	16.1	12.0

[a]Based on real plant cost from Tampa Electric, used single train includes one GE 7FA gas turbine (McDaniel 2002).
[b]The gasification plant consists of two gasifier trains. This includes two GE 7FA gas turbines, two HRSGs and one steam turbine. IG+CCS1 represented IGCC with 80 percent CO_2 capture, whereas IG+CCS2 represented IGCC with 60 percent CO_2 capture (CO_2 emission equal to that of a NGCC). Both used Selexol-based CO_2 capture process (Ordorica et al. 2004).
[c]The same plant configuration as IGCC2, except used Westinghouse 501G gas turbine (Parsons 2000).

Table 7   Economic Evaluation of IGCC with and without Capture

Cost Parameters	IGCC without Capture			IGCC with Capture	
	IGCC1	IGCC2	IGCC3[b]	IG+CCS1[b]	IG+CCS2[b]
Plant capital cost ($/KW)[a]	2535	2121	1860	2556	2336
Cost of electricity ($/MWh)[a]					
Capital	51.1	42.7	37.5	43.8	40.0
O&M	8.61	6.91	5.9	7.6	7.1
Fuel	8.84	9.5	10.5	12.6	11.9
Capture cost ($/tonne CO_2 avoided)[a]	—	—	—	15.8	12.3

[a]Cost is adjusted according to standard methodology and reported in 2004 US$ using Chemical Engineering Plant Cost Indexes (Chemical Engineering Magazine, 1990–2004) with coal price is US$1.2/GJ
[b]IGCC3 is the plant reference for IG+CCS1 and IG+CCS2 (Ordorica et al. 2004).

and new plants for base load demand, low-growth demand (1 percent growth rate), medium-growth demand (5 percent growth rate), and high-growth demand (10 percent growth rate).

7.1   Effect of CO_2 Reduction on Cost of Electricity

Figure 6 shows the effect of CO_2 reduction on the cost of electricity. As can be seen, increasing the CO_2 emission reduction would result in an increase in the cost of generating electricity. The OPG electricity cost for a 0 percent CO_2 reduction

Table 8 Plant Performance for NGCC with and without Capture

Plant Performance	NGCC without Capture			NGCC with Capture	
	NGCC1[a]	NGCC2[a]	NGCC3[b]	NG+CCS1[b]	NG+CCS2[c]
Gross plant size (MW)	334	403	517	517	
Net power generation (MW)	326.1	395	507	432	750
Heat rate, HHV(GJ/MWh)	6.32	6.74	6.37	7.48	7.8
Plant efficiency, HHV	50.6	53.4	50.2	42.8	41.0
Capacity factor, ACF	85	85	75	75	85
CO_2 emission (tonne/MWh)	0.398	0.377	0.404	0.047	0.037
CO_2 capture (%)	—	—	—	90	90
CCS energy penalty (%)	—	—	—	14.7	18.3

[a]Plant performance is gathered from Parson (2000). The plant configuration is the same as NGCC3 except NGCC1 used Westinghouse 501G gas turbine and NGCC2 used General Electric "H" gas turbine.
[b]NGCC3 equipped with two GE-PG-7241FA gas turbines, three HRSG, and feed a single two-flow, reheat condensing GE D-11 steam turbine. NG+CCS1 equipped by amine system for CO_2 capture (Rubin 2004).
[c]NG+CCS2 used two GE 9FA gas turbine, three HRSG and one steam turbine.

Table 9 Economic Evaluation of NGCC with and without Capture

Cost Parameters	NGCC without Capture			NGCC with Capture	
	NGCC1	NGCC2	NGCC3[b]	NG+CCS1[c]	NG+CCS2[c]
Plant capital cost ($/KW)[a]	617	552	442	1437	1207
Cost of electricity ($/MWh)[a]					
Capital	12.4	11.1	10.1	24.6	20.7
O&M	8.1	9.37	9.3	9.7	5.3
Fuel	28.4	26.9	25.5	29.9	31.2
Capture cost ($/tonne CO_2 avoided)[a]	—	—	—	57.1	37.5

[a]Cost is adjusted according to standard methodology and reported in 2004 US$ using Chemical Engineering Plant Cost Indexes (*Chemical Engineering Magazine*, 1990–2004) with natural gas price US$4/GJ. Gas turbine cost is based on vendor quote (*World Turbine Magazine* 2004).
[b]NGCC3 is the plant reference for NG+CCS1 and NG+CCS2.
[c]The only additional equipment requirement compare than NGCC3 is amine system that cost about $13.1/MWh for NG+CCS1 (Rubin 2004). Thus, amine system cost for NG+CCS2 is estimated using the 0.6 power law suggested by Peters et al. (2003). Thus, amine cost for 750 MW NGCC+CCS was $13.1/MWh $(750/432)^{0.6}$.

is 2.36¢/KWh. The optimization results show that fuel balancing can contribute to the reduction of CO_2 emissions by only 3 percent. Fuel balancing also results in a reduction of the cost of electricity to 2.32¢/KWh by reducing electricity generation from all four natural gas boilers by 32.1 percent and two coal-fired boilers by 33.4 percent and 59.4 percent, respectively. The electricity generation

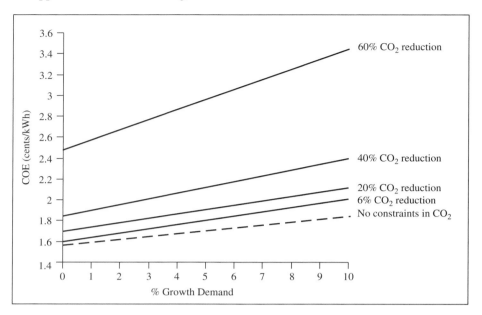

Figure 6   Effect of CO_2 reduction on the cost of electricity

from the other fossil fuel boilers and non–fossil fuel power plants were increased by 1 percent above the nominal operational level to maintain the electricity to the grid. However, if CO_2 emissions are to be reduced beyond 3 percent, more stringent measures that include fuel switching, plant retrofitting, and CO_2 capture have to be employed. The optimization results show that electricity generation from one natural gas boilers (LN4) must be reduced by 32.1 percent, one natural gas boiler (LN2) reduced by 4.1 percent and eight coal-fired boilers (L1, L2, L3, L4, N7, N8, LV3, TB1) need to be switched to natural gas, resulting in an increase of the cost of electricity by about 7.1 percent. Finally, the other coal-fired boilers and non–fossil fuel power plants increase the electricity generation by 1 percent higher than the nominal operational level to meet the electricity demand. For the case of 60 percent CO_2 reduction, the cost of electricity is increased by 28.3 percent, since four new natural gas boilers with capture are put online to compensate for the reduction in electricity output due to shutdown of all eight Lakeview boilers, two Lambton boilers (L1, L2), and two Nanticoke boilers (N7, N8), as well as to achieve greater CO_2 reductions.

7.2   Effect of CO_2 Reduction on Electricity Demand Distribution

Figures 7 to 12 illustrate the electricity distribution for base case, 3 percent, 6 percent, 20 percent, 40 percent, and 65 percent CO_2 reduction target in order to meet electricity demand. The base case represents the current operational level. As can be seen, increasing CO_2 reduction target leads to the switching of more

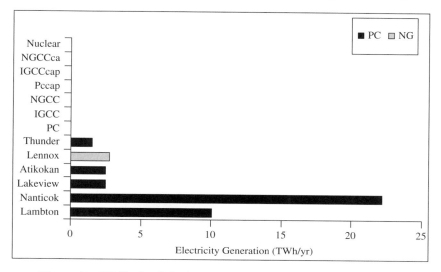

Figure 7 OPG's fossil-fuel electricity generation by plants in 2002

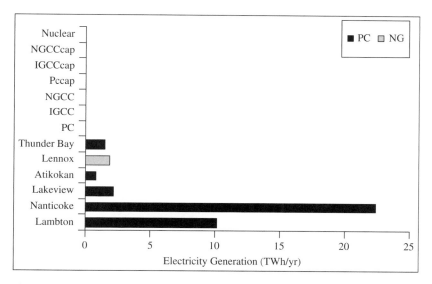

Figure 8 Optimal electricity generation for fossil fuel plants and 3 percent CO_2 reduction

coal-fired boilers to natural gas. New power plants with and without capture must also be added to the resource mix in order to meet the electricity demand requirement. Among the options, NGCC seems to be the favorable one.

Figures 13 to 15 show the electricity distribution for the base case, 1 percent, 5 percent, and 10 percent electricity demand growth rate for 20 percent, 40 percent, and 65 percent CO_2 reduction. New plants with and without CCS compete

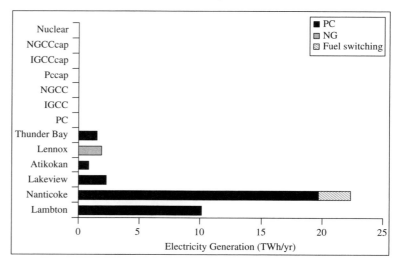

Figure 9 Optimal electricity generation for fossil fuel plants and 6 percent CO_2 reduction

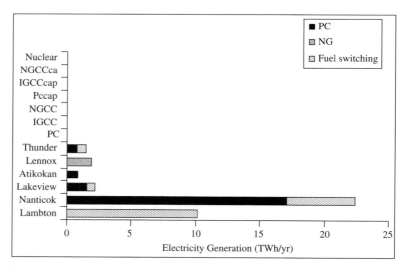

Figure 10 Optimal electricity generation for fossil fuel plants and 20 percent CO_2 reduction

with existing plants that have been paid off but remain competitive due to lower overall cost. Figure 13 shows that a 20 percent CO_2 reduction, while at the same time meeting current electricity demand, could be achieved by implementing fuel balancing and switching coal-fired boilers to natural gas. However, as demand increases (> 5.0 percent growth rate demand), the company has to switch more coal-fired boilers to natural gas and also built new NGCC and NGCC+CCS for replacing aging coal-fired power plants and increase the electricity generation of

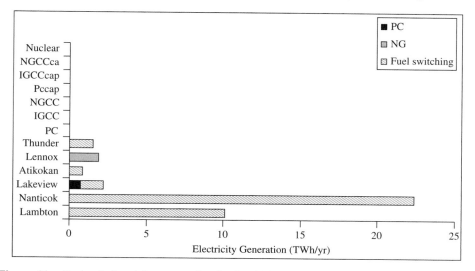

Figure 11   Optimal electricity generation for fossil fuel plants and 40 percent CO_2 reduction

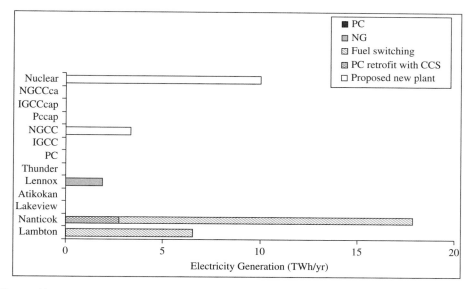

Figure 12   Optimal electricity generation for fossil fuel plants and 60 percent CO_2 reduction

non–fossil power plants by 1 percent higher than the nominal operational level to meet the electricity demand.

For the case of 40 percent CO_2 reduction, the company must switch most of the coal-fired boilers to natural gas, replacing, for example, all eight Lakeview boilers with a new NGCC and NGCC+CCS. In order to achieve higher CO_2 reductions at higher growth rates (e.g., 40 percent CO_2 reduction at 1.0 percent

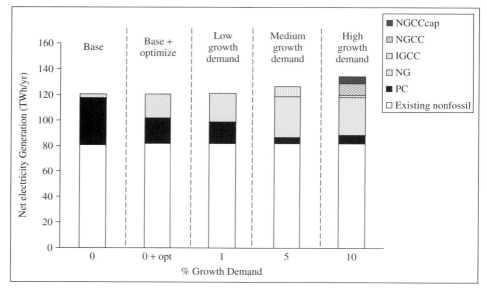

Figure 13 Effect of electricity demand growth on distribution of electricity generation for 20 percent CO_2 reduction

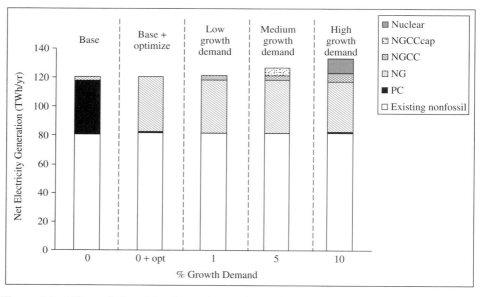

Figure 14 Effect of electricity demand growth on distribution of electricity generation for 40 percent CO_2 reduction

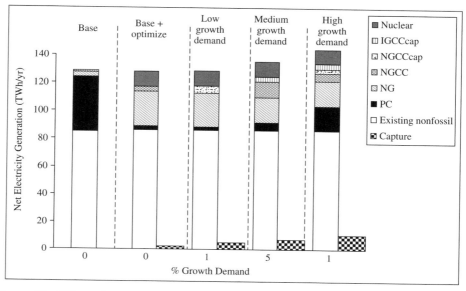

Figure 15 Effect of electricity demand growth on distribution of electricity generation for 60 percent CO_2 reduction

growth rate demand), as illustrated in Figure 14, four new plants (2 NGCC, 2 NGCC+CCS) must be built, and carbon capture retrofit on the Nanticoke (N1) coal-fired power plant (which is the largest coal-fired boiler in fleet and hence the largest CO_2 emitter) must be implemented. The electricity generation of other coal-fired power plants and non–fossil power plants will also need to increase by 1 percent higher than the nominal operational level to meet the growth in electricity demand.

If greater than 65 percent CO_2 reduction is required, new plants and carbon capture retrofits on existing coal-fired boiler must be considered. Among the attractive options are NGCC, NGCC+CCS, and IGCC+CCS, as illustrated in Figure 15. Finally, the other coal–fired boilers and non–fossil fuel power plants increase the electricity generation by 1 percent higher than the nominal operational level to meet the growth in electricity demand.

7.3 Effect of Natural Gas Price on COE and Mitigation Strategy

Future natural gas prices are highly uncertain. According to the New York Mercantile Exchange (NYMEX 2005), the average annual natural gas price for 2005 was predicted to be US$5.7/GJ. In comparison, natural gas prices are assumed to increase by 50 percent and 100 percent from the baseline price at US$4/G for base load demand, 5 percent and 10 percent growth in demand. The volatility of natural gas prices provide a significant impact on CO_2 mitigation strategies, as well as on the COE as illustrated in Figures 16 to 18. Increasing NG prices result in early aggressive CO_2 mitigation strategies, especially at higher growth rate demands.

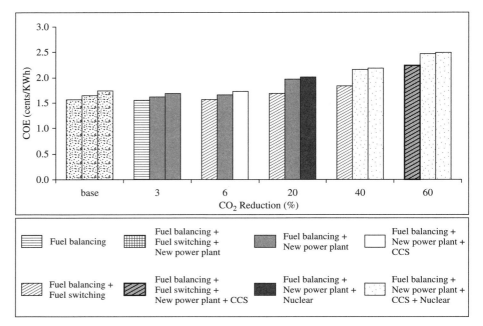

Figure 16   Effect of natural gas price on COE and the best mitigation strategy for base load demand

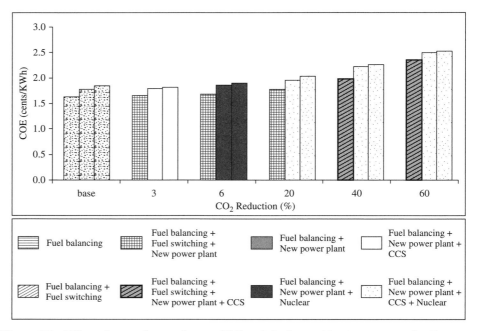

Figure 17   Effect of natural gas price on COE and the best mitigation strategy for 5 percent growth demand

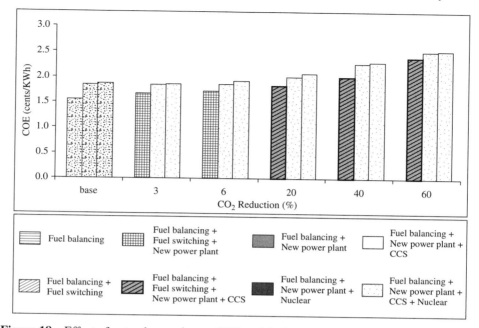

Figure 18 Effect of natural gas price on COE and the best mitigation strategy for 10 percent growth demand

8 SUMMARY

This chapter presented a fleetwide model of the electricity subsector that can be used to determine the optimal structure necessary to meet a given CO_2 reduction target while maintaining or enhancing power to the grid. The model incorporates power generation and CO_2 emissions from a mixed-fuel fleet of generating stations. The model can be used for an existing fleet and can also recommend new additional generating stations and carbon capture and store retrofit on existing generating stations to meet a CO_2 reduction target and electricity demand at minimum overall cost.

The model is illustrated on a case study that deals with the energy supply system operated by Ontario Power Generation (OPG) for the province of Ontario, Canada. The results showed that fuel balancing alone can contribute to the reduction of CO_2 emissions by only 3 percent and a slight, 1.6 percent, reduction in the COE compared to the calculated base case COE. In order to achieve a 20 percent CO_2 reduction, the results show that 8 out of 23 coal-fired boilers needed to be switched to natural gas, resulting in a 7 percent increase in the COE. Indeed, up to 40 percent CO_2 emission reductions, fuel switching from coal to natural gas was more attractive than retrofitting CCS on existing fossil-fuel stations. For the case of 60 percent CO_2 reduction, cost of electricity has increased by almost

40 percent since four new natural gas generating stations with capture were put online and a further 12 coal-fired boilers were shut down.

Four electricity demand scenarios were considered over a span of 10 years: current baseload, 1 percent, 5 percent, and 10 percent. The optimizer determined the size of new power generation capacity with and without capture that will be needed to meet growth in demand and supply supplemental energy to run capture processes. Six electricity-generating technologies have been allowed for:

1. Subcritical pulverized coal-fired (PC)
2. PC with carbon capture (PC+CCS)
3. Integrated gasification combined cycle (IGCC)
4. IGCC with carbon capture (IGCC+CCS)
5. Natural gas combined cycle (NGCC)
6. NGCC with carbon capture (NGCC+CCS)

A 20 percent CO_2 reduction at current electricity demand could be achieved by implementing fuel balancing and switching 8 out of 27 coal-fired to natural gas. However, as demand increases (> 5.0 percent growth rate demand), more coal-fired boilers were converted to natural gas and new NGCC and NGCC+CCS were built to replace aging coal-fired power plants. In order to achieve a 40 percent CO_2 reduction at 1.0 percent demand growth rate the optimizer recommends four new plants: 2 NGCC, 2 NGCC+CCS, as well as implementing carbon capture retrofit on the largest coal-fired boiler. If greater than 60 percent CO_2 reductions is required, at least one carbon capture retrofit on existing coal-fired power plants must be put online along with NGCC, NGCC+CCS, and IGCC+CCS power plants. The volatility of natural gas prices has had a significant impact on the optimal CO_2 mitigation strategy as well as the COE. Increasing the natural gas price by 50 percent and 100 percent resulted in early aggressive CO_2 mitigation strategy, especially at higher growth rate demand. All capture options recommended were located relatively close to the major CO_2 geological storage location on Ontario under Lake Erie. This will result in more than doubling the COE compare than a base case (3.37¢/kWh).

REFERENCES

Bai, H., and Wei, J. (1996), "The CO_2 Mitigation Options for the Electric Sector," *Energy Policy*, Vol. 24, pp. 221–228.

Climaco, J., et al., (1995), "A Multiple Objective Linear Programming Model for Power Generation Expansion Planning," *International Journal of Energy Research*, Vol. 19, pp. 419–432.

Costa-Pierce, B. A. (1996), "Environmental Impacts of Nutrients from Aquaculture: towards the Evolution of Sustainable Aquaculture Systems," in D. J. Baird, M. C. M. Beveridge, L. A. Kelly, and J. F. Muir (eds.), *Aquaculture and Water Resource Management. Blackwell Science*, pp. 81–113.

David, J., and Herzog, H. (2002), "The Cost of Carbon Capture," *Proceedings of the 5th International Conference on Greenhouse Gas Control Technologies* (Cairns, Australia), August 13–16.

ESCAP-UN (1995), *Energy Efficiency*. UN Publications. New York.

Genchi, Y., Saitoh, K., Arashi, N., and Inaba, A. (2002), "Assessment of CO_2 Emissions Reduction Potential by Using an Optimization Model for Regional Energy Supply Systems," *Proceedings of the 6th International Conference on Greenhouse Gas Control Technologies* (Kyoto, Japan), October 1–4.

Gielen, D., and Taylor, M. (2007), "Modeling Industrial Energy Use: The IEAs Energy Technology Perspectives," *Energy Economics*, Vol. 29, No. 4, pp. 889–912.

Herzog, H., et al. (2000)., "Update of the International Experiment on CO_2 Ocean Sequestration." Presented at the Fifth International Conference on Greenhouse Gas Control Technologies, Cairns, Australia, August 13–August 16 (2000).

Independent Electricity Market Operator (IESO) (2005), "10 Year Outlook Highlights: From January 2005 to December 2014," www.ieso.com. Retrieved January 11, 2005.

John, D., and Kelly, T. (2004), "Technologies for Capture of Carbon Dioxide," *Proceedings of the 7th International Conference on Greenhouse Gas Control Technologies* (Vancouver, BC), September 5–9.

Johnson, T. L., and Keith, D. W. (2004), "Fossil Electricity and CO_2 Sequestration: How Natural Gas Prices, Initial Conditions and Retrofits Determine the Cost of Controlling CO_2 Emissions," *Energy Policy*, Vol. 32, pp. 367–382.

Linares, P., and Romero, C. (2002), "Aggregation of Preferences in an Environmental Economics Context: A Goal Programming Approach," *Int. J. Manage. Sci.*, Vol. 30, pp. 89–95.

Loulou, R., and Kanudia A. (1999a), "Minimax Regret Strategies for Greenhouse Gas Abatement: Methodology and Application," *Operations Research Letters*, Vol. 25, pp. 219–230.

Loulou, R., and Kanudia A. (1999b), "The Kyoto Protocol, Inter-Provincial Cooperation, and Energy Trading: A Systems Analysis with Integrated MARKAL Models," *Energy Studies Review*, Vol. 9, No. 1, pp. 1–23.

Loulou R., Kanudia A., and Lavigne D. (1999), "GHG Abatement in Central Canada with inter-provincial cooperation," *Energy Studies Review*, Vol. 8, No. 2, pp. 120–129.

Mavrotas, G., Diakoulaki, D., and Papayannakis, L. (1999), "An energy planning approach based on mixed 0–1 multiple objective linear programming," *International Transaction In Operational Research*, Vol. 6, pp. 231–244.

McDaniel, J. E. (2002), "Tampa Electric Polk Power Station Integrated Gasification Combined Cycle Project," Final technical report. Tampa Electric Company, Tampa, FL.

Narula, R. G., wen, H., and Himes, K. (2002), "Incremental Cost of CO_2 Reduction in Power Plants," *Proceedings of IGTI, ASME TURBO EXPO* (Amsterdam, Netherlands), June 3–6.

Noonan, F., and Giglio, R. (1977), "Planning electric power generation: A nonlinear mixed integer model employing benders decomposition," *Manage. Sci.*, Vol. 23, pp. 946–956.

Ordorica-Garcia, G., Croiset, E., Douglas, P., and Zheng, L. (2004), "Simulation of IGCC Power Plants with Glycol-Based CO_2 Capture Using Aspen Plus," *Energy Conversion and Management*.

Parson, E. A. (2000), "Environmental Trends and Environmental Governance in Canada," *Canadian Public Policy*, Vol. 26 (SUPP), pp. S123–S143.

Rao, A. B., and Rubin, E. S. (2002). "A Technical, Economic, and Environmental Assessment of Amine-based CO_2 Capture Technology for Power Plant Greenhouse Gas Control," *Environmental Science and Technology*, Vol. 36, pp. 4467–4475.

Rao, A. B., Rubin, E. S., and Berkenpas, M. B. (2004), "An Integrated Modeling Framework for Carbon Management Technologies," Technical documentation: Amine-based CO_2 capture and storage systems for fossil fuel power plant. Carnegie Mellon University, Center for Energy and Environmental Studies, Department of Engineering and Public Policy, Pittsburgh, PA 15213–3890.

Rubin, E. S., Rao, A. B., and Chen, C. (2004), "Comparative Assessments of Fossil Fuel Power Plants with CO_2 Capture and Storage," *Proceedings of the 7th International Conference on Greenhouse Gas Control Technologies* (Vancouver, BC), September 5–9.

Simbeck, D., and McDonald, R. (2000), "Existing Coal Power Plant Retrofit CO_2 Control Options Analysis," *Proceedings of the Fifth Conference on Greenhouse Gas Control Technologies* (GHGT-5) (Cairns, Australia).

Singh, D., Croiset, E., Douglas, P.L., and Douglas, M.A. (2003), "Techno-economic study of CO_2 capture from an existing coal-fired power plant: MEA scrubbing vs. O2/CO2 recycle combustion," *Energy Conversion and Management*, Vol. 44, No. 19, pp. 3073–3091.

Vassos, Spyros, and Vlachou, Andriana (1997), "Investigating Strategies to Reduce CO_2 Emissions from the Electricity Sector: The Case of Greece," *Energy Policy*, Vol. 25, No. 3, pp. 327–336.

Wise, M. A., and Dooley, J. J. (2004), "Baseload and Peaking Economics and the Resulting Adoption of a Carbon Dioxide Capture-and-Storage System for Electric Power Plants," *Proceedings of the 7th International Conference on Greenhouse Gas Control Technologies* (Vancouver, BC), September 5–9.

CHAPTER 9

ENERGY AND EXERGY ANALYSES OF NATURAL GAS-FIRED COMBINED CYCLE POWER GENERATION SYSTEMS

K. Mohammed
Department of Mechanical
Engineering
University of New Brunswick
Fredericton, New Brunswick,
Canada

B. V. Reddy
Faculty of Engineering and
Applied Science
University of Ontario Institute of
Technology
Oshawa, Ontario, Canada

1 INTRODUCTION

The energy demand is growing worldwide in developed and developing countries. The power sector contributes significant amount of greenhouse gas emissions responsible for global warming. Various fossil fuel–based advanced energy technologies, renewable energy sources, and nuclear power technologies are developed to generate electric power with higher efficiencies. The natural gas–based combined cycle power generation systems are receiving attention because they are able to generate power with high efficiencies, thereby reduced emissions.

2 COMBINED CYCLE POWER PLANTS

A combined cycle power plant consists of two or more cycles or plants coupled in series, where the heat rejected by the topping cycle or plant is absorbed, partly or wholly, by the bottoming plant. If a higher mean temperature of heat addition and/or a lower mean temperature of heat rejection can be achieved this way, then a higher overall plant efficiency will be achieved (Nag 1994). The concept of a gas turbine combined cycle with a bottoming steam cycle, coupled through a heat recovery steam generator (HRSG) has been well established (Horlock 1995). In such a cycle, the hot exhaust gases from the gas turbine are passed through the HRSG, where they produce superheated steam in a bottoming steam cycle. The interest in combined gas turbine–steam turbine power generation systems is motivated by the defense and aircraft research and development programs. Such programs resulted in a revolution in gas turbine materials, initially intended for military use. These metallurgical accomplishments pushed the limits on gas turbine operating temperatures. Over the past two decades, combined gas and steam turbine cycle power plants have been put to operation with superior fuel utilization when compared to a separately operating gas turbine or steam turbine power plant. Among thermal power plants commercially available, the combined cycle power plant has high efficiency (Bolland 1991, Facchini et al. 2000). In addition, the mean temperature of heat rejection is lower than that of the gas turbine cycle operating separately, resulting in a higher combined cycle thermal efficiency when compared to the efficiency levels of the gas turbine and steam turbine cycles functioning individually.

The worldwide demand for combined cycle power generation systems is growing dramatically, with some experts forecasting explosive growth in the coming years (Makansi 1990). A gas turbine unit exhausting into a heat recovery steam generator (HRSG) that supplies steam to a steam turbine cycle is the most efficient combined cycle power generation system generating electricity today. There are six main advantages of a combined gas turbine and steam turbine cycle (Nag 1994):

1. Higher overall plant efficiency
2. Higher availability
3. Shorter design and construction period

4. Reduced emission levels
5. Environmentally friendly power generation
6. Rapid-load change and start up capability

3 NATURAL GAS-FIRED COMBINED CYCLE POWER PLANTS

The focus of this chapter is to discuss the thermodynamic analysis of natural gas–based combined cycle power generation systems and the related energy and exergy analyses. The major advantages of using natural gas in a combined cycle for power generation are high thermal efficiency of combined power cycles and lower emission levels characteristics. Recently, some attention has been focused on CO_2 control and/or capture in combined cycle power plants as a means of emission reduction. Chiesa and Consonni (2000) studied the feasibility of CO_2 removal by chemical absorption from the flue gases of a natural gas fired combined cycle. The work by Lozza and Chiesa (2002a, b) focused on two schemes of direct chemical treatment of the fuel (natural gas). These include partial oxidation and methane reforming. Srinivas et al. (2009) discussed the carbon dioxide emission reduction from combined cycle plant with partial oxidation of natural gas. Other studies are based on first and second laws of thermodynamics, the latter being referred to as *exergy* analysis. Exergy analysis, from which thermodynamic optimization (or entropy generation minimization, EGM) stems, is the most favorable, as it provides a deeper insight into the power plant's irreversibility and overall performance. The fundamental guideline for such a study is laid down by Bejan (2002). Various recent studies have been directed at exergy analysis of different power cycles configurations. The work by Horlock et al. (2000) took a general approach to developing terms for rational (exergetic) efficiency of modern fossil-fuel power plants. The focus is on the effect of exergy analysis, based on the gas turbine inlet temperature and the level of steam injection into the gas turbine. Facchini et al. (2000) performed an exergy analysis of a combined power cycle using high gas turbine inlet temperature. In this study, only a limited range of pressure ratios is used in the analysis. A comparison of exergy analysis between a gas turbine combined cycle and a cogeneration cycle was performed by Sue and Chuang (2004). Zwebek and Pilidis (2003a, b) conducted a more detailed analysis by demonstrating the degradation effects that combined cycle power plant components have on the overall plant performance. Dincer and Rosen (2004) discussed the need for exergy analysis and its role in achieving improved design and sustainability.

One of the options in combined cycle power plants is the use of supplementary firing (SF) in the heat recovery steam generator (HRSG) to increase steam generation in the bottoming cycle by raising the temperature of the exhaust gases from the gas turbine. With the projected increase in energy demands from existing combined cycle power plants, there is a renewed interest in studying the effects of supplementary firing on combined cycle plant operation and optimization. A study on the effects and optimization of supplementary firing was conducted by De and Nag (2000). Mohammed (2005) conducted the energy and exergy analyses for advanced combined cycle power generation systems. Reddy and Mohammed (2007) presented the exergy analysis and the role of operating conditions for a natural gas-fired combined cycle power generation plant with reheat combustor and supplementary firing.

4   THERMODYNAMIC ANALYSIS OF NATURAL GAS-FIRED COMBINED CYCLE POWER GENERATION UNIT

The schematic diagram of the natural gas-fired combined cycle power generation unit with reheat and supplementary firing is shown in Figure 1. The topping gas turbine cycle consists of staged compression of air with intercooling. The gas turbine has stages and reheating is utilized. The unit also has the supplementary

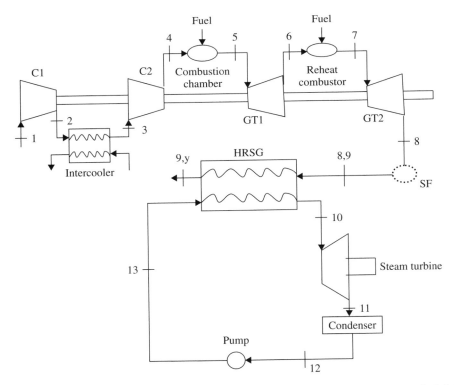

Figure 1   Combined cycle power generation unit with reheat and supplementary fuel firing options

firing combustor, which is optional. The double numbering for the states after the supplementary firing combustor represents the cases with and without supplementary firing respectively. The heat recovery steam generator (HRSG) couples the topping cycle to the bottoming steam cycle. The steam cycle consists of a basic cycle with a single stage steam turbine, a condenser and a pump. Here, it is assumed that the natural gas is pure methane (CH_4). The heating value of methane is taken from Moran and Shapiro (2000).

4.1 Energy Analysis

The conditions at the compressor No. 1 (C1) inlet are set to atmospheric conditions, which is also the dead state for the topping cycle. Based on the assigned pressure ratio, the approximate optimum intermediate pressure for the staged air compressor can be calculated (Moran and Shapiro 2000).

Based on thermodynamic details, the temperature and enthalpy are estimated for isentropic compression for compressor No. 1. From the assigned temperature at state 1, the enthalpy, relative pressure, and absolute entropy are obtained for the compressor inlet conditions. From the isentropic efficiency of the compressor, the actual enthalpy at the exit of compressor No. 1 (C1) can be calculated as follows:

$$h_2 = \frac{h_{2s} - h_1}{\eta_c} + h_1 \tag{1}$$

where η_c is the isentropic efficiency of the air compressor No. 1 (C1) and No. 2 (C2). Using the actual enthalpy, interpolation is done using the air tables, to obtain the actual temperature and entropy at the compressor No. 1 (C1) exit, giving T_2 and s_2. For the intercooler, a suitable temperature drop (of 100 K) is assumed in this analysis. Thus, all properties for state 3 can be read or interpolated from the air tables. The calculations for state 4 are conducted using this approach.

Main Combustion Chamber

Based on the assigned gas turbine No. 1 (GT1) inlet temperature, the fuel mass flow rate into the main combustion chamber (CC1) is determined through a trial and error approach. The total mass flow rate through the gas turbine No. 1 (GT1) is fixed; therefore, determining the fuel input subsequently allows for the calculation of the air mass flow rate required. The trial and error procedure begins by assigning a fuel input to the main combustion chamber. The combustion reaction is assumed to take the following form (Moran and Shapiro 2000):

$$aCH_4 + x(O_2 + 3.76N_2) \rightarrow aCO_2 + bH_2O + cN_2 + dO_2 \tag{2}$$

Knowing the fuel and air mass flow rates into the combustion chamber, the air-to-fuel mass ratio (AF) is calculated. The d term is added to ensure that there will be excess oxygen in the exhaust gases—in other words, that the condition in the combustion chamber is dilute. In addition, a and x can be computed through dividing the mass flow rates of methane and air by their corresponding molar masses. In the case for air, the molar mass is multiplied by a factor of 4.76 to account for the total number of moles of air. The variables in the chemical equations are taken to be on kmol/s basis. Now by balancing the elements on both sides of equation (2); b, c, and d can be obtained. This produces the balanced chemical equation for the main combustion chamber.

The total molar flow rate of the combustion products and their corresponding molar ratios are then calculated. This can be written as $n = a + b + c + d$, where the molar ratios of the products are given by: $y_{CO_2} = a/n$, $y_{H_2O} = b/n$, $y_{N_2} = c/n$, $y_{O_2} = d/n$ and. The molar mass of the mixture is computed using:

$$M_{mix} = \sum_i y_i M_i \tag{3}$$

An energy balance on the main combustion chamber is used to calculate the conditions at the gas turbine No. 1 (GT1) inlet. Taking into account a 3 percent heat loss in the main combustion chamber, the energy balance can be written as:

$$0.97 \left(\dot{m}_{air} h_4 + \dot{m}_{f1} CV_f \right) = \dot{m}_{mix} h_5 \tag{4}$$

where $\dot{m}_{mix} = \dot{m}_{f1} + \dot{m}_{air}$.

From equation (4), the enthalpy at the gas turbine No. 1 (GT1) inlet (h_5) is calculated. The enthalpy at the gas turbine No. 1 (GT1) inlet can also be calculated using the law of ideal gas mixtures (Moran and Shapiro 2000).

$$\overline{h}_5 = \sum_i y_i \overline{h}_i (T_5) \tag{5}$$

The temperature at the gas turbine inlet (T_5) can now be iterated, by using equation (5). If equation (5) does not iterate to give the desired temperature, the air-to-fuel ratio is varied until the desired temperature is accomplished. Knowing the gas turbine inlet temperature, the entropy at the gas turbine No. 1 (GT1) inlet can be calculated from:

$$\overline{s}_5 = \sum_i y_i \overline{s}_i (T_i, p_i) \tag{6}$$

To estimate the temperature and enthalpy at the gas turbine No. 1 (GT1) exit, it is first assumed that the gas turbine expansion is isentropic. Then equations for the entropy change of each of the exhaust gas constituents are written. Those equations are then combined and are equated to zero:

$$\Delta \overline{s} = \overline{s}_{exit} - \overline{s}_{inlet} = \sum_i y_i \Delta \overline{s}_i = 0 \tag{7}$$

$$\Delta \overline{s}_i = \left[\Delta \overline{s}^{\circ}_i (T_6) - \Delta \overline{s}^{\circ}_i (T_5) - \overline{R} \ln \left(\frac{p_6}{p_5} \right) \right] \tag{8}$$

The temperature at the gas turbine exit can then be estimated for an isentropic expansion by iterating equation (8) for T_6. From this temperature, the enthalpy of the combustion products at the gas turbine exit, based on isentropic expansion, (h_{6s}) can be calculated. Finally, the isentropic efficiency of the turbine is taken into account and the actual enthalpy at the gas turbine exit is computed using this equation:

$$h_6 = h_5 - \eta_{GT}(h_5 - h_{6s}) \tag{9}$$

From this actual enthalpy, the actual temperature of the exhaust gases leaving the first stage gas turbine (No. 1) are estimated in a manner similar to equation (5). Using the actual temperature (T_6), the entropy of the GT1 exhaust gases is calculated.

Reheat Combustor

The calculations for the reheat combustor and the gas turbine No. 2 (GT2) follow exactly the same procedure. In the reheat combustor (CC2), the reheat fuel (assigned) now reacts with the available oxygen in the exhaust gases from gas turbine No. 1 (GT1). The gas turbine No. 2 (GT2) inlet temperature is set equal to that of the gas turbine No. 1 (GT1) inlet. The chemical reaction in this case can be written as follows:

$$zCH_4 + aCO_2 + bH_2O + cN_2 + dO_2 \rightarrow pCO_2 + qH_2O + sN_2 + rO_2 \tag{10}$$

The combustion products from the reheat combustor will have a different chemical composition from that of the main combustor. As a result, the molar ratios and the molar mass for the exhaust gases from the reheat combustor are updated using the data from equation (10).

Supplementary Firing

In the case with supplementary firing, an additional combustion reaction takes place in which the fuel reacts with gas turbine (GT2) exhaust gases. Again, in this case, the changed gas composition dictates the molar ratios and molar mass. The chemical reaction for the supplementary firing combustion in with gas turbine reheat can be written as follows:

$$vCH_4 + pCO_2 + qH_2O + sN_2 + rO_2 \rightarrow eCO_2 + fH_2O + hN_2 + gO_2 \tag{11}$$

For the bottoming steam cycle, the steam conditions at the steam turbine inlet (state 10) are selected. The condenser pressure is also fixed in the analysis. To find the properties at the exit of the steam turbine, first an isentropic expansion is assumed through the steam turbine, based on which a steam quality (x_{11s}) and enthalpy (h_{11s}) are calculated. The actual enthalpy (h_{11}) is then computed by incorporating the steam turbine isentropic efficiency. The condition at the condenser exit is assumed to be saturated liquid.

Using the energy balance for heat recovery steam generator (HRSG), with an assumed 3 percent heat loss, the amount of steam generation can be written as

$$0.97\left[\dot{m}_{\text{mix}}(h_{8,9} - h_{9,y})\right] = \dot{m}_{st}(h_{10} - h_{13}) \tag{12}$$

The two subscripts for exhaust gas flow represent both cases, with and without supplementary firing, respectively.

Work output:

$$\text{Gas turbine No. 1: } \dot{W}_{\text{GT1}} = \dot{m}_{\text{mix}}(h_5 - h_6) \tag{13}$$

$$\text{Gas turbine No. 2: } \dot{W}_{\text{GT2}} = \left(\dot{m}_{\text{mix}} + \dot{m}_{f2}\right)(h_7 - h_8) \tag{14}$$

$$\text{Steam turbine: } \dot{W}_{\text{ST}} = \dot{m}_{st}(h_{10} - h_{11}) \tag{15}$$

Work input:

$$\text{Compressor No. 1: } \dot{W}_{C1} = \dot{m}_{\text{air}}(h_2 - h_1) \tag{16}$$

$$\text{Compressor No. 2: } \dot{W}_{C2} = \dot{m}_{\text{air}}(h_4 - h_3) \tag{17}$$

$$\text{Pump: } \dot{W}_p = \dot{m}_{st}(h_{13} - h_{12}) \tag{18}$$

The heat input to the topping gas turbine cycle and the bottoming steam cycle are given by:

$$\dot{Q}_{\text{in,top}} = \left(\dot{m}_{f1} + \dot{m}_{f2}\right)\text{CV}_f \tag{19}$$

$$\dot{Q}_{\text{in,bot}} = \left(\dot{m}_{\text{mix}} + \dot{m}_{f2} + \dot{m}_{SF}\right)(h_{8,9} - h_{9,y}) \tag{20}$$

where the underlined term, along with the subsequent subscripts, represent the case with supplementary firing (SF).

The thermal efficiency for the overall combined cycle is given by:

$$\eta_{\text{ovl}} = \frac{\dot{W}_{\text{GT1}} + \dot{W}_{\text{GT2}} - \dot{W}_{C1} - \dot{W}_{C2} + \dot{W}_{\text{ST}} - \dot{W}_p}{\dot{Q}_{\text{in,top}}} \tag{21}$$

4.2   Exergy Analysis

The exergy analysis is gaining prominence because it can provide the details on sources and magnitudes of irreversibilities in various power plant components and the scope for further improvement. The reference or dead state for the topping cycle exergy analysis is set at $T_0 = 298$ K, $p_0 = 1$ atm, which is the atmospheric condition for the gas turbine cycle. For the bottoming steam cycle, the dead state is taken to be the lowest thermodynamic state in the cycle, which in this case is the flow at the condenser exit.

Once the temperature, enthalpy, and entropy are calculated for each state, the availability is computed using

$$av_i = h_i - h_o - T_o(s_i - s_o) \tag{22}$$

The physical exergy at each state is the product of the availability at that state and the corresponding mass flow rate:

$$\dot{Ex}_{ph,i} = \dot{m}_i \, av_i \tag{23}$$

Another important factor in the exergy analysis, that is also dependent on the flow composition, is the chemical exergy. The chemical exergy of the fuel is given by Moran and Shapiro (2000):

$$\dot{Ex}_{fuel,ch} = \left(\frac{\dot{m}_{fuel}}{M_{CH_4}} \right) \left\{ \left(g_{CH_4} + 2g_{O_2} - g_{CO_2} - 2g_{H_2O(g)} \right) \right.$$
$$\left. + \left\{ RT_o \ln \left(\frac{(y_{O_2}^e)^2}{y_{CO_2}^e (y_{H_2O}^e)^2} \right) \right\} \right\} \tag{24}$$

In a similar manner, the chemical exergy of the exhaust gases resulting from a combustion reaction is given by:

$$\dot{Ex}_{ch,exhaust} = \dot{m}_{exhaust} \left(\frac{\overline{R}}{M_{exhaust}} \right) T_o \left\{ y_{CO_2} \ln \left(\frac{y_{CO_2}}{y_{CO_2}^e} \right) + y_{H_2O} \ln \left(\frac{y_{H_2O}}{y_{H_2O}^e} \right) \right.$$
$$\left. + y_{N_2} \ln \left(\frac{y_{N_2}}{y_{N_2}^e} \right) + y_{O_2} \ln \left(\frac{y_{O_2}}{y_{O_2}^e} \right) \right\} \tag{25}$$

where y_i^e is the molar ratio for the environmental component i, and g_i is the Gibbs function of formation for gas i (Moran and Shapiro 2000). For different combustion reactions, the mass flow rate, molar mass, and molar ratios vary—thus, causing the chemical exergy to change with composition, as is expected.

The total exergy at each state of the combined cycle is the sum of the physical and chemical exergy, neglecting the exergy associated due to potential and kinetic energy effects:

$$\dot{Ex}_i = \dot{Ex}_{ph,i} + \dot{Ex}_{ch,i} \tag{26}$$

The exergy destruction/losses and exergetic efficiency (Moran and Shapiro 2000) for each of the considered combined cycle power plant components are given next. Again here, double subscripts are used when more than one of the component under consideration is present in the combined cycle. The control volume selected for the analysis does not include the cooling water loops in the compressor intercooler or the steam cycle condenser. As a result, the exergy disappearing in those two components will be regarded as exergy losses.

Air compressor:

The exergy destruction rate for the air compressor can be written as:

$$\dot{Ex}_{dC1,2} = \dot{W}_{C1,2} - (\dot{Ex}_{2,4} - \dot{Ex}_{1,3}) \tag{27}$$

The exergetic efficiency is given by:

$$\eta_{exC1,2} = \frac{\dot{Ex}_{2,4} - \dot{Ex}_{1,3}}{\dot{W}_{C1,2}} \tag{28}$$

The exergy destruction rate for a gas turbine combustion chamber (including reheat and supplementary firing combustor) is given by this equation:

$$\dot{Ex}_{dcomb1,2,SF} = \dot{Ex}_{f1,2,SF} - (\dot{Ex}_{5,7,9} - \dot{Ex}_{4,6,8}) \tag{29}$$

The exergetic efficiency is given by

$$\eta_{ex} = \frac{\dot{Ex}_{5,7,9} - \dot{Ex}_{4,6,8}}{\dot{Ex}_{f1,2,SF}} \tag{30}$$

Gas turbine:
The exergy destruction rate is given by

$$\dot{Ex}_{dGT1,2} = (\dot{Ex}_{5,7} - \dot{Ex}_{6,8}) - \dot{W}_{GT1,2} \tag{31}$$

The exergetic efficiency is given by

$$\eta_{ex} = \frac{\dot{W}_{GT1,2}}{\dot{Ex}_{5,7} - \dot{Ex}_{6,8}} \tag{32}$$

The exergy destruction rate for a heat recovery steam generator (with and without SF) is given by this equation:

$$\dot{Ex}_{dHRSG} = (\dot{Ex}_{8,9} - \dot{Ex}_{9,y}) - (\dot{Ex}_{10} - \dot{Ex}_{13}) \tag{33}$$

$$\dot{Ex}_{lHRSG} = (\dot{Ex}_{9,y} - \dot{Ex}_1) \tag{34}$$

The exergetic efficiency is given by

$$\eta_{ex} = \frac{\dot{Ex}_{10} - \dot{Ex}_{13}}{\dot{Ex}_{8,9} - \dot{Ex}_{9,y}} \tag{35}$$

The exergy destruction rate for a steam turbine is given by this equation:

$$\dot{Ex}_{dST} = (\dot{Ex}_{10} - \dot{Ex}_{11}) - \dot{W}_{ST} \tag{36}$$

The exergetic efficiency is given by

$$\eta_{ex} = \frac{\dot{W}_{ST}}{\dot{Ex}_{10} - \dot{Ex}_{11}} \tag{37}$$

The combined cycle exergetic efficiency can be calculated as follows (Kotas 1985):

$$\eta_{ex,ovl} = \frac{\dot{W}_{GT1} + \dot{W}_{GT2} - \dot{W}_{C1} - \dot{W}_{C2} + \dot{W}_{ST} - \dot{W}_p}{\dot{Ex}_f} \tag{38}$$

where the exergy of the fuel, Ex_f, depends on the number of combustors present and the amount of fuel consumed, in other words, with and without supplementary firing.

5 EFFECT OF OPERATING PARAMETERS ON THE PERFORMANCE

All gases considered are assumed to be ideal gases, where the law of ideal gas mixtures applies. Thermodynamic properties of air, CO_2, CO, $H_2O_{(g)}$, $H_2O_{(saturated)}$, N_2, and O_2 and are taken from Moran and Shapiro (2000), whereas thermodynamic properties of CH_4, H_2, and C are obtained from JANAF (1974). The first stage air compressor (No. 1) inlet conditions are set at atmospheric pressure and 298 K. The temperature drop across the air compressor intercooler is set to be 100 K. The mass flow rate through the gas turbine is held constant. The exhaust gas temperature at the HRSG exit is fixed at $150\,^\circ$C. A 3 percent heat loss is assumed in the combustion chambers. The isentropic efficiency of all the working components (air compressor, gas turbine, and steam turbine) is set at 0.90 for the performance results. The total mass flow rate through the gas turbine (No. 1) is fixed at 300 kg/s. The steam turbine inlet conditions are fixed at 80 bar and 400°C. The condenser pressure is set at 0.08 bar. The gas turbine inlet temperature ranges from 1100 K to 1600 K. For the supplementary firing effect, the fuel inputs are considered from 1.0 kg/s to 2.0 kg/s.

5.1 Effect of Gas Turbine Inlet Temperature on Natural Gas-Fired Combined Cycle Unit Performance

The topping, bottoming, and combined cycle net work output variation with gas turbine inlet temperature is shown in Figure 2. The topping, bottoming, and

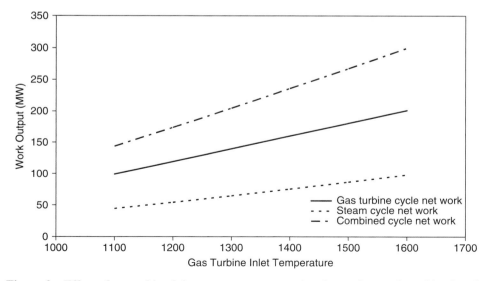

Figure 2 Effect of gas turbine inlet temperature on topping, bottoming, and combined cycle plant net work output ($rp = 16$)

combined cycle net work output increases with the gas turbine inlet temperature. At a higher gas turbine inlet temperature, the enthalpy of gases, and the enthalpy drop across the gas turbine are higher. This results in an increase in work output for a fixed mass flow rate through the gas turbine. At a fixed pressure ratio, the higher gas turbine inlet temperature results in a higher gas turbine exhaust temperature and more heat being supplied to the bottoming cycle. As a result, the combined cycle net work output increases at a fixed pressure ratio with increasing gas turbine inlet temperature. The results clearly demonstrate that the gas turbines with high inlet temperatures produce more work output.

5.2 Effect of Supplementary Fuel Firing on Combined Cycle Unit Performance

The combined cycle net work output variation with and without supplementary firing is presented in Figure 3. With more supplementary firing, the increase in bottoming cycle work output results in an increase in the combined cycle net work output as shown in Figure 3. The incremental increase in steam cycle work output for a case study of supplementary firing from SF = 1.0 kg/s to SF = 2.0 kg/s is higher than that from no supplementary firing to SF = 1.0 kg/s. The same applies to the combined cycle total net work output. This is due to a higher incremental increase in the HRSG inlet gases temperature, when the supplementary firing fuel is raised from 1.0 kg/s to 2.0 kg/s, as compared to the increase in temperature

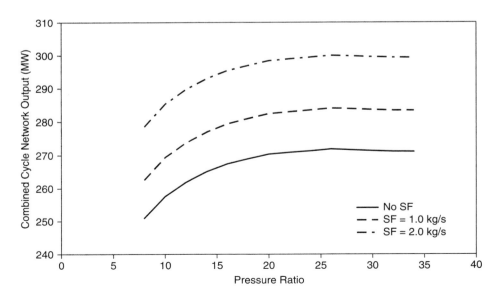

Figure 3 Effect of supplementary fuel firing with gas turbine pressure ratio on combined cycle plant net work output (T_{GT} = 1,500 K)

from the case without supplementary firing, to that of a supplementary firing input of 1.0 kg/s.

5.3 Effect of Gas Turbine Pressure Ratio on Natural Gas-Fired Combined Cycle Unit Exergy Destruction/Losses

The effect of gas turbine pressure ratio on the combined cycle rate of exergy destruction/losses, expressed in percentage of exergy input rate, is shown in Figure 4. This is reported for a gas turbine inlet temperature of 1400 K. The exergy destruction rate in the topping cycle components (SET1) is found to increase with higher gas turbine pressure ratios. A higher pressure ratio results in the air compressors requiring more work input. This causes an increase in irreversibilities in the air compressors caused by higher entropy generation. The same is true for the gas turbines: A higher pressure ratio causes an increase in entropy generation rate and more exergy destruction rate.

For the intercooler, a higher pressure ratio means a higher air temperature at the intercooler inlet. Even though the temperature drop is fixed across the intercooler, the higher that temperature at the inlet, the higher the temperature gradient across the intercooler, so more irreversibilities in the intercooler. However, based on the control volume adapted, this exergy destruction rate is looked on in terms of exergy loss rate. Another factor that causes the exergy destruction rate in the

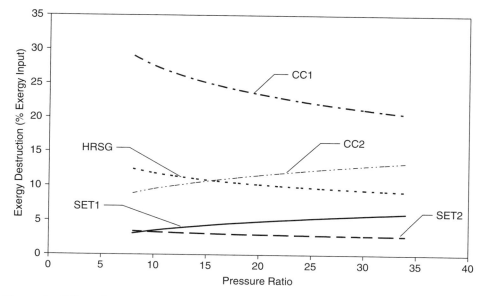

Figure 4   Effect of pressure ratio on rate of exergy destruction/losses ($T_{GT} = 1400$ K) SET1: Compressor No. 1 and No. 2, Intercooler, and Gas turbine No. 1 and No. 2 SET2: Steam turbine, condenser, and pump CC1: Gas turbine main combustion chamber (No. 1) CC2: Gas turbine reheat combustor (No. 2)

topping cycle components (SET1) to increase with pressure ratio, is the increased mass flow rate of air, which results from less fuel being required in the combustion chambers, in other words, increased air-to-fuel ratio. For gas turbine combustion chamber No. 1 (CC1), the exergy destruction rate drops with pressure ratio mainly as a result of the higher temperature of air, and lower mass flow rate of air at the combustion chamber No. 1 (CC1) inlet. This in turn results in lowering the temperature gradients within the combustion chamber, which in turn results in lower irreversibilities and lower exergy destruction rate. Increasing the gas turbine pressure ratio has the opposite effect when it comes to exergy destruction rate in the reheat combustor (CC2).

At higher pressure ratios, a fixed gas turbine inlet temperature results in a lower gas turbine exhaust temperature from gas turbine No. 1. As a result, not only do the exhaust gases enter the reheat combustor (CC2) at a lower temperature, but also, more fuel is required to achieve the desired temperature at the reheat combustor (CC2) exit. In other words, a higher exergy destruction rate in the reheat combustor (CC2) brought about by more irreversibilities and entropy generation rate. The exergy destruction rate in the HRSG drops with increasing gas turbine pressure ratio because the temperature of the exhaust gases leaving gas turbine No. 2 (GT2) reduces. This lowers the temperature gradients within the HRSG and less steam generation, therefore, a lower exergy destruction rate is noticed. The lower steam generation is the cause for the lower exergy destruction rate in the bottoming cycle components (SET2). In terms of magnitude, the gas turbine main combustion chamber is the highest source of exergy destruction rate. This is followed by the reheat combustor and the HRSG, both with relatively comparable shares of exergy destruction rates. This is also in agreement with the trends reported by Facchini et al. (2000). At low gas turbine pressure ratios, the temperature drop across the gas turbine No. 1 is lower and this results in higher flow temperatures into the reheat combustor and the HRSG. This is advantageous in terms of exergy destruction rate when it comes to the reheat combustor, whereas the opposite is true when it comes to the HRSG. The exergy destruction rate in the topping cycle components (SET1) and the bottoming cycle components (SET2) is relatively lower and considered to be less significant compared to combustion chambers.

The combined cycle energy and exergetic efficiencies, in addition to the total exergy destruction rate (expressed in percentage of exergy input) with gas turbine pressure ratio are shown in Figure 5. This is reported for a gas turbine inlet temperature of 1400 K. When compared to the thermal efficiency, the exergetic efficiency is lower. The increase in exergetic efficiency of the combined cycle can be directly related to the drop in the total exergy destruction rate of the combined cycle unit. As more of the exergy input to the cycle is utilized in work production, less exergy input is being lost/destroyed.

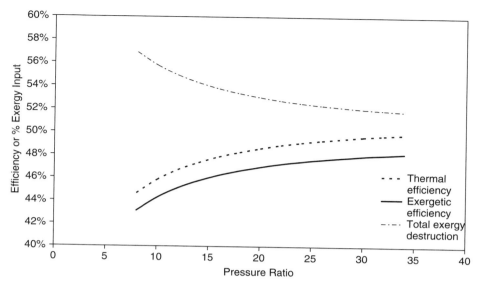

Figure 5 Effect of pressure ratio on combined cycle plant energy and exergetic efficiency, and total exergy destruction rate ($T_{GT} = 1400$ K)

5.4 Effect of Gas Turbine Inlet Temperature on Natural Gas-Fired Combined Cycle Unit Rate of Exergy Destruction/Losses

The combined cycle unit rate of exergy destruction/losses and combined cycle net work output with gas turbine inlet temperature for a particular gas turbine pressure ratio (of 14) is shown in Figure 6. The values are expressed in percentage of total exergy input rate. For the considered pressure ratio and for a particular gas turbine inlet temperature, the gas turbine main combustion chamber (CC1) is the major source of exergy destruction rate. This is due to the result of the vigorous combustion process, and the irreversibilities associated with it. With higher gas turbine inlet temperature, less air is supplied into the main combustion chamber (CC1), since the total mass flow rate through the gas turbine is fixed. This lowers the temperature gradients within the combustion chamber, and thus, the irreversibilities and the exergy destruction rate. The exergy destruction rate in the topping cycle components (SET1) is mainly due to irreversibilities, result of heat loss from the gas turbines, air compressors, and the intercooler.

The exergy destruction rate in the topping cycle components (SET1) drops with increases in the gas turbine inlet temperature. This is because there is higher work production in the gas turbines at higher gas turbine inlet temperatures; thus, more exergy is used for work production and less is destroyed by irreversibilities. Less air is required at higher gas turbine inlet temperature and this drops the air-to-fuel ratio. This results in lower exergy destruction rate in the compressor and exergy loss rate in the intercooler.

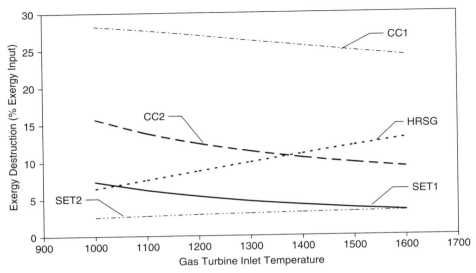

Figure 6 Effect of gas turbine inlet temperature on exergy destruction/losses ($rp = 14$) SET1: Compressor No. 1 and No. 2, Intercooler, and Gas turbine No. 1 and No. 2 SET2: Steam turbine, condenser, and pump CC1: Gas turbine main combustion chamber (No. 1) CC2: Gas turbine reheat combustor (No. 2)

The exergy destruction rate in the reheat combustor (CC2) also drops with increasing the gas turbine inlet temperature. Since the gas turbine pressure ratio is fixed, a higher gas turbine inlet temperature to gas turbine stage 1 results in a higher temperature of the gas turbine exhaust gases. This, in turn, lowers the temperature gradients within the reheat combustor and as a result lowers the exergy destruction rate. For a higher gas turbine inlet temperature, the exhaust gases enter the HRSG at a higher temperature. This causes higher temperature gradients within the HRSG, and consequently, higher exergy destruction rate. The higher steam generation rate in the bottoming cycle with increasing the gas turbine inlet temperature also contributes to the increase in exergy destruction rate in the HRSG.

The exergy destruction rate in the bottoming cycle components (SET2) increases with higher gas turbine inlet temperature. This is mainly caused by the higher steam flow rate in the bottoming cycle. The overall exergy destruction rate and losses in the combined cycle unit drops with increasing the gas turbine inlet temperature; thus, the combined cycle net work output, expressed in percentage of exergy input, increases with increasing the gas turbine inlet temperature. The results clearly demonstrate that, for a particular pressure ratio, the gas turbine with higher gas inlet temperatures will perform better from the exergy analysis point of view.

5.5 Natural Gas-Fired Combined Cycle Power Generation Unit Exergy Plots

As discussed earlier, the exergy destruction rate in the combustion chambers and the HRSG is higher in the combined cycle power plant. To give a better view of how the exergy destruction rates in these major components vary as a function of the pressure ratio and the gas turbine inlet temperature together, exergy destruction/loss rate *plots* are created. The main advantage of these plots is that they provide clear details on exergy destruction/loss rates with operating variables such as gas turbine pressure ratio and gas turbine inlet temperature.

The exergy destruction rate plot for the main combustion chamber (CC1) is shown in Figure 7. The exergy destruction rate drops with increasing the pressure ratio and increasing gas turbine inlet temperature. The point of highest exergy destruction rates occurs at the lowest pressure ratio and the lowest gas turbine inlet temperature. This is mainly as a result of the air properties entering the main combustor (CC1). At a lower pressure ratio, the air enters at a lower temperature. The plot representing the exergy destruction rate in the reheat combustor (CC2) is shown in Figure 8. In this case, the exergy destruction rate increases as the pressure ratio is raised, and drops at higher gas turbine inlet temperatures. Like the main combustion chamber, the highest exergy destruction rate occurs at the

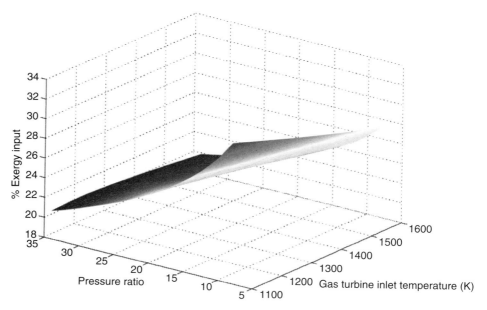

Figure 7 Exergy destruction rate plot for the main combustion chamber (CC1)

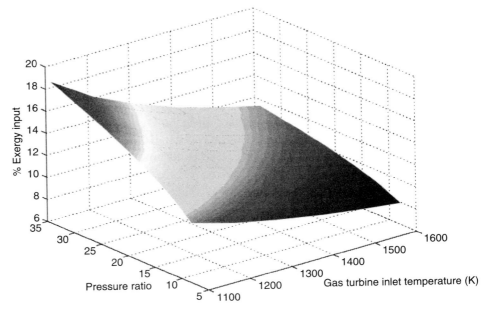

Figure 8 Exergy destruction rate plot for the reheat combustion chamber (CC2)

lowest gas turbine inlet temperature, but unlike the main combustion chamber (CC1), that simultaneously occurs at the highest pressure ratio. The point of lowest exergy destruction takes place at the highest gas turbine inlet temperature and the lowest pressure ratio. As the pressure ratio is increased, the temperature of the exhaust gases entering the reheat combustor, after expansion through the first stage gas turbine (No. 1), drops. Again in this case, the lowest gas turbine inlet temperature means that there will be more of the lower-temperature exhaust gases supplied to the reheat combustor. At the lowest pressure ratio, and the highest gas turbine inlet temperature, not only do the exhaust gases enter the reheat combustor at a higher temperature, but also at a lesser amount. Consequently, this results in the lowest exergy destruction in the reheat combustor. The exergy destruction rate plots for both combustion chambers (CC1 and CC2) are shown in Figure 9 in one plot. This assists in comparing the two in terms of exergy destruction magnitudes and trends. When it comes to the gas turbine inlet temperature, the main and reheat combustion chambers are affected in a similar manner. However, with regards to the gas turbine pressure ratio, the opposite is true.

Given that the magnitude of exergy destruction is more significant in the main combustion chamber (CC1), it would be more advantageous to minimize the exergy destruction in the main combustion chamber, as opposed to the reheat

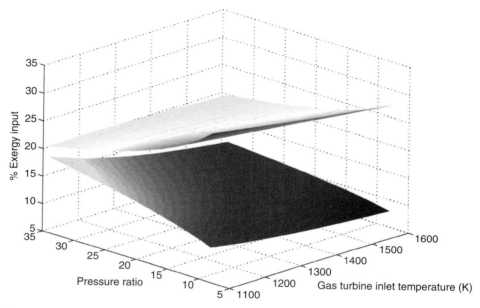

Figure 9   Exergy destruction rate plots for the main and reheat combustion chambers (CC1 and CC2)

combustor. This can be observed from Figure 10, in which the plot for the total sum of exergy destruction in the combustion chambers is shown. Minimization of the total exergy destruction in the combustion chambers can be achieved by operating at the highest gas turbine inlet temperature and pressure ratio—in other words, by reducing the exergy destruction in the main combustion chamber to the minimum. At a higher gas turbine inlet temperature, less dilution is required in the combustion chamber, which results in less irreversibilities and lower exergy destruction. Increasing the pressure ratio raises the temperature of the air entering the main combustion chamber No. 1, therefore, reducing the temperature gradients within the combustion chamber and the total exergy destruction.

The total exergy destruction rate plot for the HRSG is shown in Figure 11. The exergy destruction rate is low at low gas turbine inlet temperature and for higher gas turbine pressure ratio. At a low gas turbine inlet temperature and at high pressure ratio, the temperature of the exhaust gases entering the HRSG is low. This will diminish the temperature gradients across the HRSG, and as a result, lower the exergy destruction rate in the HRSG to the minimum, among the data range reported in the current analysis. The opposite occurs at higher gas turbine inlet temperature and lower gas turbine pressure ratio, forcing elevated exergy destruction rates in the HRSG.

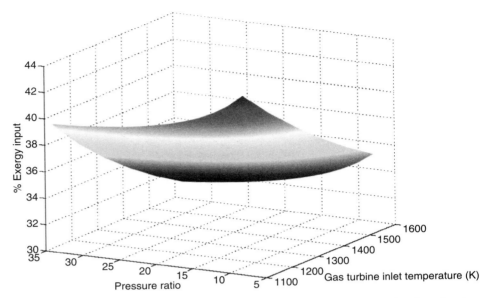

Figure 10   Total exergy destruction rate plot for the main and reheat combustion chambers together (CC1 and CC2)

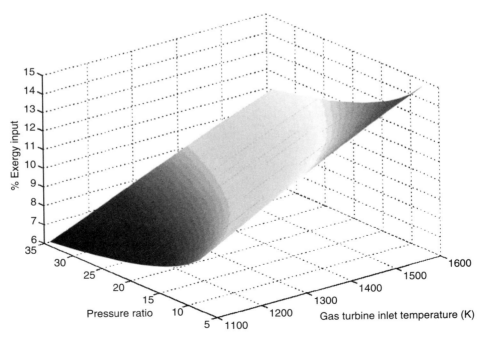

Figure 11   Exergy destruction rate plot for the heat recovery steam generator (HRSG)

6 CONCLUDING REMARKS

For the natural gas-fired combined cycle power generation unit, increasing the pressure ratio results in an increase in the combined cycle net work output, followed by a drop that is brought about by higher work demand from the compressor, and the drop in net work production from the bottoming steam cycle. With regards to the combined cycle net work output, the optimum pressure ratio increases as the gas turbine inlet temperature is raised.

The gas turbine combustion chambers (CC1, CC2) are the major source of exergy destruction in the unit. The exergy destruction in the main combustion chamber, in the HRSG, and in the steam cycle components (SET2) drops with increasing pressure ratio.

For the same gas turbine pressure ratio, increasing the gas turbine inlet temperature results in a higher combined cycle work output and thermal efficiency. The exergy destruction in the combustion chambers and the gas turbine cycle components (SET1) can be lowered by raising the gas turbine inlet temperature. For the bottoming steam cycle components (SET2) and the HRSG, the increase in gas turbine inlet temperature results in higher exergy destruction/loss rate.

The work production from the bottoming cycle can be enhanced by utilizing supplementary firing to boost steam generation in the HRSG. This, in turn, results in an increase in the combined cycle net work output, but has a negative effect on the combined cycle thermal efficiency.

The exergy destruction rate in the main combustion chamber (CC1) is more significant, and setting the conditions to lower the exergy destruction rate in the main combustion chamber, results in an overall drop in the combustion chambers total exergy destruction rate. In the HRSG, the maximum total exergy destruction rate and losses occurs at the lower pressure ratio and higher gas turbine inlet temperature. At low gas turbine inlet temperatures ($< 900°C$), increasing the pressure ratio results in a significant increase in the combined cycle thermal efficiency, with higher rates of supplementary firing fuel.

REFERENCES

American Chemical Society and the American Institute of Physics for the National Bureau of Standards, (1974), "JANAF Thermochemical Tables," Washington.

Bejan, Adrian (2002), "Fundamentals of exergy analysis, entropy generation minimization, and the generation of flow architecture," *Int. J. Energy Research*, Vol. 26, pp. 545–565.

Bolland, O. (1991), "A Comparative Evaluation of Advanced Combined Cycle Alternatives." *ASME J. Eng. Gas Turbines Power*, Vol. 113, pp. 190–197.

Chiesa, P., and Consonni, S. (2000), "Natural Gas-Fired Combined Cycles with Low CO_2 Emissions," *ASME J. Eng. Gas Turbines Power*, Vol. 122, pp. 429–436.

De, S., and Nag, P. K. (2000), "Effect of Supplementary Firing on the Performance of an Integrated Gasification Combined Cycle Power Plant," *India, Proc. Instn. Mech. Engineers*, Vol. 214, Part A.

Dincer, I., and Rosen, M. A. (2004), "Exergy as a Driver for Achieving Sustainability," *Int. J Green Energy*, Vol. 1, pp. 1–19.

Facchini, B., Fiaschi, D., and Manfrida, G. 2000. "Exergy Analysis of Combined Cycles Using Latest Generation Gas Turbines," *ASME J. Eng. Gas Turbines Power*, Vol. 122, pp. 233–238.

Horlock, J. H. (1995), "Combined Cycle Power Plants—Past, Present and Future," *ASME J. Eng. Gas Turbines Power,* Vol. 117, pp. 608–616.

Horlock, J. H., Young, J. B., and Manfrida, G. (2000), "Exergy Analysis of Modern Fossil-Fuel Power Plants," *ASME J. Eng. Gas Turbines Power*, Vol. 122, pp. 1–7.

Kotas, T. J. (1985), The Exergy Method of Thermal Plant Analysis, Krieger, Melbourne, Florida.

Lozza, G., and Chiesa, P. (1996), "Combined Cycle Power Stations Using "Clean-Coal Technologies": Thermodynamic Analysis of Full Gasification versus Fluidized Bed Combustion with Partial Gasification," *ASME J. Eng. Gas Turbines Power*, Vol. 118, pp. 737–748.

Lozza, G., and Chiesa, P. (2002a), "Natural gas Decarbonization to Reduce CO_2 Emissions from Combined Cycles—Part I: Partial Oxidation," *ASME J. Eng. Gas Turbines Power*, Vol. 124, pp. 82–88.

Lozza, G., Chiesa, P. (2002b), "Natural Gas Decarbonization to Reduce CO_2 Emissions from Combined Cycles—Part II: Methane Reforming," *ASME J. Eng. Gas Turbines Power,* Vol. 124, pp. 89–95.

Makansi, J. (1990), "Combined Cycle Power Plants," *Power* (June), pp. 91–126.

Mohammed, K. (2005), "Parametric Analysis of Advanced Combined Cycle Power Generation Systems," MScE thesis, University of New Brunswick, Fredericton, NB, Canada.

Moran, M. J., and Shapiro, H. N. (2000), *Fundamentals of Engineering Thermodynamics*, 4th ed., John Wiley, Toronto.

Nag, P. K. (1994), "Development of Combined Cycles," in the short course on combined cycle power plants, Calcutta, India, January 4–5, pp. 1–24.

Reddy, B. V., and Mohammed, K. (2007), "Exergy Analysis of a Natural Gas Fired Combined Cycle Power Generation Unit," *J of Exergy*, Vol. 4, pp. 180–196.

Srinivas, T., Gupta, A. V. S. S. K. S., and Reddy, B. V. (2009), "Carbon Dioxide Emission Reduction from Combined Cycle with Partial Oxidation of Natural Gas," *Energy for Sustainable Development*, Vol. 13, pp. 32–36.

Sue, Deng-Chern, and Chuang, Chia-Chin (2004), "Engineering Design and Exergy Analyses for Combustion Gas Turbine Power Generation System," *Energy*, Vol. 29, pp. 1183–1205.

Zwebek, A., and Pilidis, P. (2003a), "Degradation Effects on Combined Cycle Power Plant Performance—Part I: Gas Turbine Cycle Component Degradation Effects," *ASME J. Eng. Gas Turbines Power*, Vol. 125, pp. 651–657.

Zwebek, A., and Pilidis, P. (2003b), "Degradation Effects on Combined Cycle Power Plant Performance—Part II: Steam turbine cycle component degradation effects," *ASME J. Eng. Gas Turbines Power*, Vol. 125, pp. 658–663.

Index